Physics Meets Mineralogy

Physics Meets Mineralogy: Condensed-Matter Physics in Geosciences gives a state-of-the-art description of the interaction between geosciences and condensed-matter physics. Condensed-matter physics leads to a first-principles way of looking at crystals, enabling physicists and mineralogists to study the rich and sometimes unexpected behaviour that minerals exhibit under the extreme conditions (high pressure/high temperature) found deep within the Earth.

Two technological advances in recent decades have facilitated a rapid advance in our understanding of mineral structures and properties: one theoretical, the advance in computational solid-state physics with the aid of increased computing power, and the other experimental, with high-pressure mineralogy becoming possible with, for example, the aid of the diamond-anvil-cell. These have not only led to a better understanding of the formation of natural minerals deep in the Earth's interior, but also have aroused interest in condensed-matter physics.

Leading international researchers from both geosciences and condensed-matter physics discuss the state of the art of this interdisciplinary field. The volume is an excellent summary for specialists and graduate students in mineralogy and crystallography.

Hideo Aoki is a Professor in the Department of Physics at the University of Tokyo.

Yasuhiko Syono is an Emeritus Professor in Solid State Chemistry at the Institute for Materials Research of Tohoku University.

Russell J. Hemley is a Staff Scientist at the Geophysical Laboratory of the Carnegie Institution of Washington.

Physics Meets Mineralogy

Condensed-Matter Physics in Geosciences

Edited by
Hideo Aoki
Yasuhiko Syono
Russell J. Hemley

CAMBRIDGE
UNIVERSITY PRESS

CAMBRIDGE UNIVERSITY PRESS
Cambridge, New York, Melbourne, Madrid, Cape Town, Singapore, São Paulo, Delhi

Cambridge University Press
The Edinburgh Building, Cambridge CB2 8RU, UK

Published in the United States of America by Cambridge University Press, New York

www.cambridge.org
Information on this title: www.cambridge.org/9780521643429

First published 2000
This digitally printed version 2008

A catalogue record for this publication is available from the British Library

Library of Congress Cataloguing in Publication data
Physics meets mineralogy : condensed matter physics in geosciences / edited by Hideo
Aoki, Yasuhiko Syono, Russell J. Hemley.
 p.   cm.
 Includes index
 ISBN 0-521-64342-2 (hb)
 1. Solid state physics.   2. Condensed matter.   3. Mineralogy.   I. Aoki, Hideo, 1950–
 II. Syono, Yasuhiko, 1935–   III. Hemley, Russell J. (Russell Julian) 1954–
QC176.P482   2000
549′.01′53041 – dc21

                                                                              99-086299

ISBN 978-0-521-64342-9 hardback
ISBN 978-0-521-08422-2 paperback

Contents

Preface

If you look at the cosmic abundance of elements, oxygen, carbon, neon, nitrogen, magnesium, and silicon are the top six (apart from hydrogen and helium); most of them are indeed the main constituents of rock-forming minerals. If the goal of modern mineralogy is to understand classes of materials on an atomic scale, this is also the aim of modern condensed-matter physics. Therefore it is natural that condensed-matter physics should be applied to mineralogy and earth sciences, just as particle physics is a key tool in understanding cosmology.

This book is intended to give a state-of-the-art description of intensive interactions between geophysics and condensed-matter physics that recent years have witnessed. Although you might assume that, given the maturity of the solid-state physics, the crystalline structures of materials must have been readily understood in atomistic, nonempirical terms, this rather naive expectation has not been satisfied until quite recently.

Although traditional mineralogical principles had remained largely empirical, a few individuals pursued the idea that modern theoretical solid-state physics must become the foundation of mineralogy and geophysical sciences. One of the pioneers in this respect is Professor Yoshito Matsui in Japan. Following the tradition of Goldschmidt, he realized the importance of identifying and characterizing the underlying physics that controls geochemical, geophysical, and geological phenomena over the entire range of pressure and temperature relevant to the Earth. In so doing, he, in collaboration with Dr. Eiji Ito, directed the development of new experimental high-pressure facilities that would be required for understanding the rich and sometimes unexpected behaviour of minerals under conditions found deep within the Earth.

In the tradition of Pauling, he sought a fundamental understanding of the behaviour of materials from interatomic interactions. He was among the first to apply state-of-the-art molecular dynamics to understanding minerals and to predict behaviours under extreme conditions. To this end, he and his group obtained remarkably accurate interatomic potentials for silicates now used by a large number of physicists, chemists, and materials scientists, as well as mineralogists throughout the world.

An event remembered by the international community was the successful symposium "Computational Physics in Mineralogical Sciences," organised by Matsui and one of the editors of this book (Hemley) at the 1992 International Geological Congress in Kyoto. The event convinced us that the first-principles way of examining crystals enabled physicists and mineralogists to interact seriously for virtually the first time.

Two technological advances, theoretical and experimental, play major roles here: One is the advances in the computational solid-state physics with supercomputers and parallel computers, the other is high-pressure geoscience. We can even make the two compete in obtaining novel crystal structures under high pressures, which was, in Matsui's words, "the very first see-saw game between the computer science and real experiments." The impact of the works of Matsui in collaboration with Tsuneyuki and co-workers (condensed-matter physicists) on the first-principles study of crystal structures was noted by Sir John Maddox in an article entitled "Crystals from First Principles" in the News and Views column of *Nature* [*London*, **335**, 201 (1988)]. He even added that "a demonstration of success (along this line) can rank, psychologically, with the example set by those who first climbed Everest." This article in turn continues to be quoted [e.g., *Nature* (*London*) **381**, 648 (1996)] in an urge to push the progress still further.

Although the publication of the present volume was originally planned as a Festschrift to honour Professor Matsui's accomplishments at his retirement from the Institute for Study of the Earth's Interior, Misasa, Japan, in 1997, it eventually evolved into a standard book, covering recent investigations, contributed by outstanding researchers from both geosciences and condensed-matter physics, that can be referred to for coming decades. To ensure that purpose, we start with an introductory chapter. Section 1.2 of Chap. 1 highlights how major advances in the interplay between mineralogy and condensed-matter physics arose. This forms the backdrop for the articles contributed to this volume, which are summarized in Section 1.3.

We believe that the present volume will benefit a wide range of readers, including condensed-matter physicists, geophysicists/mineralogists, crystallographers, solid-state chemists, and materials scientists.

We thank all the contributors who have made the present volume rich and of high standards. We are also indebted to Shinji Tsuneyuki and Stephen Gramsch for their help in editing and to Matt Lloyd for his interest in publishing this book.

Finally, to quote John Ziman in his *Principles of the Theory of Solids*, physicists "have heard the music of the spheres; and yet they know that science is made for man, not man for science." So let us finish by paraphrasing that we wish to hear the music of the globe through the marriage of physics and mineralogy.

H. Aoki, Y. Syono, and R. J. Hemley

List of Contributors

C. A. Angell
Department of Chemistry
Arizona State University
Tempe, AZ 85287, USA

Hideo Aoki
Department of Physics
University of Tokyo
Hongo, Tokyo 113-0033, Japan

James Badro
Laboratoire de Mineralogie-
  Cristallographie
Université Paris VI
4, Place Jussieu
F-75252 Paris Cedex 05 France

M. B. Boisen, Jr.
Department of Geological Sciences and
  Materials Sciences and Engineering and
  Mathematics
Virginia Tech
Blacksburg, VA 24061, USA

R. E. Cohen
Geophysical Laboratory and
  Center for High Pressure Research

Carnegie Institution of Washington
5251 Broad Branch Road N.W.
Washington, D.C. 20015-1305, USA

Z. Fang
Joint Research Center for Atom Technology
Angstrom Technology Partnership
1-1-4 Higashi
Tsukuba, Ibaraki 305-0046, Japan

Larry W. Finger
Geophysical Laboratory and Center for
  High Pressure Research
Carnegie Institution of Washington
5251 Broad Branch Road N.W.
Washington, D.C. 20015-1305, USA

G. V. Gibbs
Department of Geological Sciences and
  Materials Sciences and Engineering and
  Mathematics
Virginia Tech
Blacksburg, VA 24061, USA

Russell J. Hemley
Geophysical Laboratory and Center for
  High Pressure Research

Carnegie Institution of Washington
5251 Broad Branch Road N.W.
Washington, D.C. 20015-1305, USA

M. Hemmati
Department of Chemistry
Arizona State University
Tempe, AZ 85287, USA

F. C. Hill
Department of Geological Sciences and
    Materials Sciences and Engineering and
    Mathematics
Virginia Tech
Blacksburg, VA 24061, USA

Raymond Jeanloz
Department of Geology and Geophysics
University of California, Berkeley
Berkeley, CA 94704, USA

H. Kagi
Laboratory for Earthquake Chemistry
Graduate School of Science
University of Tokyo
Tokyo 113-0033, Japan

Masami Kanzaki
Institute for Study of the Earth's Interior
Okayama University
Misasa, Tottori-ken 682-0193, Japan

Abby Kavner
Department of Geosciences
Princeton University
Princeton, NJ 08544, USA

Keiji Kusuba
Institute for Materials Research
Tohoku University
Sendai 980-8577, Japan

K. Kusakabe
Graduate School for Science and
    Technology
Niigata University
Ikarashi, Niigata 950-2181, Japan

J. S. Loveday
Department of Physics and Astronomy
The University of Edinburgh
Mayfield Road
Edinburgh, EH9 3JZ, U.K.

W. Marshall
Department of Physics and Astronomy
The University of Edinburgh
Mayfield Road
Edinburgh, EH9 3JZ, U.K.

Takeo Matsumoto
Department of Earth Sciences
Faculty of Science
Kanazawa University
Kakumamachi, Kanazawa 920-1192, Japan

T. Miyazaki
National Research Institute for Metals
1-2-1 Sengen
Tsukuba, Ibaraki 305-0047, Japan

R. J. Nelmes
Department of Physics and Astronomy
The University of Edinburgh
Mayfield Road
Edinburgh, EH9 3JZ, U.K.

T. Ogitsu
Institute for Solid State Physics
University of Tokyo
Roppongi, Minato-ku, Tokyo 106-8666,
Japan

J. B. Parise
Department of Geosciences
State University of New York
Stony Brook, NY 11794-2100, USA

Atul Patel
Research School of Geological and
Geophysical Sciences
University College London
Gower Street
London WC1E 6BT, U.K.

Brent T. Poe
Bayerisches Geoinstitut
Universitat Bayreuth
D-95440 Bayreuth, Germany

G. David Price
Research School of Geological and
    Geophysical Sciences
University College London
Gower Street
London WC1E 6BT, U.K.

David C. Rubie
Bayerisches Geoinstitut
Universitat Bayreuth
D-95440 Bayreuth, Germany

H. Sawada
Joint Research Center for Atom Technology
Angstrom Technology Partnership
1-1-4 Higashi
Tsukuba, Ibaraki 305-0046, Japan

I. Solovyev
Joint Research Center for Atom Technology
Angstrom Technology Partnership
1-1-4 Higashi
Tsukuba, Ibaraki 305-0046, Japan

Terry Speed
Department of Statistics
University of California, Berkeley
Berkeley, CA 94704, USA

Lars Stixrude
Department of Geological Sciences
University of Michigan
Ann Arbor, MI 48109, USA

Kazumasa Sugiyama
Department of Earth and Planetary Science
University of Tokyo
Hongo, Tokyo 113-0033, Japan

Yasuhiko Syono
Institute for Materials Research
Tohoku University
Sendai 980-8577, Japan

Osamu Tamada
Graduate School of Human and
    Environmental Studies
Kyoto University
Kyoto 606-8501, Japan

Y. Tateyama
National Research Institute for
    Metals
1-2-1 Sengen
Tsukuba, Ibaraki 305-0047, Japan

K. Terakura
Joint Research Center for Atom
    Technology
National Institute for Advanced
    Interdisciplinary Research
1-1-4 Higashi
Tsukuba, Ibaraki 305-0046, Japan

David M. Teter
Geochemistry Department 6118
Sandia National Laboratories
Albuquerque, NM 87185-0750, USA

Taku Tsuchiya
Department of Earth and Space
    Science
Graduate School of Science
Osaka University
1-16 Machikaneyama
Toyonaka, Osaka 560-8531, Japan

S. Tsuneyuki
Institute for Solid State Physics
University of Tokyo
Roppongi
Minato-ku, Tokyo 106-8666
    Japan

Lidunka Vocadlo
Research School of Geological and
    Geophysical Sciences
University College London
Gower Street
London WC1E 6BT, U.K.

Yoshio Waseda
Institute for Advanced Materials
   Processing
Tohoku University
Sendai 980-8577, Japan

Takehiko Yagi
Institute for Solid State Physics
University of Tokyo
Roppongi
Minato-ku, Tokyo 106-0032
   Japan

Masaaki Yamakata
Institute for Solid State Physics
University of Tokyo
Roppongi
Minato-ku, Tokyo 106-0032, Japan

Takamitsu Yamanaka
Department of Earth and Space Science
Graduate School of Science
Osaka University
1-16 Machikaneyama
Toyonaka, Osaka 560-8531, Japan

Part I

Introduction

Chapter 1.1

Physics and Mineralogy: The Current Confluence

Hideo Aoki, Yasuhiko Syono, Russell J. Hemley

The interplay among geosciences, mineralogy, and condensed-matter physics is briefly overviewed in the context of the chapters contained in this volume. The chapters report current developments along many lines of inquiry introduced by one of the leading figures, Yoshito Matsui, for which we start with a brief history. The diverse body of work summarized in the book includes ranges from recent advances in theoretical condensed-matter physics, crystallography and crystal chemistry, high-pressure transformations, and the behaviour of high-temperature melts.

1.1.1 Introduction

It has long been recognized that atomic interactions are key in understanding large-scale geological phenomena, which is an approach dating at least as far back as the days of Goldschmidt [1]. Conversely, a study of the materials that comprise the planets can tell us much about fundamental physics and chemistry. This line of approach is subsequently exemplified by Pauling's development of the theory of chemical bonding from the structural studies of minerals [2].

Among many facets that revolutionized modern earth science are plate tectonic theory, discoveries in planetary astronomy, and extraordinary technical advances that provide windows on our planet's deep interior. Modern condensed-matter physics, on the other hand, has acquired much impetus from the discovery of new phenomena, the development of new experimental methods, and vast improvements in computational

H. Aoki et al. (eds), *Physics Meets Mineralogy* © 2000 Cambridge University Press.

methods. Although physics and mineralogy have continued largely on distinct and disparate trajectories, many recent advances in both earth science and physics stem from science done at the crossroads of the two fields. Indeed, it has taken the vision, wisdom, and dedication of a few leading individuals, like Goldschmidt and Pauling in the earlier era, to bring together these two fields and cultures in order to answer numerous long-standing questions about the natural world.

The interplay of earth sciences and physics may indeed be best illustrated by an examination of the brief history of a key scientist, which we attempt in the next section.

1.1.2   From Mineral Assemblages to First-Principles Theory

1.1.2.1 Geochemical Beginnings

Yoshito Matsui [3] worked under the influence of Eiiti Minami, who had studied in Göttingen under Goldschmidt himself, and introduced to Japan a tradition of seeking origins of geological phenomena in atomic-level structures of minerals and their solutions in the tradition of Goldschmidt. In two of the earliest papers of Matsui, coauthored with Masuda [4, 5], he addressed the problem of the difference in abundance of lanthanide elements between the chondrites (majority of the meteorites originated from the primordial solar nebula) and the Earth's crust, thereby evoking "once completely molten Earth." The molten Earth, or the "magma ocean," was a new concept at that time.

As this work [4, 5] indicates, the rock-forming minerals occur mostly as solid solutions with a wide variety of compositions, so that their chemical thermodynamics is a central issue. Hence Matsui felt it mandatory to establish a thermodynamic description of the partitioning of elements among silicate minerals and melts. In contrast to the simple model for alloys between elements, these compound solid solutions have specific sites or crystallographic positions that accommodate the relevant ions. Through a close collaboration with Banno [6, 7], Matsui provided for the first time a thermodynamic model that could accurately describe the Fe–Mg distribution within minerals that have two energetically nonequivalent classes, I and II, of sites (Fig. 1.1.1) for cations.

If we represent the stoichiometry of minerals generically as $(A, B)Z$ (e.g., $A = Mg^{2+}$, $B = Fe^{2+}$, $Z = Si_2O_6^{2-}$ for pyroxene [8]), such solid solutions have the formula

$$(A, B)^I (A, B)^{II} Z,$$

which may be called an intracrystalline ion-exchange solid solution, and a preferential distribution of ions A and B over the nonequivalent sites is called site preference. The site preference can cause deviations from the conventional Vegard's law, which asserts that the size of a unit cell of a solid solution should be linear in its molar composition.

To detect any such anomaly, Matsui and Syono [9, 10] measured the unit-cell parameters of solid solutions of olivine and (orthorhombic) pyroxene. Olivine, an olive-coloured mineral having a stoichiometry of M_2SiO_4 (where M is a metallic element), and pyroxene ($MSiO_3$) are both common minerals. They studied the (Mg, Ni) olivine

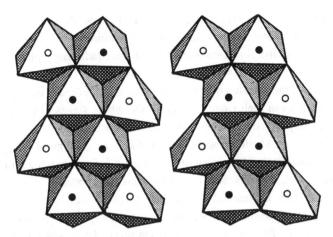

Figure 1.1.1 Two classes of sites (indicated here as solid and open circles, the latter accommodated in distorted octahedra) are shown for a pyroxene, $M_2Si_2O_6$ (enstatite with M = Mg here). A part of the crystal structure (double chain of MO_6 octahedra) is displayed.

solid solution, Mg_2SiO_4–Ni_2SiO_4 for instance, and indeed found an unexpectedly large anomaly [9]. The similar study was subsequently reported for (Mg, Zn) olivine [11] and for (Mg, Mn) olivine [12].

Matsui further pursued the thermodynamic approach [6, 7] along two lines of studies. First, silicate solid solutions were characterized thermodynamically with respect to the partitioning of elements between phases such as olivine and pyroxene or olivine and garnet $[M_3Al_2(SiO_4)_3]$ [12–14]. Second, partitioning of elements among solid silicates (minerals) and silicate melts (magma) was examined [7, 15, 16]. For instance, the partitioning of Mn and Mg between olivine (MM′SiO₄) and pyroxene [MM′(SiO₃)₂] was measured and analyzed [12] on the basis of the Matsui–Banno model. A paper by Matsui and Nishizawa on Fe^{II}–Mg partitioning [13] incorporates various approaches of Matsui, namely, a critical evaluation of existing data by application of the least-squares fit and thermodynamic analysis á la Matsui–Banno [6], later refined by Matsui with his collaborators. This work successfully resolved the long-standing controversy on Fe–Mg partitioning between olivine and pyroxene, a result that remains unchallenged to this day. The equilibrium constant for the exchange reaction,

$$Mg_2SiO_4 + 2FeSiO_3 \leftrightarrow Fe_2SiO_4 + 2MgSiO_3,$$

has turned out to be insensitive to P and T, where P is pressure and T is temperature, and this property was exploited as an indicator of nonequillibrium in volcanic rocks.

Now, ferromagnesian ions (i.e., divalent ions of Mg- and Fe-group transition elements, Ni, Co, Zn, Fe, and Mn) are important ingredients in minerals of metamorphic rocks and those of the Earth's upper mantle. A reconnaissance study on the partitioning of these ions by Matsui and Banno [7] indicated that the controlling factor is their ionic sizes, as originally introduced by Goldschmidt [1]. However, an exception was found in the distribution of Zn, which exhibits a strong preference for tetrahedral coordination by O ions. This finding later led to a study on crystal structures of Zn silicates by

Morimoto and co-workers jointly with Akimoto and co-workers [17]. Also realized was the fact that both Goldschmidt's and Pauling's definitions of ionic radii were too crude for geochemical use. A self-consistent set of ionic radii that replaced the earlier ones was finally published in 1969 by Shannon and Prewitt [18].

The geochemical fractionation of every element is the outcome of the crystallization of silicate minerals from melt (magma) during the whole stages of Earth evolution. Goldschmidt [1] proposed that the major fractionation occurred during the primordial stage of Earth's history. Masuda and Matsui [5] followed this line of thought and revealed the factors that control fractionation of elements during crystallization of minerals from magma, a topic that has remained of fundamental importance in geochemistry.

An ambitious effort, nicknamed the "all-ium project," was launched a few years after the 1966 paper by Masuda and Matsui [5] in order to put the approach in a wider scope. The goal of this effort, which was led by Onuma and Nagasawa, was to measure concentrations of all technically detectable elements in major rock-forming minerals and in liquid magma (groundmass in geological terms) from which they crystallized. Matsui recognized the significance of the project and served as its supervisor [15] and co-worker [16]. The final paper by Matsui et al. [16], dedicated "to the memory of the late Professor Minami," contains partition coefficient versus ionic radius diagrams for groundmass versus all the four classes (olivine, pyroxene, hornblende [8], and mica) of ferromagnesian silicates, namely, (1) olivine–groundmass, (2) orthorhombic (calcium-poor) pyroxene–groundmass, (3) monoclinic (calcium-rich) pyroxene–groundmass, (4) hornblende–groundmass, and (5) mica–groundmass, covering more than 30 elements with the ionic radii defined by Shannon and Prewitt [18]. The diagrams clearly endorse the idea that the partitioning of elements between crystal and magma is indeed controlled by both the size of available sites in relevant crystal and the size (and charge) of the ions to be accommodated. This serves as a final, negative answer to the well-known Goldschmidt postulate [19] on element distribution in magmatic processes.

1.1.2.2 Mineral Physics and High-Pressure Frontier

The late 1960s and early 1970s witnessed major advances in experimental high-pressure methods. With these techniques, Matsui and co-workers demonstrated that pressure could induce extensive polymorphism in silicates at upper-mantle pressures [11, 20]. This was followed by a series of pioneering high-pressure studies of mantle silicates at high-pressure facilities developed for and built at the Institute for Study of the Earth's Interior at Misasa, Japan. The synthesis of spinel-type γ-Mg_2SiO_4 in 1974 was a landmark as the first reliable structure determination from powder-x-ray-diffraction data, including d spacings and intensities [21]. This was followed by detailed crystal chemistry and phase equilibria of postspinel phases (i.e., phases that would appear when the spinel was subject to even higher pressures) of ferromagnesian silicates, including those having the ilmenite structure [22].

High-pressure studies in the mid-1970s of postspinel phases led to the discovery of dense silicates that have the perovskite structure. Most important was the discovery of

$MgSiO_3$ perovskite, a revolutionary result as it was soon realized that this could be the Mg end member of the dominant mineral of the Earth's lower mantle. Determination of the structure of $MgSiO_3$ perovskite was therefore a major problem in deep-earth mineralogy and geophysics. Matsui carried out one of the first x-ray-diffraction structure refinements of $MgSiO_3$ perovskite based on powder-x-ray data. The crystal structure was determined from powder-diffraction data in combination with a simulation of crystal structures based on ionic radii (distance-least-squares method) [23]. The results agreed with the powder-diffraction study carried out at the same time by Yagi et al. [24].

High-pressure experiments were further used to address major questions regarding the mineralogy and composition of the lower mantle [25]. The maximum solubility of Fe in perovskite, which had been discrepant in earlier studies, was determined at 26 GPa, a result that has recently been extended to higher temperatures [26] and pressures [27]. The composition of the lower mantle inferred by comparison of density and bulk modulus of the minerals with seismological data was close to that of Ringwood's "pyrolite" [28], which is a hypothetical composition (pyroxene + olivine) that has a stoichiometry close to that of M_2SiO_4 (orthosilicate). The results, however, suggested that the lower mantle could be rich in perovskite (i.e., $MSiO_3$ stoichiometry, close to the silicate portion of enstatite chondrites), which is consistent with later work; this is still an active area of research in mineral physics [29]. Later synthesis by Ito of single-crystal $MgSiO_3$ perovskite provided crucial characterization of these deep-mantle materials. The analysis covered full crystal structure refinements [30], Brillouin scattering [31], Raman scattering [32], infrared absorption [33, 34], NMR [35], and calorimetry [36].

Matsui also played a direct role in the refinements of important crystallographic structures (see, e.g., Ref. [17]). Intriguing high-pressure phases such as the high-pressure phase of $NaAlSiO_4$ with the $CaFe_2O_4$ structure were later found and characterized in terms of crystal chemistry [37]. These crystal-chemical studies included the use of Mössbauer spectroscopy, a technique that is widely used in mantle mineralogy [38] because of the information it can provide about oxidation state and local bonding properties of Fe in materials. Matsui et al. [39] used Mössbauer spectroscopy to examine the crystal chemistry of Fe^{2+} and Fe^{3+} in clinopyroxenes. They showed the importance of combining local-probe techniques with complementary information on long-range structure obtained by diffraction techniques.

Later, a systematic study of the high-pressure transformations of ABO_3-type silicates, germanates, and titanates [40] provided a basis for further insight into the polymorphism in silicates; this also set the stage for work along these lines that has continued to this day (see, e.g., Ref. [41]). Materials in this class can possess unusual physical properties, including exceptionally high dielectric constants, and may find use in various applications. The earlier high-pressure study of Zn silicates [11] has had an unanticipated spin-off: The bonding topology of the high-pressure phase of Zn_2SiO_4 (so-called willemite II) was used as a candidate structure for a hypothetical cubic high-pressure phase of C_3N_4, which has been predicted by first-principles calculations to have a bulk modulus exceeding that of diamond [42].

1.1.2.3 Computational Theory

Matsui was one of the first to introduce the computational technique of molecular dynamics (MD) in the geological sciences, his third major accomplishment. This effort eventually led to the introduction of nonempirical MD; that is, following the report of Woodcock et al. [43] on MD calculations on liquid and glassy SiO_2, Matsui and co-workers proceeded, with Kawamura, to apply the technique to Mg silicates for various pressures for $MgSiO_3$ [44, 45]. Motivation came from the fact that the pioneering work [43] had a serious shortcoming in that the Si atom remains fourfold coordinated by O atoms even at very high pressures. The 1980 paper of Matsui and Kawamura [44] showed, for a certain realistic choice of the interatomic potentials, that conversion between fourfold and sixfold coordinations does occur under compression. This result was later supported by shock-wave equation-of-state studies [46] and high-pressure spectroscopic studies of glasses [47, 48]. The technique was also applied to crystals. A previously found instability of SiO_2 in the hypothetical fluorite structure was shown to lead to the α-PbO_2-type phase [49]. These simulations also showed that a variety of structures closely related to that of α-PbO_2 have similar energies and densities; this finding was recently examined and confirmed by accurate first-principles calculations for SiO_2 [50].

In these early calculations, however, limitations existed principally from uncertainties inherent in approximate interatomic (pairwise) potentials that were used. These potentials were typically obtained by the fitting of selected experimental data to simple functions. To go beyond the empirical approach, Tsuneyuki et al. launched a programme to develop accurate pair potentials systematically from first-principles methods. The success of this effort for SiO_2 [51] was heralded as not only a major breakthrough for the geosciences but also as a conceptual milestone for condensed-matter physics.

In contrast to previous potentials, the new ones were the first to reproduce accurately the dynamical stabilities and physical properties of four principal silica polymorphs. Tsuneyuki et al. [52] went one step further with these potentials to predict new high-pressure structures (as discussed in Aoki's paper, Chap. 5.1 of this book). These polymorphs have a mixture of fourfold and sixfold coordination of Si, and interpretations were made of the high-pressure transformations found experimentally in quartz and coesite [53]. A transition in quartz, further observed experimentally [54], continues to be the focus of theoretical study (see, e.g., Ref. [55]).

Tsuneyuki et al. [52] also showed that cristobalite transforms into mixed fourfold and sixfold coordinations before it becomes stishovite at higher pressures. This prediction promoted Yagi and Yamakata to undertake an experimental test. Initially, there seemed to be a discrepancy between theory and experiment, but this was later resolved: For a (quasi-)hydrostatic pressure the stishovitelike phase is indeed obtained in experiments [56] as mentioned by Yagi et al. in Chap. 4.4 of this book.

MD calculations were also applied to the long-standing and controversial problem of the temperature-induced α–β transition in quartz near 850 K [57]. The calculations indicated that the transition has what may be called a dynamically order–disorder character: The atoms in the high-temperature β structure hop between two equivalent sites anharmonically rather than vibrating around an idealized (i.e., higher-symmetry) structure. This is in agreement with the interpretation of transmission electron microscopy studies [58].

The MD method was also applied to examine the properties of novel low-pressure, open structures, which include new zeolitelike phases [59]. The approach pioneered by Tsuneyuki et al. [51, 52] has been extensively used by other groups [60–63]. The many recent studies include investigations of low-frequency floppy modes in β-cristobalite [64] and soft-mode behaviour and elastic anomalies in the α–β transition in quartz [65].

A wider range of transformations of AO_2-type compounds in general was also explored. The transformation from the tetragonal rutile-type to the orthorhombic $CaCl_2$-type phase was first pointed out by Nagel and O'Keeffe [66]. Calculations carried out by Kusaba, Syono, and Matsui [67] indicated that the transition in TiO_2 depends on the nature of the compression, with compression in different directions ([100] and [110]) leading to the CaF_2-type structure (or a twinned type), whereas [001] compression leads to the $CaCl_2$-type structure. Preliminary evidence for the stability of the $CaCl_2$-type phase in silica was obtained by powder-x-ray diffraction [68]. The problem was examined by Matsui and Tsuneyuki [69], who used the first-principles-derived pair potential described above [51]; their calculation showed that the system at the transition fluctuates between two equivalent configurations (having $a > b$ or $b > a$ unit cells, respectively) of the $CaCl_2$ structure. Subsequently the transition was identified experimentally near 50 GPa, first by Raman spectroscopy [70] and then by x-ray diffraction [71,72]. Although the calculated transition pressure is now known to be too high, indicating the need for the extended calculations described below [73, 74], the result did reveal important aspects of anharmonic dynamics of the system in the vicinity of the transition. This anharmonicity appears to be responsible for complexities observed spectroscopically at the phase transition [70].

Armed with these new theoretical techniques, researchers reexamined the anomalous properties of high-temperature silicate melts; in particular, the remarkable property that ion mobilities increase with pressure in some liquid silicates was explored. This was inferred from the reported decrease in viscosity of some melts such as jadeite ($NaAlSi_2O_6$) [75]. Seeking a physical understanding for this anomalous behaviour, Tsuneyuki and co-workers used first-principles-derived pair potentials [51] to examine the predicted pressure enhancement of ion mobilities in liquid silica [76]. They proposed a model involving coordination changes, which is identified earlier as a key process in crystal-to-crystal transitions in high pressures. They proposed that the anomalous diffusion results from fluctuations in the melt between fourfold and sixfold coordinations and showed that this could account quantitatively for the predicted diffusivity maximum as a function of pressure.

The earlier simulations for SiO_2 in the fluorite structure [49] also found that a new modified fluorite that has space group $Pa3$ could be formed. The $Pa3$-type phase was then characterized in more detail [77] and its stability and physical properties calculated with first-principles methods, i.e., the linearized augmented plane-wave (LAPW) method [73]. The predicted stability of $Pa3$-type silica was subsequently confirmed by additional calculations and the transition pressure from the rutile (or $CaCl_2$ phase) was predicted to be above 100 GPa [74].

1.1.3 Physics Meets Mineralogy: An Overview of the Articles in this Book

The above background sets the stage for the articles collected in this book. In view of the development described above, we start with theory, including both fundamental principles and key applications. We then move on to recent studies in crystallography and crystal chemistry, including both systematics and new structural techniques. This naturally leads to a series of studies of temperature- and pressure-induced transformations, tackled from both theoretical and experimental points of view. Finally, we turn to high-temperature melts and crystal–melt interactions.

1.1.3.1 Advances in Theoretical and Experimental Techniques

Major advances have been made in the application of increasingly accurate theoretical methods for calculating electronic structures of crystalline solids. Stixrude (Chap. 2.1) reviews the applications of density functional theory and, in particular, local density approximation (LDA) methods to Earth materials. These methods, in standard use in condensed-matter physics for years, have found their way into mineralogy only recently and have become prevalent. Particular attention is paid to the LAPW technique and its use in calculating total energies, equations of state, crystal structures, phase stability, and elasticity. This and related methods are particularly accurate for a number of systems, including the geophysically important oxides and silicates cited above [73]. Matsumoto in Chap. 2.2 describes the technique of crystallographic orbits [78] in classifying space-group hierarchies (of which he is a pioneer), with particular emphasis placed on the fluorite structures.

   Experimentally, there have been remarkable recent developments in x-ray-diffraction techniques. New synchrotron-radiation methods have improved the precision of structural studies. Finger (Chap. 2.3) examines sources of errors in powder-diffraction measurements, particularly those obtained in energy-dispersive x-ray-diffraction experiments, which have been used extensively for high-pressure studies. As described by Finger, new developments include the use of three-dimensional techniques (e.g., measuring x rays collected radially as well as axially relative to the principal stresses produced by a high-pressure vessel). With these advances, elasticity, strength, texture, and rheology can be studied in situ at high pressures [79].

   We have seen that an important area of deep-mantle mineralogy involves the accurate determination of phase relations in high-pressure minerals and their analogues.

The accurate determination of phase boundaries continues to be a major concern in high-pressure research on such systems, including the determination of boundaries in situ under extreme $P - T$ conditions. Kavner et al. (Chap. 2.4) examine the use of a novel statistical method for the treatment of phase-boundary observations. The results show promise for the correct treatment of data under extreme deep-earth conditions and may find important utility in materials science problems.

1.1.3.2 New Findings in Oxides and Silicates

The correlation between chemical bond strength and bond length dates back to Pauling [2]. Using first-principles molecular cluster calculations, Gibbs et al. (Chap. 3.1) explore the more recent treatment of this correlation for metal–oxygen bonds in crystals and show that the bond strength correlates with covalency. Also of interest here is the fact that the molecular orbital calculations are of the same general type as that used in the development of the Tsuneyuki potentials [51] for SiO_2.

Cohen (Chap. 3.2) applies first-principles theory to MgO, a prototypical oxide material that has served as a testing ground for intensive theoretical and experimental studies, including numerous high-pressure investigations. The electronic structure, bonding, thermal equation of state, elasticity, melting, thermal conductivity, and diffusivity have all been calculated from first principles, including ab initio models. The results provide first-order guidance as to the behaviour with respect to these properties of oxides and silicates deep within the Earth.

Extensions of the LDA method represent a major current effort in condensed-matter physics. This is particularly important for the treatment of so-called Mott insulator systems, such as FeO and MnO, in which electron correlation causes the metal–insulator transition. These are important as end members and model systems for deep-earth minerals as well. Fang et al. (Chap. 3.3) discuss applications of methods that go beyond LDA, i.e., the generalized gradient approximation and LDA $+ U$ methods (U is a Coulomb interaction parameter). The results show unusual polymorphism, including the possible stability of an inverse NiAs structure of FeO at high temperatures and pressures for which there is experimental evidence as well as additional theoretical support [80]. Also, the discovery of the high-pressure metallic phase of MnO with the normal NiAs structure above 90 GPa under both dynamic and static compression [81,82] favourably compares with the results of first-principles calculations [83]. More recently, a series of new high-pressure phase transformations in CoO above 80 GPa was shown to occur within the framework of the rock-salt structure [84] and could be explained by the magnetic collapse concept proposed by Cohen et al. [85].

Continuing on this theme of computer simulations, Patel et al. (Chap. 3.4) provide a review for recent theoretical studies of $MgSiO_3$ and MgO, crucial lower-mantle end-member minerals, by using approximate interatomic potentials such as those originally developed and applied by Matsui. These authors examine the microscopic (atomistic) origin of thermoelastic and transport properties of perovskite and MgO phases, including important defect behaviour and dynamical issues, through the use of both lattice dynamics and MD. The results allow an independent evaluation of the conflict-ing experimental data and, more important, provide predictions for properties (e.g., diffusion) well outside the range of current experimental techniques.

1.1.3.3 Transformations in Silica

As described in Subsection 1.3.2, the combination of theory and experiment has played a key role in elucidating high-pressure polymorphism in crystalline and glassy SiO_2, a geologically and technologically important material, at high pressures. Surprisingly, new discoveries continue to be made in this well-studied material, helped notably by the application of new synchrotron-radiation methods and new theoretical techniques, so the fascination continues. Hemley et al. (Chap. 4.1) review these developments by using recent calculations for understanding the new experimental results as well as sorting out inconsistencies. The sequence of phases at higher phases is of continuing interest. This transformation is also the subject of the paper by Kusaba and Syono (Chap. 4.2), which further describes calculations of the orientational dependence (anisotropy) of the shock-induced transition in TiO_2.

Crystalline silica was the first ceramic material and mineral to exhibit pressure-induced amorphization, as mentioned in the paper by Aoki (Chap. 5.1). This is the direct transformation from a crystal to an amorphous (or significantly disordered) solid, which was first shown for ice I [86].

Yamanaka and Tsuchiya (Chap. 4.3) address the origin of this class of transfor-mations, now observed in a growing number of materials of interest to mineralogy, condensed-matter physics, and materials science. The authors examine both reversible and irreversible amorphization processes from the point of view of both x-ray diffrac-tion and MD simulations. They assess the relationship between finite crystallite size and x-ray coherence lengths as well as the role of the frustrated crystal growth in reported observations of amorphization. Melting and instability of amorphization models are contrasted; this contrast is directly related to a specific debates about the connection between crystal instabilities and melting in general (see also Chap. 3.2 by Cohen).

Simulations that use the potential developed by Tsuneyuki et al. showed that in-triguing metastable crystal phases of SiO_2 could be formed under pressure [52] (see subsection 1.1.2.3). There is growing awareness of the effect of nonhydrostatic stresses on phase transformations. Just as Kusaba and Syono in Chap. 4.2 discern the effect of nonhydrostatic compression on the polymorphism of TiO_2 from computer simu-lations, Yagi and Yamakata (Chap. 4.4) present experimental evidence for unusual transformation behaviour in SiO_2 cristobalite under nonhydrostatic stress conditions (although in the case of TiO_2, shock compression produces such anisotropic com-pression within the shock front and quickly switches to hydrostatic). Phases other than stishovite are formed when the material is subjected to nonhydrostatic conditions at room (or relatively low) temperature. One is a new phase that occurs at 15 GPa and room temperature but is not yet identified. On the other hand, a stishovitelike phase appears on quasi-hydrostatic compression, in agreement with the theoretical predictions (as noted in subsection 1.1.2.3). The unexpected results for nonhydrostatic compression show the need to extend the calculations to the nonhydrostatic case as well.

1.1.3.4 Novel Structures and Materials

Aoki (Chap. 5.1) examines the diversity of crystal structures from the standpoint of condensed-matter theory. He combines crystal-chemical systematics based largely on structural motifs found in mineral systems to consider new classes of materials with possibly novel electronic properties. In doing so, he places the above-mentioned concepts such as surprisingly rich polymorphism or pressure-induced amorphization in silica in a wider scope. For the systems in which electron correlation is essential, he develops the concept of electron correlation engineering by considering electron correlation for diverse crystal structures. Here again we discover the important role that the phases of silica play as model structures, or more generally, the emerging input of concepts from mineralogy and crystallography in theoretical condensed-matter physics.

The application of new theoretical methods to the search for new materials recurs as a theme in the article by Tsuneyuki et al. (Chap. 5.2). The authors focus on a specific class of materials – graphite intercalation compounds under pressure. Dense, exotic diamondlike phases formed from LiC_6 and LiC_{12} with interesting electronic properties are predicted. These results suggest important implications for the high-pressure chemistry involving C and alkali metals (e.g., the incorporation of Li atoms in ultrahigh-pressure diamond anvils).

Pressure-amorphized material, as described above, represents only one type of metastable state induced in crystals by pressure. Other metastable states involving more delicate types of order–disorder may occur. Parise et al. (Chap. 5.3) examine the question of pressure-induced order–disorder transition. They propose sublattice amorphization in the H sublattice of simple hydroxides, in which the role of possible H . . . H interactions is discussed. Using β-Co(OD)$_2$ as an example, they report high-pressure neutron-diffraction measurements that show evidence of a competition between H . . . H repulsion and H-bond formation under pressure. These mineral hydroxides under pressure are important model systems for understanding more complex hydrous minerals that may reside deep within the Earth. These materials are also prototypes for examining the behaviour of the H bond as a function of interatomic distances, a problem of fundamental interests in physics, chemistry, and biology.

1.1.3.5 Melts and Crystal–Melt Interactions

The nature of silicate and oxide melts continues to be a challenge to both theory and experiment. In the tradition of Matsui, research on these high-temperature liquids is motivated by geoscience problems – understanding the behaviour of melts within the Earth, including implications for the evolution of the planet – as well as condensed-matter physics problems – understanding the liquid state of matter, including its anomalous properties. Hemmati and Angell (Chap. 6.1) carry out a detailed comparative study of various interatomic potentials for predicting thermodynamic, structural, and dynamic properties of liquid silica. The comparison exhibits, in particular, progress in modelling the unusual properties of liquid silica such as the density maximum as a function of temperature. The existence of such maxima and other anomalous transport properties of liquid silicates that are formed deep within the Earth, now or early in Earth's history, plays a crucial role in governing magma ascent and crystal settling, which would in turn determine fractionation and differentiation of the planet.

The diffusivity of silicate liquids at high pressures and temperatures is also the theme of the article by Poe and Rubie (Chap. 6.2), who report state-of-the-art experimental studies up to 15 GPa and 2800 K. They find a significant increase in O_2 and Si self-diffusivities in $Na_2Si_4O_9$ liquid, which suggests that the viscosity decreases with pressure. However, the result sensitively depends on composition, where diffusivities reach a maxima below 10 GPa as the liquid becomes more polymerized. The increase in viscosity with depth in strongly depolymerized melts points towards major effects on fractionation processes within the magma ocean that may have existed early in Earth's history.

Properties such as density, viscosity, surface tension, and electrical conductivity of oxide melts at very high temperatures are relevant to fields ranging from geological or materials sciences to fundamental high-temperature physics. As a result of important developments in x-ray-diffraction techniques, including angle-dispersive, energy-dispersive, and anomalous-scattering methods, the local structure of oxide melts can now be studied directly. Waseda and Sugiyama (Chap. 6.3) review the development of these techniques in recent years. So far, the techniques have been applied to local structures (radial distribution functions) as functions of composition and temperature. Extensions of these techniques include anomalous-scattering with new intense synchrotron sources that will allow identification of local structures within a wider compositional range and as functions of pressure as well as temperature.

In the final paper, Kanzaki (Chap. 6.4) examines melt–crystal systems and, in particular, trace-element partitioning, a principal focus of Matsui's early research (e.g., the "all-ium" project mentioned in Subsection 1.1.2.1). The origin of the distribution coefficients lies in interatomic interactions. Because the quantitative prediction requires comparison of energetics between two phases, accurate calculations of partition coefficients are extremely difficult. Kanzaki applies simulation techniques such as those used by Matsui's group to this problem. The approach involves calculating the elastic-energy changes associated with ion substitution. The calculations qualitatively reproduce the partition coefficient against the ionic radii diagrams described above [16]. Because the relative sizes of ions change with pressure, such an approach based on reliable interatomic potentials has an advantage in that it can be used in principle to predict partitioning to extreme P–T conditions deep within the Earth.

1.1.4 **Conclusion**

The articles in this book reflect the rich variety of multidisciplinary problems addressed at the intersection of mineralogy and physics. As such, the articles exemplify a tradition in modern science that Yoshito Matsui has helped to establish. Problems once relegated to disparate fields are now common to many. Understanding the behaviour of complex ceramic materials forms a class of problems in solid-earth geophysics and geochemistry as well as in materials science and condensed-matter physics. The advance is an outcome of the combination of the state-of-the-art theoretical and experimental studies, including those under extreme conditions. Such lines of approach, on the one hand, are providing accurate constraints on the composition and mineralogy of the Earth and other planets. On the other hand, they are leading to the discovery of new phenomena in materials subject to diverse $P-T$ conditions, including the high $P-T$ behaviours of metal oxides (e.g., Mott transition), novel transformations such as amorphization and changes in order–disorder in crystals, and the anomalous properties of oxide melts. Finally, the implications for materials technology may be profound: New calculations and experiments are uncovering a rich variety of materials that may possess conceptually novel, and possibly useful, physical and chemical properties. All of this leads us to predict that the crossroads of physics and mineralogy mapped out by Matsui will be the site of continuing scientific advances in the new century.

References

[1] V. M. Goldschmidt, Skr. Nor. Vidensk. Akad. Oslo I. Mat.-Naturv. Kl. No. 2 (1926) 1.

[2] L. Pauling, J. Am. Chem. Soc. **51**, 1010 (1929).

[3] E. Ito et al., eds., *Selected Papers of Professor Yoshito Matsui* (Terra Scientific, Tokyo, 1997).

[4] Y. Matsui and A. Masuda, Geochim. Cosmochim. Acta **27**, 547 (1963).

[5] A. Masuda and Y. Matsui, Geochim. Cosmochim. Acta **30**, 239 (1966).

[6] Y. Matsui and S. Banno, Proc. Jpn. Acad. **XLI** (6), 461 (1965).

[7] Y. Matsui and S. Banno, Chem. Geol. **5**, 259 (1969/1970).

[8] A classic description of pyroxene and amphibole (of which hornblende most commonly occurs in magmatic rocks) silicates may be found in W. L. Bragg, *Atomic Structure of Minerals*, Oxford Univ. Press, 1937, Chap. 12.

[9] Y. Matsui and Y. Syono, Geochem. J. **2**, 51 (1968).

[10] Y. Matsui, Y. Syono, S. Akimoto, and K. Kitayama, Geochem. J. **2**, 61 (1968).

[11] Y. Syono, S. Akimoto, and Y. Matsui, J. Solid State Chem. **3**, 369 (1971).

[12] O. Nishizawa and Y. Matsui, Phys. Earth Planet. Inter. **6**, 377 (1972).

[13] Y. Matsui and O. Nishizawa, Bull. Soc. Fr. Mineral. Cristallogr. **97**, 122 (1974).

[14] T. Kawasaki and Y. Matsui, Earth Planet. Sci. Lett. **37**, 159 (1977).

[15] H. Higuchi, N. Onuma, H. Nagasawa, H. Wakita, Y. Matsui and S. Banno, in *Proceedings of Int. Mineralogical Association-Int. Association of*

Geology and Ore Deposits Meeting 70 (IMA, location, 1970), Vol. I, pp. 254–255.

[16] Y. Matsui, N. Onuma, H. Nagasawa, H. Higuchi, and S. Banno, Bull. Soc. Fr. Mineral. Cristallogr. **100**, 315 (1977).

[17] N. Morimoto, Y. Nakajima, Y. Syono, S. Akimoto and Y. Matsui, Acta Crystallogr. Sect. B **31**, 1041 (1975).

[18] R. D. Shannon and C. T. Prewitt, Acta Crystallogr. Sect. B **25**, 925 (1969); **26**, 1046 (1970).

[19] See B. Mason, *Principles of Geochemistry*, 3rd ed. (Wiley, New York, 1966). Goldschmidt's rules are not found in V. M. Goldschmidt's main monographs, *"Geochemische Verteilungsgesetze der Elemente,"* I–IX (Skrifter Utgitt ar dev Norske Videnskaps-Akademii, Oslo, 1926–1937).

[20] E. Ito and Y. Matsui, Phys. Earth Planet. Inter. **9**, 344 (1974).

[21] E. Ito, Y. Matsui, K. Suito, and N. Kawai, Phys. Earth Planet. Inter. **8**, 342 (1974).

[22] E. Ito and Y. Matsui, in *High-Pressure Research: Applications in Geophysics*, M. H. Manghnani and S. Akimoto, eds. (Academic, New York, 1977), p. 193.

[23] E. Ito and Y. Matsui, Earth Planet. Sci. Lett. **38**, 443 (1978).

[24] T. Yagi, H. K. Mao, and P. M. Bell, Phys. Chem. Miner. **3**, 97 (1978).

[25] E. Ito, E. Takahashi, and Y. Matsui, Earth Planet. Sci. Lett. **67**, 238 (1984).

[26] Y. Fei, Y. Wang, and L. W. Finger, J. Geophys. Res. **101**, 11 (1996).

[27] H. K. Mao, G. Shen and R. J. Hemley, Science **278**, 2098 (1997).

[28] A. E. Ringwood, *Composition and Petrology of the Earth's Mantle* (McGraw-Hill, New York, 1975).

[29] R. J. Hemley, ed., *Ultrahigh-Pressure Mineralogy*, Vol. 37 of *Reviews in Mineralogy* (Mineralogical Society of America, Washington, D.C., 1998).

[30] H. Horiuchi, E. Ito, and D. J. Weidner, Am. Mineral. **72**, 357 (1987).

[31] A. Yeganeh-Haeri, D. J. Weidner, and E. Ito, Science **243**, 787 (1989).

[32] R. J. Hemley, R. E. Cohen, A. Yeganeh-Haeri, H. K. Mao, D. J. Weidner, and E. Ito, in *Perovskite: A Structure of Great Interest to Geophysics and Materials Science*, A. Navrotsky and D. J. Weidner, eds. (American Geophysical Union, Washington, D.C., 1989), p. 35.

[33] R. Lu, A. Hofmeister, and Y. Wang, J. Geophys. Res. **99**, 11795 (1994).

[34] C. Meade, J. A. Reffner, and E. Ito, Science **264**, 1558 (1994).

[35] R. J. Kirkpatrick, D. Howell, B. L. Phillips, X. D. Cong, E. Ito, and A. Navrotsky, Am. Mineral. **76**, 673 (1991).

[36] E. Ito, M. Akaogi, L. Topor, and A. Navrotsky, Science **249**, 1275 (1990).

[37] H. Yamada, Y. Matsui, and E. Ito, Mineral. Mag. **47**, 177 (1983).

[38] C. A. McCammon, in *Mineral Physics and Crystallography: A Handbook of Physical Constants*, Vol. 2 of AGU Reference Shelf Series, T. J. Ahrens, ed. (American Geophysical Union, Washington, D.C., 1995), p. 332.

[39] Y. Matsui, Y. Syono, and Y. Maeda, Tech. Rep. ISSP, Ser. A, No. 478 (University of Tokyo, Tokyo, Japan, 1971).

[40] E. Ito and Y. Matsui, Phys. Chem. Miner. **4**, 265 (1979).

[41] J. Linton, A. Navrotsky, and Y. Fei, Phys. Chem. Miner. **25**, 591 (1998).

[42] D. M. Teter and R. J. Hemley, Science **271**, 53 (1996).

[43] L. V. Woodcock, C. A. Angell, and P. Cheeseman, J. Chem. Phys. **65**, 1563 (1976).

[44] Y. Matsui and K. Kawamura, Nature (London) **285**, 648 (1980).

[45] Y. Matsui, K. Kawamura, and Y. Syono, in *High-Pressure Research in Geophysics*, S. Akimoto and M. H. Manghnani, eds. (Terra Scientific, Tokyo, 1982), p. 511.

[46] S. M. Rigden, T. J. Ahrens, and E. M. Stolper, Science **226**, 1071 (1984).

[47] R. J. Hemley, H. K. Mao, P. M. Bell, and B. O. Mysen, Phys. Rev. Lett. **57**, 747 (1986).

[48] Q. Williams and R. Jeanloz, Science **239**, 902 (1988).

[49] Y. Matsui and K. Kawamura, in *High-Pressure Research in Mineral Physics*, M. H. Manghnani and Y. Syono, eds. (Terra Scientific, Tokyo; American Geophysical Union, Washington, D.C., 1987), p. 305.

[50] D. M. Teter, R. J. Hemley, G. Kresse, and J. Hafner, Phys. Rev. Lett. **80**, 2145 (1998).

[51] S. Tsuneyuki, M. Tsukada, H. Aoki, and Y. Matsui, Phys. Rev. Lett. **61**, 869 (1988).

[52] S. Tsuneyuki, Y. Matsui, H. Aoki, and M. Tsukada, Nature (London) **339**, 209 (1989).

[53] R. J. Hemley, A. P. Jephcoat, H. K. Mao, L. C. Ming, and M. H. Manghnani, Nature (London) **334**, 52 (1988).

[54] K. J. Kingma, R. J. Hemley, H. K. Mao, and D. R. Veblen, Phys. Rev. Lett. **70**, 3927 (1994).

[55] R. M. Wentzcovitch, C. da Silva, J. R. Chelikowsky, and N. Binggeli, Phys. Rev. Lett. **80**, 2149 (1998).

[56] M. Yamakata and T. Yagi, Proc. Jpn. Acad. **73b**, 85 (1997).

[57] S. Tsuneyuki, H. Aoki, M. Tsukada, and Y. Matsui, Phys. Rev. Lett. **64**, 776 (1990).

[58] G. van Tendeloo, J. van Landuyt, and S. Amelinckx, Phys. Status Solidi A **33**, 723 (1976).

[59] S. Tsuneyuki, H. Aoki, and Y. Matsui in *Computer Aided Innovation of New Materials*, M. Doyama, T. Suzuki, J. Kihara and R. Yamamoto, eds. (Elsevier, New York, 1991), p. 381.

[60] J. S. Tse and D. D. Klug, J. Chem. Phys. **95**, 9167 (1991); Phys. Rev. Lett. **67**, 3559 (1991).

[61] E. R. Cowley and J. Gross, J. Chem. Phys. **95**, 8357 (1991).

[62] R. G. Della Valle and H. C. Andersen, J. Chem. Phys. **97**, 2682 (1992).

[63] N. R. Keskar and J. R. Chelikowsky, Phys. Rev. B **46**, 1 (1992).

[64] I. P. Swainson and M. T. Dove, Phys. Rev. Lett. **71**, 193 (1993).

[65] M. B. Smirnov and A. P. Mirgorodsky, Phys. Rev. Lett. **78**, 2413 (1998).

[66] L. Nagel and M. O'Keeffe, Mater. Res. Bull. **6**, 1317 (1971).

[67] K. Kusaba, Y. Syono and Y. Matsui, in *Shock Compression of Condensed Matter – 1989*, by S. C. Schmidt, J. N. Johnson, and L. W. Davidson, eds. (Elsevier, New York, 1990), p. 135.

[68] Y. Tsuchida and T. Yagi, Nature (London) **340**, 217 (1989).

[69] Y. Matsui and S. Tsuneyuki, in *High-Pressure Research: Application to Earth and Planetary Sciences*, Y. Syono and M. H. Manghnani, eds. (Terra Scientific, Tokyo; American Geophysical Union, Washington, D.C., 1992), p. 433.

[70] K. J. Kingma, R. E. Cohen, R. J. Hemley, and H. K. Mao, Nature (London) **374**, 243 (1995).

[71] H. K. Mao and R. J. Hemley, Philos. Trans. R. Soc. London A **354**, 1315 (1996).

[72] D. Andrault, G. Fiquet, F. Guyot, and M. Hanfland, Science **282**, 720 (1998).

[73] K. T. Park, K. Terakura, and Y. Matsui, Nature (London) **336**, 670 (1988).

[74] R. E. Cohen, in *High-Pressure Research: Application to Earth and Planetary Sciences*, Y. Syono and M. H. Manghnani, eds. (Terra Scientific, Tokyo; American Geophysical Union, Washington, D.C., 1992), p. 425.

[75] I. Kushiro, J. Geophys. Res. **76**, 619 (1976).

[76] S. Tsuneyuki and Y. Matsui, Phys. Rev. Lett. **74**, 3197 (1995).

[77] Y. Matsui and M. Matsui, in *Structural and Magnetic Phase Transitions in Minerals*, S. Ghose, J. M. D. Coey, and E. Salje, eds. (Springer-Verlag, New York, 1988), p. 129.

[78] T. Engel, T. Matsumoto, G. Steinman, and H. Wondratschek, Z. Krist., Suppl. No. 1 (Ordenburg, Munich, 1984).

[79] A. K. Singh, H. K. Mao, J. F. Shu, and R. J. Hemley, Phys. Rev. Lett. **80**, 2157 (1998).

[80] I. I. Mazin, Y. Fei, R. Downs, and R. E. Cohen, Am. Mineral. **83**, 451 (1998).

[81] Y. Noguchi, K. Kusaba, K. Fukuoka, and Y. Syono, Geophys. Res. Lett. **23**, 1469 (1996).

[82] T. Kondo, T. Yagi, Y. Syono, T. Kikegawa, and O. Shimomura, Rev. High Pressure Sci. Technol. **7**, 148 (1998).

[83] Z. Fang, K. Terakura, H. Sawada, T. Miyazaki, and I. Solovyev, Phys. Rev. Lett. **81**, 1027 (1998).

[84] Y. Noguchi, T. Atou, T. Kondo, T. Yagi, and Y. Syono, Jpn. J. Appl. Phys. **38**, L7 (1999).

[85] R. E. Cohen, I. I. Mazin, and D. G. Isaak, Science **275**, 654 (1997).

[86] O. Mishima, L. D. Calvert, and E. Whalley, Nature (London) **310**, 393 (1984).

Part II

Advances in Theoretical and Experimental Techniques

Chapter 2.1

Density Functional Theory in Mineral Physics

Lars Stixrude

Density functional theory of the electronic structure of condensed matter is reviewed with an emphasis on its application to geophysics. The review is placed within the context of our attempts to understand planetary interiors and the unique features of these regions that lead us to use band-structure theory. The foundations of density functional theory are briefly discussed, as are its scope and limitations. Special attention is paid to commonly used approximations of the theory, including those of the exchange-correlation potential and the structure of the electronic core. Some of the important computational methods are reviewed, including the linearized augmented plane-wave method and the plane-wave pseudopotential method. Examples of applications of density functional theory to the study of the equation of state, crystalline structure, phase stability, and elasticity of earth materials are described. Some critical areas for further development are identified.

2.1.1 Introduction

Planetary interiors represent a unique environment in the universe in which the behavior of condensed matter presents a considerable challenge. The nature and the evolution of planetary interiors, even that of our own Earth, are complex, poorly understood, and difficult to predict with current theoretical understanding. In contrast, we have a much better understanding in many ways of the interiors of distant stars. For example, we are

H. Aoki et al. (eds), *Physics Meets Mineralogy* © 2000 Cambridge University Press.

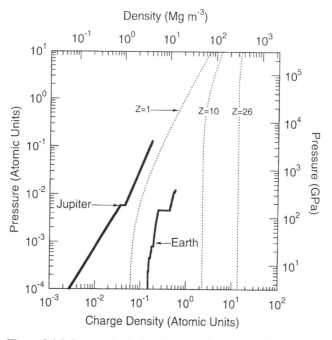

Figure 2.1.1 Pressure in the interior of Jupiter [3] and Earth [4] as a function of mass density (top) and charge density (bottom). The charge density has been calculated from the observed mass density with the assumption that the number of electrons is one-half the number of nucleons. Planetary structures are compared with limiting high-density equation of state (2.1.1) for three values of the atomic number Z.

able to calculate the structures and evolutionary history of stars with some certainty, an exercise that is not yet possible for the Earth.

There are sound physical reasons for this apparent anomaly. The stellar interior is extraordinarily simple from a condensed-matter physicist's point of view. Because the pressure is so high, the electrons obey an almost trivial limiting behavior, the uniform electron gas [1]. The fundamental reason is that the kinetic energy of electrons increases with the charge density ρ as $\rho^{2/3}$, whereas the potential energy binding the electrons to the nuclei increases only as $\rho^{1/3}$; the kinetic energy dominates at high pressure and the electrons become unbound (see Ref. [2] for an extended discussion).

The contrast with planetary interiors can be illustrated when the pressure is expressed in terms of atomic units, 1 atomic unit (29.4 TPa) being comparable with the pressure required for complete ionization and the formation of a degenerate electron gas. The structures of planets are such that pressures are much less than unity (Fig. 2.1.1). The behavior of planetary materials will be far from plasmalike and therefore much more complex.

A more useful pressure scale in the context of planetary interiors is formed from an energy typical of the spacing between electronic bands (1 eV) and a volume typically

occupied by a valence electron in a mineral (25 bohr$^3 \approx 4$ Å^3). This pressure scale
($\sim$100 GPa) is characteristic of the Earth's interior and the bulk modulus of typical
earth-forming constituents. We must then expect to find in planetary interiors not only
significant compression and phase transitions, but also electronic transitions (e.g.,
insulator to metal) and substantial changes in the mechanisms of bonding, all of which
complicate our picture of planetary structure and evolution.

From the computational point of view, the recognition that planetary interiors are
characterized by complex multiphase behavior places tremendous demands on the
required accuracy of theoretical methods. They must be general, applicable to essentially all classes of elements, and must not make any assumptions regarding the
nature of bonding. Energies and volumes must be accurate to well within typical heats
and volumes of solid–solid transformations. In the case of the Earth, these requirements rule out essentially all weak screening approaches that treat condensed matter
in perturbative fashion, beginning with the free-electron gas. Indeed, early calculations for which one such approach was used [5] incorrectly predicted that iron is
substantially lighter than the Earth's core [6]. In the case of Jupiter, weak screening
approximations have proved fruitful for investigations of planetary structure but are
unlikely to successfully capture a priori important details such as the structure of the
molecular-to-atomic transition [1, 3].

The following sections review in some detail modern methods of first-principles
theory as they have been applied in the geophysics literature. The fundamental approximations on which current implementations of density functional theory are based are
discussed. I then discuss applications of the theory to the derivation of observable
quantities of geophysical interest. In Section 2.1.3 I review computational methods
for solving the equations. Finally, I discuss some important unsolved problems in the
behavior of earth materials and possible future directions.

2.1.2 **Theory**

From the point of view of any first-principles theory, solids are composed of nuclei
and electrons; atoms and ions are constructs that play no primary role. This departure
from our usual way of thinking about minerals and solids is essential and has the
following important consequences. We may expect our theory to be equally applicable to the entire range of conditions encountered in planets (and even stars), the entire
range of bonding environments encompassed by this enormous range of pressures and
temperatures, and all elements of the periodic table.

To illustrate this way of thinking about solids and to introduce some important
concepts, consider first the properties of the simplest system, the uniform electron
gas with embedded nuclei. The total energy consists of the kinetic energy of the electrons and three distinct contributions to the potential energy: (1) Coulomb interactions
among nuclei and electrons, (2) electron exchange, and (3) electron correlation. The

first contribution is straightforward and involves only sums over point charges and/or integrals over the (uniform) electronic charge density.

Exchange and correlation account for local deviations from uniform charge that arise from the tendency of electrons to avoid each other. Correlation accounts for the mutual Coulomb repulsion, whereas exchange embodies the Pauli exclusion principle and the resulting tendency of electrons of the same spin to avoid each other. The net effect is that each electron can be thought of as digging a hole of reduced charge density about itself. Certain properties of the exchange-correlation hole are well understood; it is known for instance that its integrated charge must exactly balance that of the electron. Exchange and correlation reduce the total energy by reducing the Coulomb repulsion between electrons.

The total energy of our simple system is readily evaluated as a function of charge density; the equation of state then follows from differentiation. With the assumption that the nuclei are in a close-packed arrangement and with only the leading-order high-density contributions to exchange and correlation included, the equation of state is [7]

$$P = 0.176 r_s^{-5}[1 - (0.407 Z^{2/3} + 0.207) r_s], \tag{2.1.1}$$

where P is the static (athermal) pressure, Z is the nuclear charge, and the Wigner–Seitz radius,

$$r_s = \left(\frac{3}{4\pi\rho}\right)^{1/3}, \tag{2.1.2}$$

is a measure of the average spacing between electrons. The first term in Eq. (2.1.1) is the kinetic contribution, the second is due to the Coulomb attraction of the nuclei for the electrons and mutual repulsion of the electrons, and the third is due to exchange. Correlation, which is smaller than exchange at high density, has been neglected, as has the mutual Coulomb repulsion of the nuclei.

Comparisons with the structure of planetary interiors reveal some fundamentally important aspects of planetary matter (Fig. 2.1.1). First, the net Coulomb attraction provided by the nuclei plays an essential role at planetary densities – different mean nuclear charges account to first order for the difference in mean charge (and mass) density between Jupiter and Earth. Second, screening also has a first-order effect on the equation of state, accounting for the much lower densities of planets at a given pressure than those predicted by Eq. (2.1.1). In planetary matter, the charge density is substantially enhanced in the vicinity of the nucleus, reducing the ability of the point charges to attract the remaining (valence) electrons. Screening is weaker in the case of Jupiter because it contains dominantly lighter elements and because the pressures are much greater. Nevertheless, for all the planets screening is so strong that it must be accounted for. In the case of the terrestrial planets, the charge density near the nuclei is so much higher than in the interstitial region that this difference plays a central role in the design of modern computational methods.

2.1.2.1 Density Functional Theory

We turn now from simple to real systems and at the same time from analytically expressible results to necessarily elaborate computations. Although the electronic structure will be nontrivial, we retain the charge density as a central concept. This is appealing because this quantity is experimentally observable; it is precisely what is measured by an x-ray-diffraction experiment.

The general problem we are faced with in a nonuniform, nondegenerate electron system is formidable. Given a periodic potential set by the positions of the nuclei, we must solve the Schrödinger equation for the total wave function $\Psi(r_1, r_2, \ldots, r_N)$ of a system of N interacting electrons, where N is of the order of Avogadro's number. Density functional theory [8, 9] is a powerful and, in principle, exact method of dealing with this problem in a tractable way (see Ref. [10] for reviews).

The essence of this theory is the proof that the ground-state properties of a material are a unique functional of the charge density $\rho(\mathbf{r})$. Among these properties are the ground-state total energy,

$$E = T + U[\rho(\mathbf{r})] + E_{xc}[\rho(\mathbf{r})], \qquad (2.1.3)$$

and its derivatives (pressure, elastic constants, etc.), where T is the kinetic energy of a system of noninteracting electrons with the same charge density as that of the interacting system, U is the electrostatic (Coulomb) energy, including the electrostatic interaction between the nuclei, and E_{xc} is the exchange-correlation energy. A variational principle leads to a set of single-particle, Schrödinger-like, Kohn–Sham equations,

$$[-\nabla^2 + V_{KS}]\psi_i = \epsilon_i \psi_i, \qquad (2.1.4)$$

where ψ_i is now the wave function of a single electron, ϵ_i is the corresponding eigenvalue, and the effective potential is

$$V_{KS}[\rho(\mathbf{r})] = \sum_{i=1}^{N} \frac{2Z_i}{|\mathbf{r} - \mathbf{R}_i|} + \int \frac{2\rho(\mathbf{r}')}{|\mathbf{r} - \mathbf{r}'|}\, d\mathbf{r}' + V_{xc}[\rho(\mathbf{r})], \qquad (2.1.5)$$

where the first two terms are Coulomb potentials that are due to the nuclei and the other electrons, respectively, the last is the exchange-correlation potential, and the units are Rydberg atomic units: $\hbar^2/2m = 1$, $e^2 = 2$, energy in Ry and length is in bohrs.

The power of density functional theory is that it allows us to calculate, in principle, the exact many-body total energy of a system from a set of single-particle equations. The solution to the Kohn–Sham equations is that of the set of coupled generalized eigenvalue equations:

$$H_{ij}(\mathbf{k})\psi_j(\mathbf{r}, \mathbf{k}) = \epsilon_j(\mathbf{k})O_{ij}(\mathbf{k})\psi_j(\mathbf{r}, \mathbf{k}),$$ (2.1.6)

$$H_{ij}(\mathbf{k}) = \int \psi_i^*(\mathbf{r}, \mathbf{k})(-\nabla^2 + V_{KS})\psi_j(\mathbf{r}, \mathbf{k})\, d\mathbf{r},$$ (2.1.7)

$$O_{ij}(\mathbf{k}) = \int \psi_i^*(\mathbf{r}, \mathbf{k})\psi_j(\mathbf{r}, \mathbf{k})\, d\mathbf{r},$$ (2.1.8)

where $\mathbf{H}$ and $\mathbf{O}$ are the Hamiltonian and the overlap matrices, respectively, and $\mathbf{k}$ is a vector in reciprocal space. Because the Kohn–Sham potential is a functional of the charge density, the equations must be solved self-consistently together with the definition of the charge density in terms of the wave functions:

$$\rho(\mathbf{r}) = \int \sum_i n[E_F - \epsilon_i(\mathbf{k})]\psi_i^*(\mathbf{r}, \mathbf{k})\psi_i(\mathbf{r}, \mathbf{k})\, d\mathbf{k},$$ (2.1.9)

where n is the occupation number and E_F is the Fermi energy.

2.1.2.2 Approximations

Exchange-Correlation Potential

The Kohn–Sham equations are exact. That only approximate solutions have been possible to date is a limitation imposed only by our current ignorance of the exact exchange-correlation functional. If the exact exchange-correlation functional were known, we would be able to obtain exact solutions. All other terms in the Kohn–Sham equations are straightforward and readily evaluated.

The exchange-correlation functional is known precisely only for simple systems such as the uniform electron gas (Fig. 2.1.2). The exchange portion is known analytically, as are the leading-order contributions to correlation in the limit of high density [12]:

$$V_{xc} = \frac{\partial}{\partial \rho}(\rho E_{xc}),$$

$$E_{xc} = -\frac{3}{4\pi}\left(\frac{9\pi}{4}\right)^{1/3} r_s^{-1} + A \ln r_s + B,$$ (2.1.10)

where the first contribution to E_{xc} is exchange and the constants $A = (1 - \ln 2)/\pi$ and $B = -0.046644$ [13]. At other densities, accurate values of the exchange-correlation potential are known from quantum Monte Carlo calculations [14], which have been represented in a parametric form that obeys the high-density limiting behavior [Eqs. (2.1.10)] [15].

The precision of modern condensed-matter computations has made the accurate representation of the exchange-correlation potential of the uniform electron gas an important issue. In this context, it is important to be aware that approximate representations of V_{xc} have appeared frequently in the geophysical literature and are still

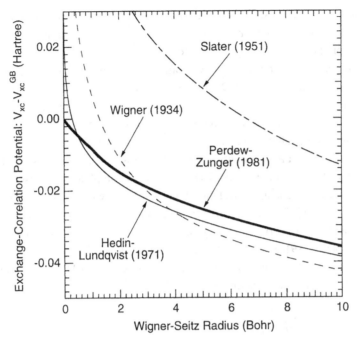

Figure 2.1.2 Difference between the exchange-correlation potential and its high-density limit [Eqs. (2.1.10)] in (bold curve) the local-density approximation and (other curves) three commonly used approximations to the local-density approximation. For the Slater [11] result, I have used $\alpha = 2/3$, which yields the pure exchange potential.

in use. Of these, the Hedin–Lundqvist [16] expression is most similar to the accurate Perdew–Zunger parameterization, that of Wigner [17] the least. The Wigner approximation shows a much stronger dependence on density than the accurate potential and leads to significant errors in density functional computations for solids. None of the commonly used approximate expressions satisfy the correct high-density limiting behavior [Eqs. (2.1.10)].

The charge density in real materials is highly nonuniform, and the exchange-correlation potential cannot be evaluated. Fortunately, simple approximations of the exchange-correlation potential have been very successful. The local-density approximation (LDA) is based on the uniform electron gas, taking into account nonuniformity to lowest order by setting V_{xc} at every point in the crystal to that of the uniform electron gas with a density equal to the local charge density [10].

The success of the LDA can be understood at a fundamental level in terms of the satisfaction of exact sum rules for the exchange-correlation hole [18]. For example, the LDA correctly predicts an exchange-correlation hole of unit charge. Ultimately, the appropriateness of the LDA can be judged only by comparison of its predictions with observation. Here the LDA has been remarkably successful. The LDA has been shown to yield excellent agreement with experiment for a wide variety of insulators, metals, and semiconductors and for bulk, surface, and defect properties. The LDA shows some general failures, such as a tendency to underpredict bandgaps, and, from a geophysical point of view, more serious failures for certain materials. For example, LDA fails to predict the correct ground state of iron.

These failures have prompted the development of new exchange-correlation functionals. One shortcoming of the LDA may be its local character, that is, its inability to distinguish among electrons of different angular momenta or energy. Generalized gradient approximations (GGAs) partially remedy this by including a dependence on local charge-density gradients in addition to the density itself. Some care must be taken in constructing gradient approximations; a straightforward Taylor series expansion in the charge-density gradient about the LDA result fails completely because it violates the sum rule for the exchange-correlation hole. The most widely used GGA satisfies the sum rules exactly [13]. This approximation and its forerunners have been shown to yield agreement with experimental data that is usually as good as the LDA and often substantially better [19]. For example, GGAs correctly predict the bcc phase as the ground state of iron [20]. The relationship of the GGA to the LDA can be expressed in terms of the enhancement factor,

$$F(r_s, s) = \frac{V_{xc}^{GGA}(r_s, s)}{V_x(r_s)}, \tag{2.1.11}$$

where V_x is the exchange potential and F is a function of the charge density and the nondimensional charge-density gradient, $s = (24\pi^2)^{-1/3}|\nabla\rho/\rho^{4/3}|$ (Fig. 2.1.3).

Frozen-Core Approximation

The physical motivation for this approximation is the observation that only the valence electrons participate in bonding and in the response of the crystal to most perturbations of interest. Unless the perturbation is of very high energy (comparable with the binding energy of the core states), the tightly bound core states remain essentially unchanged. The frozen-core approximation is satisfied to a high degree of accuracy for many applications, for example in the case of finite strains of magnitudes typically encountered in the Earth's interior.

Within this approximation, the charge density of the core electrons is just that of the free atom, which can be found readily. We then need solve for only the valence electrons in Eq. (2.1.4), often a considerable computational advantage. An important technical point is that, although in many cases the choice is obvious, there is no fundamentally sound way to decide a priori which electrons are core and which are valence. Some care is required; for example, the $3p$ electrons in iron must be treated as valence electrons as they are found to deform substantially at pressures comparable with those in the Earth's core [21].

Pseudopotential Approximation

This approximation goes one step beyond the frozen-core approximation. It replaces the nucleus and the core electrons with a simpler object, the pseudopotential, which

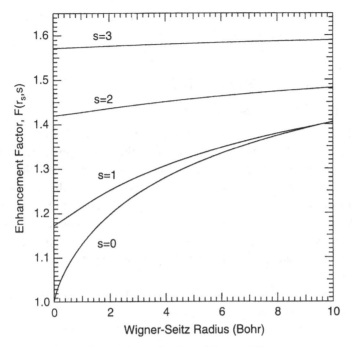

Figure 2.1.3 Effect of charge-density gradients on the exchange-correlation potential according to the GGA. The enhancement factor for zero gradient ($s = 0$) reflects the contribution of correlation to V_{xc}.

has the same scattering properties [22]. The pseudopotential is chosen such that the valence wave function in the free atom is the same as the all-electron solution beyond some cutoff radius, but nodeless within this radius. The advantages of the pseudopotential method are that (1) spatial variations in the pseudopotential are much less rapid than the bare Coulomb potential of the nucleus and (2) we need solve for only the (pseudo-) wave functions of the valence electrons, which show much less rapid spatial variation than the core electrons or the valence electrons in the core region. This means that in the solution of the Kohn–Sham equations, potential and charge density can be represented by a particularly simple, complete, and orthogonal set of basis functions (plane waves) of manageable size; with this basis set, evaluation of total energies, stresses, and forces acting on the atoms is particularly efficient.

The pseudopotential is an approximation to the potential that the valence electrons "see" and its construction is nonunique; different pseudopotentials may yield significantly different predictions of bulk properties. Several different methods for constructing pseudopotentials have been developed [23–25]. Care must be taken to demonstrate the transferability of the pseudopotentials generated by a particular method and to compare with all-electron calculations where these are available. When these conditions are met, the error that is due to the pseudopotential is generally small (a few percent in volume for earth materials).

2.1.2.3 Derivation of Observables

Total Energy and Band Structure

For a given arrangement of nuclei (crystal structure) we may solve the equations of density functional theory under one or more of the above approximations to determine the total energy, charge density, and the quasiparticle eigenvalue spectrum (electronic band structure). By examining the dependence of the total energy on perturbations to the volume V or shape of the crystal (described by the deviatoric strain tensor ϵ'_{ij}) or to the positions of the atoms, we can, in principle, deduce the Helmholtz free energy F as a function of V, ϵ'_{ij}, and T. For example, the pressure and the equation of state are simply given by the variation of the total energy with volume.

We may determine the elastic constants from total-energy calculations. For small deviatoric strains under hydrostatic stress [26],

$$F(V, \epsilon'_{ij}, T) = F_0(V) + F_{\text{TH}}(V, T) + \frac{1}{2} c_{ijkl}(V, T) \epsilon'_{ij} \epsilon'_{kl}, \qquad (2.1.12)$$

where F_0 is the static (zero-temperature) contribution, F_{TH} is due to the thermal excitation of electrons and phonons, and c_{ijkl} is the elastic-constant tensor. This equation shows that combinations of elastic constants are related to the difference in total energy between a strained and an unstrained lattice.

It is possible in principle to calculate thermal contributions to the thermodynamic and thermoelastic properties of crystals. Calculating thermal properties is much more difficult than calculating static properties, however. The reason is simple: The atomic vibrations induced by finite temperature break the symmetry of the crystal so that it is now periodic in only a time-averaged sense. In the context of total-energy calculations, our task is then to evaluate the partition function, an integral over all atomic configurations realized by a crystal at high temperature. This is essentially impossible. More efficient ways of evaluating thermal free energies from first principles are required. Some future directions are indicated in Section 2.1.5.

Forces, Stresses, and Structures

The Hellman–Feynman theorem allows us to calculate first derivatives of the total energy directly in terms of the ground-state wave functions. The application of this theorem allows us to determine the forces acting on every atom and the stresses acting on the lattice.

This is important for two related reasons. First, it allows us to determine ground-state crystal structures very effectively. This has become possible only recently for relatively complex structures such as $MgSiO_3$ perovskite [27]. The key innovation has been the development of a structural optimization strategy based on a pseudo-Lagrangian that treats the components of the strain tensor and the atomic positions as dynamical variables [28]. The optimization is performed at constant pressure. At each step of the dynamical trajectory, the Hellman–Feynman forces and stresses [29] acting on the nuclei and the lattice parameters, respectively, are evaluated and used to generate the next configuration. The optimization is complete when the forces on the nuclei vanish and the stress is hydrostatic and balances the applied pressure.

Second, once the equilibrium structure at a given pressure is determined, we can calculate the elastic constants. We do this in a straightforward way by applying a deviatoric strain to the lattice and calculating the resulting stress tensor. The elastic constant c_{ijkl} is then given by the ratio of stress σ_{ij} to strain ϵ_{ij}:

$$\sigma_{ij} = c_{ijkl}\epsilon_{kl}. \tag{2.1.13}$$

Care must be taken to reoptimize the positions of the atoms in each strained configuration, as vibrational modes typically couple with lattice strains in silicate structures.

2.1.3 Computation

2.1.3.1 Methods

First-principles methods solve Eqs. (2.1.6)–(2.1.8) by expanding the wave functions in a basis

$$\psi_i(\mathbf{r}, \mathbf{k}) = \sum_{j=1}^{N} c_{ij}(\mathbf{k})\phi_j(\mathbf{r}, \mathbf{k}), \tag{2.1.14}$$

where N is the number of basis functions ϕ_j and c_{ij} are the coefficients to be determined by solution of the Kohn–Sham equations.

The linearized augmented plane-wave (LAPW) method is the current state of the art in density functional theory computations. It makes no essential approximations beyond that to the exchange-correlation functional, allowing us to solve routinely for all electrons, both core and valence. For example, it makes no approximations of the shape of the charge density or potential. The accurate representation of the potential and the core states means that the LAPW method is equally applicable to all elements of the periodic table and over the entire range of densities of interest in planetary or astrophysical studies.

The LAPW method differs from its forerunner, the APW method, in that the APW method assumes a spherically symmetric potential near the nuclei [30, 31]. Because of its precise representation of the potential, the LAPW method is sometimes referred to as the full-potential LAPW (FLAPW). LAPW shares the ability to represent precisely the full potential and the charge density with the full-potential linear muffin-tin orbital

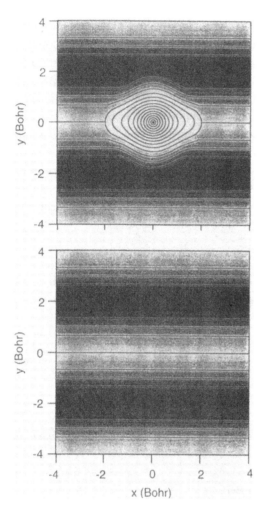

Figure 2.1.4 (Top) One LAPW basis function in the vicinity of a hydrogen nucleus located at the origin: $G = (0, \pi/2, 0)$, $l_{max} = 6$, $E_l(\mathrm{Ry}) = -(l+1)^{-2}$; (bottom) the plane wave $G = (0, \pi/2, 0)$.

(FP-LMTO) method [32]. The FP-LMTO method is very similar to the LAPW method in its capabilities and level of accuracy, differing primarily in the details of the basis functions.

The accuracy and the flexibility of the LAPW method are derived from its basis, which explicitly treats the first-order partitioning of space into near-nucleus regions, where the charge density and its spatial variability are large, and interstitial regions, where the charge density varies more slowly (Fig. 2.1.4) [33–35]. These two regions are delimited by the construction of so-called muffin-tin spheres of radius R_{MT}^{α} centered on each nucleus α. A dual-basis set is constructed, consisting of plane waves in the interstitial regions that are matched continuously to more rapidly varying functions inside the spheres. Within the muffin-tin spheres ($r' < R_{\mathrm{MT}}^{\alpha}$),

$$\phi^{\mathbf{k+G}}(\mathbf{r}) = \left[a_{lm}^{\alpha} u_l \left(E_l^{\alpha}, r' \right) + b_{lm}^{\alpha} \dot{u}_l \left(E_l^{\alpha}, r' \right) \right] Y_{lm}(\mathbf{r}'/r'), \qquad (2.1.15)$$

and for $r' > R_{MT}^\alpha$,

$$\phi^{k+G}(r) = \exp[i(k+G) \cdot r], \qquad\qquad (2.1.16)$$

where $r' = r - R_\alpha$, R_α are the positions of the nuclei, G is a wave vector, u_l and $\dot{u}_l$ are the solution to the radial part of the Schrödinger equation and its energy derivative, respectively, for the spherically symmetric portion of the potential inside the muffin-tin sphere at energy E_l, and the coefficients a and b are determined by requiring continuity of the basis function and its first radial derivative on the muffin-tin sphere.

With this basis set, all-electron calculations for silicates or transition metals typically require of the order of 100 basis functions per atom. The primary disadvantage of the LAPW method is that the complexity of the basis functions makes it relatively intensive computationally. In practice, this limits the size of the system that can be studied. Even so, LAPW computations for structures as complex as that of $MgSiO_3$ perovskite (20 atoms in the unit cell) have been performed [36].

Basis sets consisting solely of plane waves, because of their analytical simplicity, have some advantages over the LAPW basis. However, all-electron calculations are virtually impossible with a plane-wave basis set; the number of basis functions needed to represent the rapid spatial oscillations of the core region is much too large to be practical. For this reason, the plane-wave basis is generally linked in practice to the pseudopotential approximation, in which the Fourier content of charge density and potential are limited by design.

2.1.3.2 Convergence

There are two primary convergence issues: the size of the basis and the integrations over reciprocal space [e.g., Eq. (2.1.9)]. Both basis sets have the property of smooth convergence; this means that convergence of the computations is readily assessed; quantities of interest vary smoothly as the basis-set size is increased. In the LAPW method, the size of the basis set is described by the dimensionless quantity $R_{MT}K_{max}$, where K_{max} is the maximum wave number of the plane waves included in the basis set. In the pseudopotential method, the size of the basis is set by the maximum kinetic energy of the plane waves $E_{cut} = K_{max}^2$ in atomic units. Typical values for computations of silicates are $R_{MT}K_{max} = 7$ and $E_{cut} = 40-80$ Ry, depending on the pseudopotential that is used.

Sampling of the Brillouin zone is treated with the special-points method, which has been shown to yield rapid convergence [37]. For insulators, only a few points (1–10) are typically needed to achieve fully converged total energies; metals require denser sampling because of the often complex structure of the Fermi surface. The special-points method constructs a uniform grid of k points of specified resolution in the first Brillouin zone. The resulting set of k points is divided into subgroups (stars) of symmetrically equivalent points. The Kohn–Sham equations are solved for only one

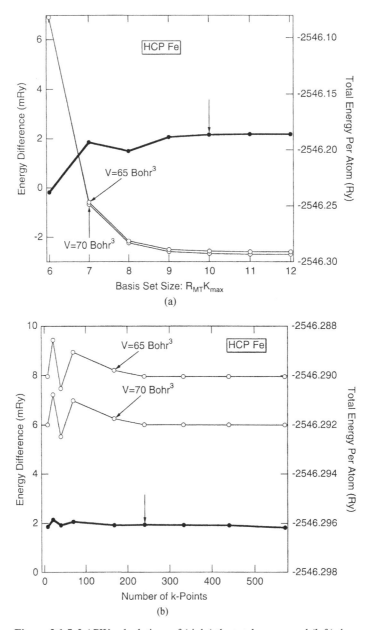

Figure 2.1.5 LAPW calculations of (right) the total energy and (left) the difference in total energy of hcp iron at two different volumes as a function of (a) the size of the basis and (b) the number of k points in reciprocal space.

member of each star, and the wave functions at other points in the star are reconstructed with the appropriate symmetry operations, weighting the contribution of each star by its degeneracy.

The results of convergence tests typical of a transition metal are shown in Fig. 2.1.5. Convergence of the total energy to within a few tenths of a millirydberg ($\sim$50 parts per 10^9 in the case of iron) are routinely achieved. This level of convergence is essential for making accurate predictions – typical solid–solid heats of transformation are of the order of a few millirydbergs. A general feature of convergence in either LAPW or pseudopotential methods is that energy differences – between two volumes of one structure or between two different structures – converge much faster than total energies.

2.1.4 Some Applications

2.1.4.1 Equation of State

The error that is due to the LDA can be evaluated by a comparison of the results of LAPW calculations, which make no further essential approximations beyond the LDA, with experiment (Fig. 2.1.6). In investigations of silicates and oxides of geophysical interest, it has been found that errors in volumes are typically 1%–4%, with theoretical volumes being uniformly smaller than experimental [36, 39–41]. Part of this small difference is due to the higher temperatures of experiments (300 K) compared with the athermal calculations. This is a highly satisfactory level of agreement for a theory that is parameter free and independent of experiment. All-electron LDA computations of transition metals show errors of similar magnitude in the zero-pressure volume; for the 3d and the 4d metals, the calculations uniformly underestimate the experimental volumes, whereas for the 5d metals, the situation is more complex [42]. For the heaviest materials, additional effects such as spin-orbit coupling, often neglected in computations, may become important and contribute to the discrepancy between theory and experiment.

The GGA improves the agreement between theory and experimental equations of state for most materials, including the 3d transition metals. In the case of iron, LAPW and FP-LMTO calculations differ from the experimentally measured room-temperature equation of state by 3% at zero pressure and by less than 1% at core pressures [21, 32, 43]; agreement with high-temperature Hugoniot data is equally good [44].

Pseudopotential calculations make additional approximations that lead to additional errors (Fig. 2.1.6). These are small in magnitude and comparable in size with the LDA error. At this level of detail, different pseudopotentials yield results that differ from each other and from the all-electron LDA result from LAPW. Because the pseudopotential method is nearly as accurate as the much more elaborate LAPW method, it is often

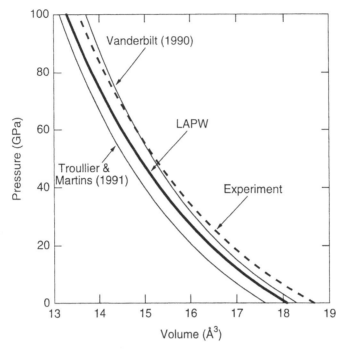

Figure 2.1.6 Equation of state of MgO from (dashed curve) experiment [38], (bold curve) all-electron LDA (LAPW) calculations, (thin curves) pseudopotential calculations based on the indicated potentials.

preferred for many applications, as its computational advantages allow much larger and more complex systems to be studied.

2.1.4.2 Structural Optimization

The structure and compression mechanisms of a number of complex silicates have been studied with density functional theory at high pressure, including $MgSiO_3$ enstatite and perovskite, Mg_2SiO_4 forsterite, and SiO_2 in the quartz, stishovite, $CaCl_2$, and columbite structures [27, 40, 41, 45–50]. These investigations (1) provide an important test of the approximations on which first-principles methods are based (2) illustrate in detail often not obtainable by experiment the nature of compression mechanisms, and (3) provide a sensitive test of the hypothesis that some minerals undergo high-order symmetry-invariant phase transformations.

When the method of Wentzcovitch [28] is used, the optimization of complex crystal structures such as forsterite is an efficient procedure. Typically, of the order of 10–20 iterations are required for full structural convergence in this mineral with 7 internal degrees of freedom and three lattice parameters [48]. The results of first-principles calculations show that volume compression is primarily accommodated by nearly isotropic compression of the MgO_6 octahedra, which are much softer than the SiO_4

tetrahedra. Changes in bond lengths and angles are smooth and monotonic, unlike the results of x-ray diffraction experiments. In particular, density functional theory does not support the proposal that compression mechanisms change suddenly at pressures near 9 GPa. Brodholt et al. [51], by using different pseudopotentials, also found no evidence for sudden changes in compression mechanisms.

2.1.4.3 Phase Stability

In many ways, phase stability provides the most stringent test of first-principles methods. The reason is that we are comparing total energies computed for two different structures with different basis sets and Brillouin zones at the level of heats of transformations, generally a miniscule fraction of the total energy (less than 1 part per 10^6).

First-principles LDA results for transformations in oxides and silicates have shown excellent agreement with experiment [39–41, 50, 52–54]. In the case of transition metals, the form of the exchange-correlation potential is critical. LDA fails to predict the correct ground state of iron, finding incorrectly that the hcp phase has a lower total energy than the bcc. The GGA correctly recovers the bcc ground state (Fig. 2.1.7). Moreover, it accurately predicts the pressure of the phase transition from bcc to hcp

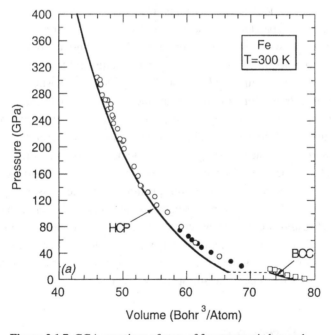

Figure 2.1.7 GGA equations of state of ferromagnetic bcc and nonmagnetic hcp (solid curve) phases of iron compared with the experimental data of Jephcoat et al. [55] (bcc, open squares; hcp, filled circles) and Mao et al. [56] (hcp, open circles). The dashed line indicates the predicted phase-transition pressure of 11 GPa. (From Ref. [44].)

near 11 GPa. This is an important result because the energetics are particularly subtle in the case of this transition as it involves a ferromagnetic and a nonmagnetic phase.

2.1.4.4 Elastic Constants

The elastic constants are of central importance geophysically because they govern the passage of seismic waves, our primary source of information on the structure of the Earth's interior. Despite their importance, density functional calculations of the elastic constants of earth materials have appeared only recently. The key development has been that of an efficient structural optimization scheme [27] and calculation of stresses from the Hellman–Feynman theorem. Elastic constants are determined by calculation of the stress generated by deviatoric strains applied to the equilibrium structure. It is straightforward to demonstrate that we are within the linear regime by performing the calculation at a variety of values of the strain magnitude and extrapolating to the limit of zero strain [54]. These calculations show that strains of the order of 1% are appropriate for silicates and oxides. By carefully choosing the symmetry of the applied strain, we can calculate all elements of the elastic-constant tensor with a small number of different strains. For example, the three elastic constants of a cubic mineral can be determined from a single strain; four different strains have been used for orthorhombic minerals (nine independent elastic constants) [57].

The full elastic-constant tensors of a number of silicates and oxides have been determined with the plane-wave pseudopotential method, including that of MgO periclase, $MgSiO_3$ perovskite, Mg_2SiO_4 forsterite and ringwoodite, and SiO_2 in the stishovite, $CaCl_2$, and columbite structures [27, 47, 54, 57–59]. Once the elastic-constant tensor is determined, it is straightforward to calculate the elastic-wave (seismic) velocities in any direction, the elastic anisotropy, and the seismic-wave velocities of isotropic aggregates. Results for $MgSiO_3$ perovskite [47] (Fig. 2.1.8) show several geophysically important features: (1) athermal longitudinal and shear-wave velocities of isotropic aggregates of this mineral nearly parallel those of the lower mantle and are uniformly larger, (2) perovskite remains highly anisotropic throughout the pressure regime of the lower mantle, and (3) its anisotropy changes qualitatively with increasing pressure. The theoretical results are consistent with the hypothesis that perovskite is the most abundant mineral in the lower mantle [66]; they further require that it exist in approximately randomly oriented aggregates, except possibly in the D'' region, where it may partially account for the seismic anisotropy observed there.

2.1.5   Future Directions

This review has only given an indication of the realm of geophysical application of density functional theory. I have reviewed only a subset of the important calculations that have been performed and have not touched on possible future directions.

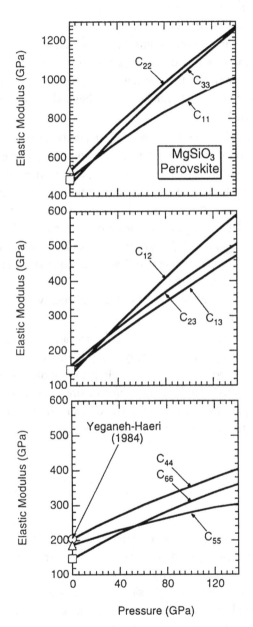

Figure 2.1.8 Elastic constants of MgSiO₃ perovskite according to theory [47] (lines) and experiment [60] symbols: c_{11}, c_{12}, c_{44} (○); c_{22}, c_{13}, c_{55} (△); c_{33}, c_{23}, c_{66} (□).

Other applications of density functional theory include the following:

1. The investigation of magnetism in metal alloys, silicates, and oxides. Studies of transition-metal compounds have shown that earth materials display remarkably rich magnetic behavior that has important implications for our understanding of bonding, equations of state, elasticity, and phase stability [61]. This is a challenging area that requires new theoretical developments because transition-metal oxides such as hematite and magnetite are Mott insulators that owe their electronic properties to strong localization of the d electrons. Mott-insulating behavior is fundamentally beyond the scope of band-structure theory as outlined here.

2. First-principles computation of phonon spectra [62, 63]. In addition to making contact with experimental observation of zone-center vibrational frequencies, these predictions allow us to investigate phase stability and, to the extent that thermal properties are quasi harmonic, high-temperature properties [67]. The computation in polar substances is subtle and necessarily involves not only the calculation of force constants, but also that of the dielectric constant and Born effective charge tensors so that coupling to the macroscopic field at zone center is properly accounted for. Efficient calculation involves a technique known as linear response, in which computation of the linear response of the charge density to a perturbation allows second derivatives of the total energy to be computed directly [64].

3. First-principles molecular dynamics simulations at high temperatures. These have now been performed for the first time in a study of liquid iron at core conditions [65]. The method is general and relies on only the principles of the plane-wave pseudopotential method and the computation of stresses and forces outlined here. The first-principles investigation of other solid and fluid earth materials by this technique represents an exciting future direction.

2.1.6   Conclusions

Modern first-principles methods are now capable of realistic predictions of many experimentally observable and geophysically important properties such as the equation of state, phase stability, crystal structure, and elasticity. Parameter free and completely independent of experiment, these methods have been shown to reproduce observations of even subtle features such as phase transitions and the elastic anisotropy with good accuracy. Density functional theory represents the ideal complement to the experimental approach toward studying the behavior of earth materials under extreme conditions. Accelerated progress is predicted on a number of fronts in this challenging field, resulting from the continued interplay of theory and experiment.

Acknowledgment

This work was supported by the U.S. National Science Foundation under grant EAR-9628199.

References

[1] S. Ichimaru, Rev. Mod. Phys. **54**, 1017 (1982).

[2] M. S. T. Bukowinski, Annu. Rev. Earth Planet. Sci., **22**, 167 (1994).

[3] G. Chabrier, Astrophys. J. **391**, 817 (1992).

[4] A. M. Dziewonski and D. L. Anderson, Phys. Earth Planet. Inter. **25**, 297 (1981).

[5] H. Jensen, Z. Phys. **111**, 373 (1938).

[6] F. Birch, J. Geophys. Res. **57**, 227 (1952).

[7] W. B. Hubbard, *Planetary Interiors* (Van Nostrand Reinhold, New York, 1984).

[8] P. Hohenberg and W. Kohn, Phys. Rev. **136**, B864 (1964).

[9] W. Kohn and L. J. Sham, Phys. Rev. **140**, A1133 (1965).

[10] S. Lundqvist and N. H. March, *Theory of the Inhomogeneous Electron Gas* (Plenum, London, 1987).

[11] J. C. Slater, Phys. Rev., **81**, 385 (1951).

[12] M. Gell-Mann and K. A. Brueckner, Phys. Rev. **106**, 364 (1957).

[13] J. P. Perdew, K. Burke, and M. Ernzerhof, Phys. Rev. Lett. **77**, 3865 (1996).

[14] D. M. Ceperley and B. J. Alder, Phys. Rev. Lett. **45**, 566 (1980).

[15] J. P. Perdew and A. Zunger, Phys. Rev. B **23**, 5048 (1981).

[16] L. Hedin, B. I. Lundqvist, 1971. J. of Phys., **4**, 2064 (1971).

[17] E. Wigner, Phys. Rev., **46**, 1002 (1934).

[18] O. Gunnarsson and B. I. Lundqvist, Phys. Rev. B **13**, 4274 (1976).

[19] J. P. Perdew, J. A. Chevary, S. H. Vosko, K. A. Jackson, M. R. Pederson, D. J. Singh, and C. Fiolhais, Phys. Rev. B **46**, 6671 (1992).

[20] P. Bagno, O. Jepsen, and O. Gunnarson, Phys. Rev. B **40**, 1997 (1989).

[21] L. Stixrude, R. E. Cohen, and D. J. Singh, Phys. Rev. B **50**, 6442 (1994).

[22] W. E. Pickett, Comput. Phys. Rep. **9**, 114 (1989).

[23] D. Vanderbilt, Phys. Rev. B **41**, 7892 (1990).

[24] N. Troullier and J. L. Martins, Phys. Rev. B **43**, 1993 (1991).

[25] J. S. Lin, Q. Qteish, M. C. Payne, and V. Heine, Phys. Rev. B **47**, 4174 (1993).

[26] D. C. Wallace, *Thermodynamics of Crystals*, 1st ed. (Wiley, New York, 1972).

[27] R. M. Wentzcovitch, J. L. Martins, and G. D. Price, Phys. Rev. Lett. **70**, 3947 (1993).

[28] R. M. Wentzcovitch, Phys. Rev. B **44**, 2358 (1991).

[29] O. H. Nielsen and R. Martin, Phys. Rev. B **32**, 3780 (1985).

[30] M. S. T. Bukowinski, Phys. Earth Planet. Inter. **14**, 333 (1977).

[31] M. S. T. Bukowinski, Geophys. Res. Lett. **12**, 536 (1985).

[32] P. Söderlind, J. A. Moriarty, and J. M. Willis, Phys. Rev. B **53**, 14063 (1996).

[33] O. K. Anderson, Phys. Rev. B **12**, 3060 (1975).

[34] S.-H. Wei and H. Krakauer, Phys. Rev. Lett. **55**, 1200 (1985).

[35] D. J. Singh, *"Planewaves, Pseudopotentials, and the LAPW Method*, 1st ed. (Kluwer Academic, Norwall, MA, 1994).

[36] L. Stixrude and R. E. Cohen, Nature (London) **364**, 613 (1993).

[37] H. J. Monkhurst and J. D. Pack, Phys. Rev. B **13**, 5188 (1976).

[38] T. S. Duffy, R. J. Hemley, and H. K. Mao, Phys. Rev. Lett. **74**, 1371 (1995).

[39] M. J. Mehl, R. E. Cohen, and H. Krakauer, J. Geophys. Res. **93**, 8009 (1988).

[40] R. E. Cohen, Am. Mineral. **76**, 733 (1991).

[41] R. E. Cohen, in *High-Pressure Research: Applications to Earth and Planetary Sciences*, Y. Syono and M. H. Manghnani, eds. (TERRAPUB, Tokyo, 1992), pp. 425–431.

[42] M. Sigalas, D. A. Papaconstantopoulos, and N. C. Bacalis, Phys. Rev. B **45**, 5777 (1992).

[43] D. M. Sherman, Earth Planet. Sci. Lett. **153**, 149 (1997).

[44] L. Stixrude, E. Wasserman, and R. E. Cohen, J. Geophys. Res. **102**, 24729 (1997).

[45] R. M. Wentzcovitch, D. A. Hugh-Jones, R. J. Angel, and G. D. Price, Phys. Chem. Minerals **22**, 453 (1995).

[46] R. M. Wentzcovitch, N. L. Ross, and G. D. Price, Phys. Earth Planet. Inter. **90**, 101 (1995).

[47] B. B. Karki, L. Stixrude, S. J. Clark, M. C. Warren, G. J. Ackland, and J. Crain, Am. Mineral. **82**, 635 (1997).

[48] R. E. Wentzcovitch and L. Stixrude, Am. Mineral. **82**, 663 (1997).

[49] R. M. Wentzcovitch, C. da Silva, J. R. Chelikowsky, and N. Binggeli, Phys. Rev. Lett. **80**, 2149 (1998).

[50] B. B. Karki, M. C. Warren, L. Stixrude, G. J. Ackland, and J. Crain, Phys. Rev. B **55**, 3465 (1997).

[51] J. Brodholt, A. Patel, and K. Retson, Amer. Mineral., **81**, 257 (1996).

[52] D. G. Isaak, Phys. Rev. B **47**, 7720 (1993).

[53] K. Kingma, H. K. Mao, and R. J. Hemley, High Pressure Res. **14**, 363 (1996).

[54] B. B. Karki, L. Stixrude, S. J. Clark, M. C. Warren, G. J. Ackland, and J. Crain, Am. Mineral. **82**, 51 (1997).

[55] A. Jephcoat, H. K. Mao, and P. M. Bell, J. Geophys. Res., **91**, 4677 (1986).

[56] H. K. Mao, Y. Wu, L. C. Chen, J. F. Shu, A. P. Jephcoat, J. Geophys. Res., **95**, 21737 (1990).

[57] C. da Silva, L. Stixrude, and R. M. Wentzcovitch, Geophys. Res. Lett. **24**, 1963 (1997).

[58] B. Kiefer, L. Stixrude, and R. M. Wentzcovitch, Geophys. Res. Lett. **24**, 2841 (1997).

[59] B. B. Karki, L. Stixrude, M. C. Warren, G. J. Ackland, and J. Crain, Geophys. Res. Lett. **24**, 3269 (1997).

[60] A. Yeganeh-Haeri, Phys. Earth Planet. Inter. **87**, 111 (1994).

[61] R. E. Cohen, I. I. Mazin, and D. G. Isaak, Science **275**, 654 (1997).

[62] C. Lee and X. Gonze, Phys. Rev. B **51**, 8610 (1995).

[63] L. Stixrude, R. E. Cohen, R. C. Yu, and
     H. Krakauer, Am. Mineral. **81**, 1293
     (1996).

[64] X. Gonze and C. Lee, Phys. Rev. B **55**,
     10355 (1997).

[65] G. A. de Wijs, G. Kresse, L. Vocadlo,
     D. Pobson, D. Aifé, M. J. G. Slan and

     G. D. Price, Nature **392**, 805
     (1998).

[66] B. B. Karki, and L. Stixrude,
     J. Geophys. Res. **104**, 13025 (1999).

[67] B. B. Karki, R. M. Wentzcovitch, S. de
     Gironcoli, S. Baroni, Science **286**,
     1705 (1999).

Chapter 2.2

Crystallographic Orbits and Their Application to Structure Types

Takeo Matsumoto

Noncharacteristic (crystallographic) orbits of space groups have been listed by Engel, Matsumoto, Steinmann, and Wondratschek. These orbits have been derived for each Wyckoff position of all space groups within the same crystal family. Noncharacteristic orbits of plane groups were studied in a more general form without limitations of crystal systems by Matsumoto and Wondratschek. Definitions of orbits and their applications to structure types of fluorite are examined. In addition, with lists of Engel et al. the structure types of fluorite for cubic and hexagonal families are derived. Currently there are no general lists for crystallographic orbits in three-dimensional space; therefore use is restricted to the separate lists by Engel et al. for each crystal family or each crystal system.

2.2.1 Introduction

In many crystal structures, several kinds of atoms form substructures with a higher symmetry; this symmetry is not what we would normally expect from the crystal structures. The following example demonstrates the crystal structure of fluorite (CaF_2) in a simplified form.

> The Ca atoms form the face-centered cubic packing that we expect in the space group $Fm\bar{3}m$ of CaF_2. The F atoms, however, form a primitive cubic packing with a cell edge that is half that of CaF_2, which follows the space-group symmetry $Fm\bar{3}m$. However, the inherent symmetry of F atoms is higher than that of the space group CaF_2.

H. Aoki et al. (eds), *Physics Meets Mineralogy* © 2000 Cambridge University Press.

F atoms do not form a lattice because they are not translationally equivalent in the space group of CaF_2; they form an apparent translation lattice. The higher symmetry of the F substructure in CaF_2 is based on the superposition in which the shape and the behaviour of the particles are symmetric enough so as not to break the symmetry of the cubic primitive packing. For spherical particles, this superposition is always fulfilled. It is also fulfilled in cases in which an atom or a group of atoms (molecules) is replaced with a point, namely, its center of gravity. If, however, the F atoms have a low symmetry, neighbouring F atoms are not translationally equivalent.

In this paper I assume that particles may be replaced with points. At this point in this CaF_2 structure, the space group is $Fm\bar{3}m$, the Ca position $4a$ 0, 0, 0 has eigensymmetry $E1 = Fm\bar{3}m$, and the F position $8c$ 1/4, 1/4, 1/4 has eigensymmetry $E2 = Pm\bar{3}m$, with $a' = a/2$. In this case $G = E1 < E2$.

2.2.2 Definitions

We start with some definitions for the space groups [1–4]. Almost all definitions are quoted from the book by Engel, Matsumoto, Steinmann, and Wondratschek (EMSW) [1] and the paper by Matsumoto and Wondratschek [2].

Definition: The set of all points that are symmetrically equivalent to point X under all motions of space group G, namely, $\{gx \mid g \in G\}$, is called the crystallographic orbit or the G orbit GX of X. The space group G is called the generating space group.

Point X is called the representing point of the crystallographic orbit. Any point of GX can be taken as the representing point, as in the *International Tables for Crystallography* (abbreviated hereafter as IT, with the publication year given in parentheses) [5]. The standard representation is restricted by conditions $0 \leq |x|, |y|, |z| \leq 1$ for the coordinates of X.

The isometries $g \in G$ leave the G orbit GX invariant, but there may be other isometries that also leave GX invariant. In any case, the set of all isometric mapping GX onto itself is a group. It is a space group because its subgroup of all translation is three dimensional and discrete.

Definition: The set of all isometries that leave the G orbit GX invariant is called the eigenstabilizer or the eigensymmetry space group E of GX. Clearly, $E \geq G$ holds.

Definition: The G orbit GX is called characteristic if $G = E$ holds. Otherwise, $G < E$, and this G orbit is called noncharacteristic.

Problem: For any space group or plane group, find all noncharacteristic G orbits.

The solutions of this problem can partly be taken from the lists of EMSW [1] for space groups and lists in Ref. [6] for planes. However, there are more solutions. For the two-dimensional case, namely for plane groups, all noncharacteristic G orbits of the plane, i.e., all those G orbits whose Euclidean stabilizer E exceeds G ($E > G$), have been derived and listed [2].

A G orbit may be noncharacteristic for the following reasons:

1. Because of the generating space group G. In two dimensions the plane groups of types $p1$, $p1m1$, $p1g1$, and $c1m1$ cannot be the eigensymmetry plane groups; they become $p2$, $p2mm$, $p2gm$, and $c2mm$, respectively. All G orbits of these plane groups are noncharacteristic. *Example:* $G = p1$; the lattice of $p1$ will have oblique symmetry and, not accidentally, a higher one. For $X = x, y$, $E = p2$ with the same lattice constants as $p1$ is a supergroup of index 2 of G. One example in three dimensions is space group $P1$, which cannot be an eigensymmetry space group: $G = P1$, $X = x, y, z$. $E = P\bar{1}$ with the same lattice constants as $P1$ is a supergroup of index 2 of G.

2. Because of the particular coordinates of the point X, referred to as a conventional coordinate system of G. *Example:* $G = p2mg$, $X = 1/3, 0$. $E = p2mm$ with lattice constants $a' = 1/2\,a$, $b' = b$ is a supergroup of index 2 of G. Another example is $G = p4$, $X = x, 0$. $E = p4mm$ with the same lattice constants of $p4$ is a supergroup of index 2 of G. An example in three dimensions is $G = P432$, $X = x, x, z$. $E = Pm\bar{3}m$ is a supergroup of index 2 of G.

3. Because of special lattice constants. *Example:* $G = p2$, $a \neq b$, $\gamma = 90°$, $X = 0, 0$. $E = p2mm$ is a supergroup of index 2 of G. In this case, G belongs to an oblique system and E belongs to a rectangular system; that is, they are different crystal systems.

4. Because of both special lattice constants and particular coordinates. *Example:* $G = p2mm$, $a = b$, $X = x, x$ or $x, x + 1/2$. $E = p4mm$ is a supergroup of index 2 of G. If either $a \neq b$ or $y \neq x$ and $y \neq x + 1/2$, then $E = G$.

There are different kinds of noncharacteristic G orbits. Some examples of these four kinds are shown for the plane groups in Fig. 2.2.1. A complete list of noncharacteristic G-orbits of the types given in points 1 and 2 above has been published by EMSW [1] for space groups. Tables of noncharacteristic G orbits of the types given in points 3 and 4 above are not yet available in the literature. However, for the plane groups, all noncharacteristic G orbits of planes have been listed in Ref. [2].

Although the space group or the plane group G describes the global symmetry of the G-orbit GX, the local symmetries at points $Y = gx$ are defined by the site-symmetry groups as follows.

Definition: The set of all isometries $g_i \in G$ that leave the point $Y \in gX$ invariant is called the site-symmetry group $S_G(Y)$. The set of all isometries of E that leave the point $Y \in gX$ invariant is called the site-eigensymmetry group $S_E(Y)$.

If the site-symmetry group $S_G(Y)$ is identity operation 1, the crystallographic orbit is called a general orbit. Otherwise, if $S_G(Y) > 1$ it is called a special orbit.

As the eigensymmetry space group E of a noncharacteristic orbit (NCO) is a supergroup of the generating space group, G is a subgroup of E and belongs to one of

(Ⅰ) Due to the generating plane group G

pl, 1 a 1   plml, 2 c 1   plgl, 2 a 1   clml, 4 b 1

p2, 1 a 2   p2mm, 2(e-h).m.   p2gm, 2(c)..m   c2mm, 4(d,e)..m

(Ⅱ) Due to particular
coordinates of the point x

p2mg
x : 1/3,0

p4, 4 d 1
x : x,0

(Ⅲ) Due to special
lattice constants
p2, a≠b, γ =90°
x : 0.0

p2mm,2(e-h)..m
a'=a/2, b'=b

p4mm,4(d,e).m.

p2mm,1(a-d)2mm

(Ⅳ) Due to special lattice constants &
particular coordinates

p2mm, 4 (i)1

a=b

x,x or
x, x+1/2

x,x or
x, x+1/2

p4mm, 4(f)..m

a=b

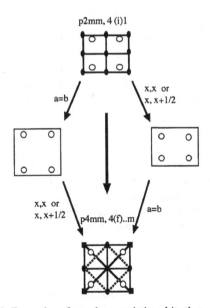

Figure 2.2.1 Examples of noncharacteristic orbits that are
due to (I) the generating plane group; (II) particular
coordinates of point X, referred to as a conventional
coordinate system of G; (III) special lattice constants; (IV)
both special lattice constants and particular coordinates.

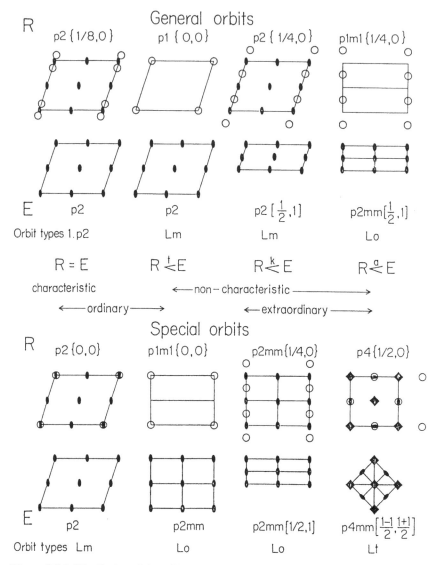

Figure 2.2.2 Distribution of the orbits among their different types. A general or special orbit for which the generating space group R (G is used in text) and the eigensymmetry space group E are the same is a characteristic orbit of R. Otherwise, if $E > R$, it is a noncharacteristic orbit. If E has additional translations, it is an extraordinary orbit.

the following types of subgroup [7]:

1. G and E belong to the same crystal class (point group), $G^k < E$. According to Hermann [7], G is classified as a klassengleiche (class-equivalent) subgroup of E or a K subgroup of E.

2. G and E have the same set of translations $LG = LE$; then $G^t < E$. G is classified as a translationengleiche (translation-equivalent) subgroup of E or a T subgroup of E.

3. G is a general subgroup of E, $G^a < E$, i.e., G is a proper subgroup of E, but neither class equivalent nor translation equivalent.

In Fig. 2.2.2, the distribution of the orbits among their different types is illustrated. In this figure, instead of G, R (for raum Gruppe or space group) is used. General orbits and special orbits for the plane groups are shown in the upper part and the lower part of the figure, respectively.

Necessarily any crystallographic orbit is always characteristic in eigensymmetry space or plane group E. As a special case, GX is called an extraordinary orbit or extraorbit, if in E there are additional translations that are not contained in G [3]. This corresponds to subgroups 1 and 3 described above.

Extraorbits are distinguished by their additional translations with respect to the generating space group. Extraorbits cause additional integral extinctions that are not required by the generating space group G but affect the whole diffraction patterns [8]. The resulting integral extinction laws are unusual in some cases. A superlattice, $P(1, 1, 1/6)$ in $P3_121$, for example, could occur in the extraorbit $\{0, 0, 1/12\}$. In this case, structure factor $F \neq 0$ for hkl if $l = 6n$ only. This extinction rule [8] in $P3_121$ is not represented in the *International Tables for (X-Ray) Crystallography* [5].

For the order, o, of site-symmetry groups, both $S_E(Y)$ and $S_G(Y)$, and the index of G in E, $i(E, G)$, the following relation holds [9, 10]:

$$o[S_E(Y_0)]/o[S_G(Y_0)] = i(E, G).$$

To be independent of the particular point, it is useful to consider the sets $\{S_G(Y)\} = \{S_G(Y) \mid Y \in GX\}$ and $\{S_E(Y)\} = \{S_E(Y) \mid Y \in EX\}$ of all site-symmetry groups and all site-eigensymmetry groups connected with the points of GX, i.e., the conjugacy classes of $S_G(Y)$ and $S_E(Y)$ relative to G.

Definition: The pair $[G, \{S_G(Y)\}]$ of the group G and the G conjugacy class $\{S_G(Y)\}$ is called the generating aspect: the pair $[E, \{S_E(Y)\}]$ is called the eigenaspect of the G orbit GX of X.

The generating aspects form the basis of the space-group or the plane-group description in IT(1935), IT(1952), and IT(1983) [5]. In EMSW [1, 2], for each plane or space group G, those G orbits are listed under the same Wyckoff letter that belong to the same generating aspect. The symmetry of a G orbit is completely determined by its eigenaspect.

Example: Let G be a space group of type $P12/m1$ with a monoclinic lattice, and consider the G orbit of $X : x = 0$, $y = 0$, $z = 0$. Its symmetry is designated by $1a2/m$, where 1 is multiplicity, i.e., the number of points in the conventional unit cell, a is the Wyckoff letter, and $2/m$ is the site-symmetry group S_G. If the conventional origin had been chosen in another $2/m$ (center-of-inversion) point, the Wyckoff letter would be b, c, d, e, f, g, or h. In this way, the symbol for the eigenaspect of GX depends on the choice of origin (and the basis). This ambiguity can be avoided if the eigensymmetry of the G orbit is described by its eigenaspect but the type of aspects to which its symmetry belongs, according to the following definition:

Definition: If $(G, \{S_G\})$ is some aspect and A is the group of all invertible affine mappings, then the affine conjugacy class $\{a^{-1}(G, \{S_G\})a \mid a \in A\}$ is called the type of aspect to which this aspect belongs.

In this way, the infinite number of individual aspects is classified into a finite number of types analogous to the infinite number of space groups, which is classified into a finite number of space-group types (230 types). There are types of aspects that do not contain any eigenaspect, e.g., $p1$, $1(a)1$; $p1m1$, $1(a, b).m.$; $p6$, $2(b)3 \ldots$; $P121$, $2(e)1$; $Pna2_1$, $4(a)1$, etc.

Definition: A type of aspects that contains at least one eigenaspect is called an Ξ type of aspect.

There are 51 types of aspects of the plane; 30 of them are Ξ types of aspects (Table 2.2.1). They are essentially already contained in the fundamental paper of Sohncke [11]. There is a close relation between the Ξ types of aspects and the lattice complexes of IT(1987) [12]. The data of the Ξ types of aspects contain the complete 'lattice complex–limiting complex' [9]. The diagram of the mutual relationship for 30 Ξ types of aspects has been shown in Ref. [2]. Here the diagram and the corresponding orbit diagram are shown in Figs. 2.2.3(a) and 2.2.3(b), respectively. For space groups, there may be 402 Ξ type of aspects.

2.2.3   Application of Noncharacteristic Orbits to the Derived Fluorite-Type Structures

The $Pa\bar{3}$ type of SiO_2 in high-pressure form has been simulated by a molecular dynamics calculation or first-principles method and compared with the fluorite type by several authors [13–16].

In relation to these problems, the derivative CaF_2-type structures are of interest. If we have the complete list of NCO or the mutual relationship of Ξ types of aspects for space, it is not so difficult to derive the structure type. However, at present we do not have general NCO lists for space groups, so we use EMSW lists to derive the derivative

Table 2.2.1. *List of Ξ types of aspects of the plane* (after Ref. [2])*

$p2$, $2(e)1$	$p2$, $1(a–d)2$	$p2mm$, $4(i)1$	$p2mm$, $2(e–h)m$
$p2mm$, $1(a–d)2mm$	$p2mg$, $4(d)1$	$p2mg$, $2(c).m.$	$p2gg$, $4(c)1$
$c2mm$, $8(f)1$	$c2mm$, $4(d, e)m$	$c2mm$, $2(a, b)2mm$	$p4$, $4(d)1$
$p4mm$, $8(g)1$	$p4mm$, $4(f)..m$	$p4mm$, $4(d, e).m.$	$p4mm$, $1(a, b)4mm$
$p4gm$, $8(d)1$	$p4gm$, $4(c)..m$	$p3$, $3(d)1$	$p3m1$, $6(e)1$
$p3m1$, $3(d).m.$	$p31m$, $6(d)1$	$p31m$, $3(c)..m$	$p6$, $6(d)1$
$p6mm$, $12(f)1$	$p6mm$, $6(e).m.$	$p6mm$, $6(d)..m$	$p6mm$, $3(c)2mm$
$p6mm$, $2(b)3m.$	$p6mm$, $1(a)6mm$		

*The symbol for a type of eigenaspect is composed of the symbol for the plane-group type and the symbol for the Wyckoff sets in its plane groups, separated by a comma.

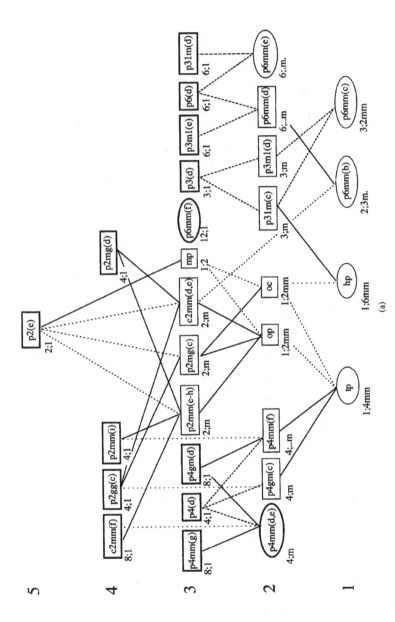

Figure 2.2.3 (a) Diagram of the relations among the 30 Ξ types of aspects of the plane [2]. Each Ξ type of aspect is represented by a frame with a plane-group type (Wyckoff letters), except five types of point lattices: mp, op, oc, tp, hp. The multiplicity and the site symmetry are given in the lower left-hand corner. The number of free parameters (x,y,a,b, γ) is given on the left-hand side. Ellipsoidal frame: not specialized to other types. Bold frame: not obtained from another type. Two Ξ types of aspects are connected by a line, if one can be obtained by specialization from the other. Solid: a lattice of higher density. broken: different site symmetry, same crystal system. Dotted: different site symmetry, different crystal system. (b) Orbits of the 30 Ξ types of aspects of the plane that correspond to (a). Each aspect is shown by plane-group (Wyckoff letter) site symmetry. (After [19].)

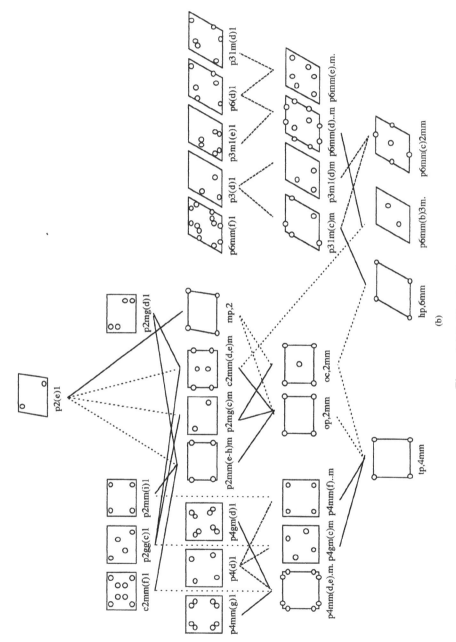

Figure 2.2.3 (Continued)

(b)

structures separately for each crystal family. In the following two subsections some examples of CaF_2-type derivations for cubic and hexagonal families made with the NCO lists are described [17].

2.2.3.1 Derivation

CaF_2 was analyzed by Bragg in 1914 [18]; it is one of the simplest structures found in compounds with the formula structure AB_2. The structure parameters are as follows:

- Space group (No. 225): $Fm\bar{3}m$, cubic face centered; unit cell $a = 5.463$ Å, $Z = 4$.
- Atomic parameters: Ca at $0, 0, 0$, $4am\bar{3}m$; F at $\pm(1/4, 1/4, 1/4)$, $8c\bar{4}3m$.

The unit cell contains four Ca and eight F atoms, i.e., four formula units of CaF_2. Each Ca atom is surrounded by eight equidistant F atoms. Each F atom, however, is four coordinated.

The Ca atoms form a cubic-face-centered packing that we would expect in the space group $Fm\bar{3}m$ of CaF_2. They belong to the characteristic orbit. The lattice of Ca atoms is $F(1, 1, 1)$, and their orbit is $Fm\bar{3}m$ ($4a, b$) (this belongs to a characteristic orbit).

The F atoms, however, form a primitive cubic packing with a cell edge that is half that of CaF_2, as described in the Introduction. The inherent symmetry of F atoms is higher than that of the space group of CaF_2. This is a NCO; the eigensymmetry is not $Fm\bar{3}m$, but $Pm\bar{3}m$, with a lattice expressed as $P(1/2, 1/2, 1/2)$. This type of eigenaspect (NCO) is $Pm\bar{3}m$ ($1a, b$), with superlattice $P(1/2, 1/2, 1/2)$, that is, $\{P(1/2, 1/2, 1/2), Pm\bar{3}m(1a, b)\}$; these can be found in the EMSW [1] lists in the Appendix at the end of this section.

- Space group No. 225: $Fm\bar{3}m$
- Superlattices: $SL\ 1 = P(1/2, 1/2, 1/2)$
- Density: 2
- Additional generators: $1/2, 0, 0$
- Wyckoff letter: $8c\bar{4}3m$
- Crystallographic NCOs: $SL\ 1$ $1/4, 1/4, 1/4$ $Pm\bar{3}m$ ($1a, b$)

The crystallographic orbit of Ca atoms is not written in the lists of EMSW, because it belongs to characteristic orbit in $Fm\bar{3}m$. This aspect of the Ca orbit should be written as follows:

- Space group No. 225: $Fm\bar{3}m$
- Superlattices: $SL\ 0 = F(1, 1, 1)$
- Density: 1
- Additional generators:
- Wyckoff letter: $4b\ m\bar{3}m$ or $4a\ m\bar{3}m$
- Crystallographic NCOs: $SL\ 0$ $1/2, 1/2, 1/2$ $Fm\bar{3}m$ ($1a, b$) or
                         $SL\ 0$ $0, 0, 0$ $Fm\bar{3}m$ ($1a, b$).

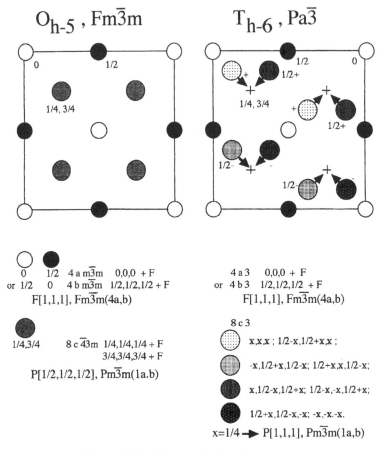

Figure 2.2.4 CaF$_2$ types in the cubic system.

Therefore the aspects of orbits for CaF$_2$-type structures are

$\{F(1, 1, 1),\ Fm\bar{3}m\ (4a, b)\ m\bar{3}m\}$; $\{P(1/2, 1/2, 1/2),\ Pm\bar{3}m\ (1a, b)\ m\bar{3}m\}$.

Both orbits of one space group in a cubic family (cubic system) are suitable for fluorite types.

We currently have no general lists for NCOs, without limitations of crystal families or systems. Therefore it is necessary to derive the structure types separately for each crystal family with the lists of EMSW [1] and IT [5].

2.2.3.2 Examples in the Cubic Family (Cubic System)

The CaF$_2$-type structure consists of two aspects of orbits, which I have already discussed. These two aspects of orbits can be found in the EMSW [1] lists and in the IT [5] in each space group of the cubic system (also see the Appendix at the end of this Section and Fig. 2.2.4).

The following possibilities that are suitable for CaF$_2$ structures are easily found.

First, the CaF_2-type structure: $Fm\bar{3}m \cdots Fm\bar{3}m$, $F432$, $Fm\bar{3}$ ($4a$ or $4b$ and $8c$), $F\bar{4}3m$, $F23$ ($4a$ or $4b$ and $4c + 4d$).

Here we must remember that the orbit type of the composite arrangement of two fcc's with a difference of $1/2, 1/2, 1/2$ for origin is $\{P(1/2, 1/2, 1/2), Pm\bar{3}m (1a, b)$ $m\bar{3}m\}$. For the above example of $F\bar{4}3m$ or $F23$, the orbit of $4c + 4d$ corresponds to $8c$ in $Fm\bar{3}m$. Also for the above five space groups, the coordinates of two types of orbits are nonvariable; consequently these belong to one type, the CaF_2 type $Fm\bar{3}m$.

Second, the pyrite- or CO_2-type structure $Pa\bar{3}$ ($4a$ $0, 0, 0$ and $8c$ x, x, x). In this type, we find the following orbits in the lists of EMSW [1]:

- $8c$ 3 $SL 3 = P(1/2, 1/2, 1/2)$ $1/4, 1/4, 1/4$ $Pm\bar{3}m$ $(1a, b)$
- $4b$ 3 $SL 2 = F(1, 1, 1)$ $1/2, 1/2, 1/2$ $Pm\bar{3}m$ $(4a, b)$ or
- $4a$ 3 $SL 2 = F(1, 1, 1)$ $0, 0, 0$ $Pm\bar{3}m$ $(4a, b)$

This pyrite structure is the derived CaF_2-type structure. From the orbit of $8c$ 3, if x, x, x becomes $1/4, 1/4, 1/4$, then the true CaF_2 structure will appear. Therefore the structure with orbits of $8c$ 3, x, x, x and $4a$ or b 3, are then one of the structures derived from the CaF_2 type (pyrite or CO_2 type).

Thus, in a cubic system, there are two CaF_2-type structures.

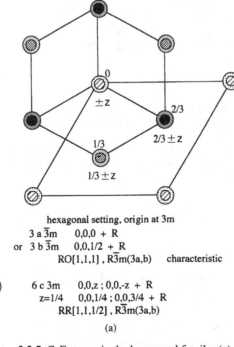

$$D_{3d}\text{-}5 \ , \ R\bar{3}m$$

hexagonal setting, origin at 3m

    ○ $3a\,\bar{3}m$ $0,0,0 + R$
    or $3b\,\bar{3}m$ $0,0,1/2 + R$
           $RO[1,1,1]$, $R\bar{3}m(3a,b)$ characteristic

    ○ $6c\,3m$ $0,0,z\,;0,0,\text{-}z + R$
    $z=1/4$ $0,0,1/4\,;0,0,3/4 + R$
           $RR[1,1,1/2]$, $R\bar{3}m(3a,b)$

(a)

Figure 2.2.5 CaF_2 types in the hexagonal family: (a) $R\bar{3}m$, (b) $P3_121$, (c) $P3_221$.

D3-4, P3$_1$21

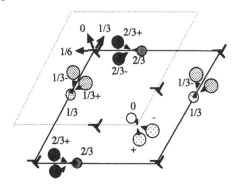

○   3 a .2.   x,0,1/3; 0,x,2/3; -x,-x,0   or   3 b .2.   x,0,5/6; 0,x1/6; -x,-x,1/2

x=1/3→→1/3,0,1/3; 0,1/3,2/3; 2/3,2/3,3,0 or 1/3,0,5/6; 0,1/3,1/6; 2/3,2/3,1/2
        RO[1,1,1], R$\overline{3}$m(3a,b)

○   6 c 1   x,y,z; -y,x-y,1/3+z; y-x,-x,2/3+z; y,x,-z; -x,y-x,1/3-z; x-y,-y,2/3-z

x=y=2/3, z=1/4 →→2/3,2/3,1/4;1/3,0,7/12; 0,1/3,11/12; 2/3,2/3,3/4; 1/3,0,1/12;0,1/3,5/12
        RR[1,1,1/2], R$\overline{3}$m(3a,b)

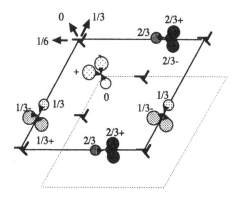

○   3 a .2.   x,0,1/3; 0,x,2/3; -x,-x,0   or   3 b .2.   x,0,5/6; 0,x1/6; -x,-x,1/2

x=2/3→→2/3,0,1/3; 0,2/3,2/3; 1/3,1/3,0 or 2/3,0,5/6; 0,2/3,1/6; 1/3,1/3,1/2
        RR[1,1,1], R$\overline{3}$m(3a,b)

○   6 c 1   x,y,z; -y,x-y,1/3+z; y-x,-x,2/3+z; y,x,-z; -x,y-x,1/3-z; x-y,-y,2/3-z

x=y=1/3, z=1/4 →→1/3,1/3,1/4;2/3,0,7/12; 0,2/3,11/12; 1/3,1/3,3/4; 2/3,0,1/12;0,2/3,5/12
        RO[1,1,1/2], R$\overline{3}$m(3a,b)

(b)

Figure 2.2.5 (Continued)

2.2.3.3 Examples in the Hexagonal Family (Hexagonal and Trigonal Systems)

For systems other than cubic, we should choose the corresponding unit cell of CaF$_2$
types. In the hexagonal family, the rhombohedral axes of $a/2 + b/2$, $b/2 + c/2$, and
$c/2 + a/2$, where a, b, and c are cubic axes, can be obtained as shown in Fig. 2.2.5(a).

D3-6, P3$_2$21

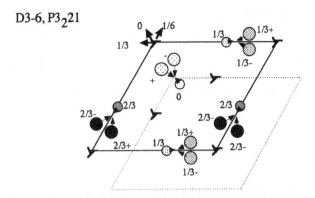

○ 3 a .2. x,0,2/3; 0,x,1/3; -x,-x,0   or   3 b .2. x,0,1/6; 0,x,5/6; -x,-x,1/2
  x=2/3→→2/3,0,2/3; 0,2/3,1/3; 1/3,1/3,0   or   2/3,0,1/6; 0,2/3,5/6; 1/3,1/3,1/2
  RO[1,1,1], R3m(3a,b)
○ 6 c 1   x,y,z; -y,x-y,z+2/3; -x+y,-x,z+1/3; y,x,-z; x-y,-y,-z+1/3; -x,-x+y,-z+2/3
  x=y=1/3, z=1/4→→1/3,1/3,1/4; 2/3,0,11/12; 0,2/3,7/12; 1/3,1/3,3/4; 0,2/3,1/12;2/3,0,5/12
  RR[1,1,1/2], R3m(3a,b)

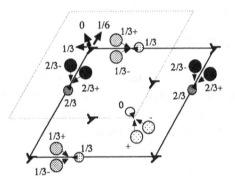

○ 3 a .2. x,0,2/3; 0,x,1/3; -x,-x,0   or   3 b .2. x,0,1/6; 0,x,5/6; -x,-x,1/2
  x=1/3→→1/3,0,2/3; 0,1/3,1/3; 2/3,2/3,0   or   1/3,0,1/6; 0,1/3,5/6; 2/3,2/3,1/2
  RR[1,1,1], R3m(3a,b)
◠ 6 c 1   x,y,z; -y,x-y,z+2/3; -x+y,-x,z+1/3; y,x,-z; x-y,-y,-z+1/3; -x,-x+y,-z+2/3
  x=y=2/3, z=1/4→→2/3,2/3,1/4; 1/3,0,11/12; 0,1/3,7/12; 2/3,2/3,3/4; 0,1/3,1/12; 1/3,0,5/12
  RO[1,1,1/2], R3m(3a,b)

(c)

Figure 2.2.5 (Continued)

The rhombohedral structure of CaF$_2$ is expressed as

$$R\bar{3}m, \ Ca\{Ro(1.1.1), \ R\bar{3}m(3a, b)\}, \quad F\{Rr(1, 1, 1/2), R\bar{3}m(3a, b)\}$$

in a hexagonal setting, and its reverse structure is structurally the same.

From the $R\bar{3}m$ structure, two additional structures are derived with the EMSW [1] lists (see Appendix). Thus there are three structures in the hexagonal family:

$R\bar{3}m$ type 1: $R\bar{3}m$, R32, $R\bar{3}$ (3 a   0,0,0   or 3 b   0,0,1/2,   and   6 c   0,0, z).

In this structure, if $z = 1/4$ and axial angle $= 60°$, then it becomes $Fm\bar{3}m$.

$P3_121$, $P3_221$ type 2 : $P3_121$ (3 a $x, 0, 1/3$ and   6 c x, y, z)   [Fig. 2.2.5(b)],

                3 : $P3_221$ (3 a $x, 0.2/3$ and   6 c x, y, z)   [Fig. 2.2.5(c)].

We can apply the same procedure to other crystal systems to derive the CaF_2-type structure.

2.2.4   Summary

Noncharacteristic crystallographic orbits of the plane groups have been listed in a general form without the limitations of crystal systems [2]. For three-dimentional space, NCOs of space groups also have been listed by EMSW [1]. However, these orbits have been derived within the same crystal family, not in the general form. Therefore, to derive the structure types we must use the separate lists by EMSW [1] for each crystal family or crystal system, as indicated in this paper. In the future, a more general form of NCOs of space groups should be derived.

Acknowledgments

The investigation of crystallographic orbits has been done in collaboration with H. Wondratschek, P. Engel, and G. Steinmann. I especially thank H. Wondratschek for his help and advice. His comments and discussion were extremely valuable, but it should be emphasized that I alone am responsible for the views expressed in this paper.

References

[1] P. Engel, T. Matsumoto, G. Steinmann, and H. Wondratschek, *The non-characteristic orbits of the space groups*, Z. Kristallogr. Suppl. No. 1, (1984).

[2] T. Matsumoto and H.Wondratschek, Z. Kristallogr. **179**, 7 (1987).

[3] H. Wondratschek, Z. Kristallogr. **143**, 460 (1976).

[4] H. Wondratschek, in *International Tables for Crystallography* (Reidel, Dordrecht, The Netherlands/Boston, 1983), Vol. A.

[5] C. Hermann, ed., *Internationale Tabellenzur Bestimmung von*

Kristallstrukturen. Borntrager. Berlin, (1935). N. F. M. Henry and K. Lonsdale, eds., *International Tables for X-ray Crystallography*, Vol. 1. Kynoch Press, Birmingham, England (1952). [revised ed.: 1965, 1969, and 1977]. Th. Hahn, ed., *International Tables for Crystallography* Vol. A. Reidel, Dordrecht, Holland/Boston, USA (1983). Kluwer Academic, Dordrecht, The Netherlands / Boston / London (1995).

[6] J. E. Lawrenson and H. Wondratschek, Z. Kristallogr. **143**, 471 (1976).

[7] C. Hermann, Z. Kristallogr. **69**, 533 (1929).

[8] T. Matsumoto and H. Wondratschek, Z. Kristallogr. **150**, 181 (1979).

[9] W. Fischer and E. Koch, Z. Kristallogr. **147**, 255 (1978).

[10] P. Engel, Z. Kristallogr. **163**, 243 (1983).

[11] L. Sohncke, *Entwickelung einer Theorie der Kristallstruktur* (Teubner, Leipzig, Germany 1879).

[12] W. Fischer and E. Koch, Z. Kristallogr. **139**, 268 (1974).

[13] Y. Matui and K. Kawamura, in *High-Pressure Research in Mineral Physics*, M. H. Manghnani and Y. Syono, eds., (Terra Scientific, Tokyo, and the American Geophysical Union, Washington, D.C., 1987), p. 305.

[14] Y. Matsui and M. Matsui, in *Structural and Magnetic Phase Transitions in Minerals*. Vol. 7 of Advances in Physical Chemistry Series, E. S. Ghose and J. M. D. Coey, eds. (Springer-Verlag, New York, 1988), p. 129.

[15] K. T. Park, K. Terakura, and Y. Matsui, Nature (London) **336**, 670 (1988).

[16] S. Tsuneyuki, Y. Matsui, H. Aoki and M. Tsukada, Nature (London) **339**, 209 (1989).

[17] T. Matsumoto, Acta Crystallogr. A, **46**, Suppl. C-447 (1990).

[18] W. L. Bragg, Proc. R. Soc. London A **89**, 468 (1914).

[19] T. Matsumoto, (in Japanese). *Kesshou Kaiseki Handbook (Structure Analysis Handbook)*, Kyouritu, Tokyo, 41, (1999).

Appendix: Noncharacteristic Orbits of the Space Groups (after EMSW [1], 1984).

Cubic system: Space group No. 225 $Fm\bar{3}m$ (added $4a, b\ m\bar{3}m$, characteristic orbits).

Space group No. 205 $Pa\bar{3}$.

Trigonal system: Space group No. 166 $R\bar{3}m$ (added $4\,a, b\bar{3}m$, characteristic orbits).

Space group No. 152 $P3_121$.

Space group No. 154 $P3_221$.

Short explanation of tables.

No. and Hermann-Mauguin symbol of the space group are written according to IT (1983). For each superlattice No. corresponding to the numbering in Table 2 in EMSW[1] monograph, and the Bravais letter indicating the lattice type together with the transformation of the basis vectors are given. Density is index of the original lattice in the superlattice. Additional generators (add. gen.): $g1, g2, g3$ are short symbols for those vectors $g1\mathbf{a} + g2\mathbf{b} + g3\mathbf{c}$ which generate the superlattice together with the basis of the space group.

First column: multiplicity, Wyckoff letter, and site symmetry of the Wyckoff of the space group.

Second column: *SL* number of the superlattice.

The orbits in one or two columns, followed by the eigensymmetry space group and the multiplicity and Wyckoff letters of the appropriate Wyckoff set.

Appendix: NCOs of the Space Groups [1]

Cubic system

Space group No. 225 $Fm\bar{3}m$
Superlattices

SL 1 $P(1/2, 1/2, 1/2)$ Density 2 Additional generators 1/2, 0, 0
SL 2 $I(1/2, 1/2, 1/2)$ Density 4 Additional generators 1/4, 1/4, 1/4

Wyckoff letter   noncharacteristic crystallographic orbits

192	*l*	1	*SL* 1	1/4, *y*, *z*	$Pm\bar{3}m$	$(24k, l)$
96	*k*	*m*	*SL* 1	*x*, *x*, 1/4	$Pm\bar{3}m$	$(12i, j)$
96	*j*	*m*	*SL* 1	0, 1/4, *z*	$Pm\bar{3}m$	$(12h)$
			SL 2	0, 1/4, 1/8	$Im\bar{3}m$	$(12d)$
48	*g*	*mm2*	*SL* 1	*x*, 1/4, 1/4	$Pm\bar{3}m$	$(6e, f)$
24	*e*	*4mm*	*SL* 1	1/4, 0, 0	$Pm\bar{3}m$	$(3c, d)$
24	*d*	*mmm*	*SL* 1	0, 1/4, 1/4	$Pm\bar{3}m$	$(3c, d)$
8	*c*	$\bar{4}3m$	*SL* 1	1/4, 1/4, 1/4	$Pm\bar{3}m$	$(1a, b)$
4	*b*	$m\bar{3}m$	*SL* 0	1/2, 1/2, 1/2		
4	*a*	$m\bar{3}m$	*SL* 0	0, 0, 0		

$\left.\begin{array}{c}4\ b\ m\bar{3}m\ SL\ 0\ 1/2,1/2,1/2\\4\ a\ m\bar{3}m\ SL\ 0\ 0,0,0\end{array}\right\}$ $Fm\bar{3}m$ $(4a, b)$ characteristic

Space group No. 205 $Pa\bar{3}$
Superlattices

SL 1 $I(1, 1, 1)$ Density 2 Additional generators 1/2, 1/2, 1/2
SL 2 $F(1,1,1)$ Density 4 Additional generators 0, 1/2, 1/2 1/2, 0, 1/2
SL 3 $P(1/2, 1/2, 1/2)$, Density 8 Additional generators 1/2, 0, 0 0, 1/2, 0 0, 0, 1/2

Wyckoff letter   noncharacteristic crystallographic orbits

24	*d*	1	*SL* 1	1/4, *y*, 0	$Ia\bar{3}$	$(24d)$	1/4, 1/8, 0	$Ia\bar{3}d$	$(24c)$
				1/4, 3/8, 0	$Ia\bar{3}d$	$(24d)$			
			SL 2	0, 0, *z*	$Fm\bar{3}m$	$(24e)$			
			SL 3	0, 1/4, 0	$Pm\bar{3}m$	$(3c, d)$	1/4, 1/4, 0	$Pm\bar{3}m$	$(3c, d)$
8	*c*	3	*SL* 3	1/4, 1/4, 1/4	$Pm\bar{3}m$	$(1a, b)$			
4	*b*	$\bar{3}$	*SL* 2	1/2, 1/2, 1/2	$Fm\bar{3}m$	$(4a, b)$			
4	*a*	$\bar{3}$	*SL* 2	0, 0, 0	$Fm\bar{3}m$	$(4a, b)$			

Trigonal System

Space group No. 166 $R\bar{3}m$
Superlattices

SL 1 $RR(1, 1, 1/2)$ Density 2 Additional generators 0, 0, 1/2
SL 2 $P(1^x, 1^x, 1/3)$ Density 3 Additional generators 0, 0, 1/3
SL 5 $P(1^x, 1^x, 1/6)$ Density 6 Additional generators 0, 0, 1/6

Wyckoff letter   noncharacteristic crystallographic orbits

36	i	1	SL 1 $x, 0, 1/4$	$R\bar{3}m$ $(18f, g)$
			SL 2 $1/3, 1/3, z$	$P6/mmm$ $(4h)$
			SL 5 $1/3, 1/3, 1/12$	$P6/mmm$ $(2c, d)$
18	h	m	SL 1 $1/2, 1/2, 1/4$	$R\bar{3}m$ $(9d, e)$
18	g	2	SL 2 $1/3, 0, 1/2$	$P6/mmm$ $(2c, d)$
18	f	2	SL 2 $1/3, 0, 0$	$P6/mmm$ $(2c, d)$
6	c	$3m$	SL 1 $0, 0, 1/4$	$R\bar{3}m$ $(3a, b)$
3	b	$\bar{3}m$	SL 0 $0, 0, 1/2$	
3	a	$\bar{3}m$	SL 0 $0, 0, 0$	$\left.\right\}$ $R\bar{3}m$ $(3a, b)$ characteristic

Space group No. 152 $P3_121$

Superlattices

SL	0	$P(1, 1, 1)$	Density 1	Additional generators
SL	1	$P(1, 1, 1/2)$	Density 2	Additional generators $0, 0, 1/2$
SL	2	$P(1, 1, 1/3)$	Density 3	Additional generators $0, 0, 1/3$
SL	4	$RO(1, 1, 1)$	Density 3	Additional generators $2/3, 1/3, 1/3$
SL	5	$RR(1, 1, 1)$	Density 3	Additional generators $1/3, 2/3, 1/3$
SL	8	$P(1, 1, 1/6)$	Density 6	Additional generators $0, 0, 1/6$
SL	10	$RO(1, 1, 1/2)$	Density 6	Additional generators $2/3, 1/3, 1/6$
SL	11	$RR(1, 1, 1/2)$	Density 6	Additional generators $1/3, 2/3, 1/6$

Wyckoff letter   noncharacteristic crystallographic orbits

6 c 1	SL	0	$x, -x, 3/12$	$P6_122$ $(6b)$	$x, -x, 1/2$	$P6_422$ $(6i, j)$
			$x, -x, 0$	$P6_4, 22$ $(6i, j)$	$1/2, 0, z$	$P6_4, 22$ $(6f)$
	SL	1	$x, 0, 1/12$	$P3_221$ $(3a, b)$	$1/2, 0, 1/12$	$P6_222$ $(3c, d)$
	SL	2	$2/3, 1/3, z$	$P\bar{3}m1$ $(2d)$	$2/3, 1/3, 1/6$	$P6/mmm$ $(2c, d)$
			$2/3, 1/3, 0$	$P6/mmm$ $(2c, d)$	$2/3, 1/3, 1/12$	$P6_3/mmc$ $(2c, d)$
			$2/3, 1/3, 3/12$	$P6_3/mmc$ $(2c, d)$	$0, 0, z$	$P6/mmm$ $(2e)$
	SL	4	$2/3, 2/3, z$	$R\bar{3}m$ $(6c)$		
	SL	5	$1/3, 1/3, z$	$R\bar{3}m$ $(6c)$		
	SL	8	$0, 0, 1/12$	$P6/mmm$ $(1a, b)$		
	SL	10	$1/3, 1/3, 1/4$	$R\bar{3}m$ $(3a, b)$		
	SL	11	$2/3, 2/3, 1/4$	$R\bar{3}m$ $(3a, b)$		
3 b 2	SL	0	$1/2, 0, 5/6$	$P6_422$ $(3c, d)$		
	SL	2	$0, 0, 5/6$	$P6/mmm$ $(1a, b)$		
	SL	4	$1/3, 0, 5/6$	$R\bar{3}m$ $(3a, b)$		
	SL	5	$2/3, 0, 5/6$	$R\bar{3}m$ $(3a, b)$		
3 a 2	SL	0	$1/2, 0, 1/3$	$P6_422$ $(3c, d)$		
	SL	2	$0, 0, 1/3$	$P6/mmm$ $(1a, b)$		
	SL	4	$1/3, 0, 1/3$	$R\bar{3}m$ $(3a, b)$		
	SL	5	$2/3, 0, 1/3$	$R\bar{3}m$ $(3a, b)$		

Space group No. 154 $P\,3_2 21$

Superlattices

SL 0 $P(1, 1, 1)$ Density 1 Additional generators
SL 1 $P(1, 1, 1/2)$ Density 2 Additional generators 0, 0, 1/2

SL 2 $P(1, 1, 1/3)$ Density 3 Additional generators 0, 0, 1/3
SL 4 $RO(1, 1, 1)$ Density 3 Additional generators 2/3, 1/3, 1/3
SL 5 $RR(1, 1, 1)$ Density 3 Additional generators 1/3, 2/3, 1/3
SL 8 $P(1, 1, 1/6)$ Density 6 Additional generators 0, 0, 1/6
SL 10 $RO(1, 1, 1/2)$ Density 6 Additional generators 2/3, 1/3, 1/6
SL 11 $RR(1, 1, 1/2)$ Density 6 Additional generators 1/3, 2/3, 1/6

Wyckoff letter   noncharacteristic crystallographic orbits

6 c 1	SL	0	$x, -x, 1/4$	$P6_522\ (6b)$	$x, -x, 0$	$P6_222\ (6i, j)$	
			$x, -x, 1/2$	$P6_222\ (6i, j)$	$1/2, 0, z$	$P6_222\ (6f)$	
	SL	1	$x, 0, 5/12$	$P3_121\ (3a, b)$	$1/2, 0, 5/12$	$P6_422\ (3c, d)$	
	SL	2	$2/3, 1/3, z$	$P\bar3m1\ (2d)$	$2/3, 1/3, 1/6$	$P6/mmm\ (2c, d)$	
			$2/3, 1/3, 0$	$P6/mmm\ (2c, d)$	$2/3, 1/3, 3/4$	$P6_3/mmc\ (2c, d)$	
			$2/3, 1/3, 1/4$	$P6_3/mmc\ (2c, d)$	$0, 0, z$	$P6/mmm\ (2e)$	
	SL	4	$1/3, 1/3, z$	$R\bar3m\ (6c)$			
	SL	5	$2/3, 2/3, z$	$R\bar3m\ (6c)$			
	SL	8	$0, 0, 1/12$	$P6/mmm\ (1a, b)$			
	SL	10	$2/3, 2/3, 1/4$	$R\bar3m\ (3a, b)$			
	SL	11	$1/3, 1/3, 1/4$	$R\bar3m\ (3a, b)$			
3 b 2	SL	0	$1/2, 0, 1/6$	$P6_222\ (3c, d)$			
	SL	2	$0, 0, 1/6$	$P6/mmm\ (1a, b)$			
	SL	4	$2/3, 0, 1/6$	$R\bar3m\ (3a, b)$			
	SL	5	$1/3, 0, 1/6$	$R\bar3m\ (3a, b)$			
3 a 2	SL	0	$1/2, , 2/3$	$P6_222\ (3c, d)$			
	SL	2	$0, 0, 2/3$	$P6/mmm\ (1a, b)$			
	SL	4	$2/3, 0, 2/3$	$R\bar3m\ (3a, b)$			
	SL	5	$1/3, 0, 2/3$	$R\bar3m\ (3a, b)$			

Chapter 2.3

Accuracy in X-Ray Diffraction

Larry W. Finger

The career and achievements of Yoshito Matsui have emphasized the attainment
of accuracy in experimental studies. To honor him, this contribution presents an
analysis of several sources of systematic errors in powder-diffraction experiments
and evaluates their effects on the results. In particular, the error arising from
uncorrected axial divergence in an energy-dispersive experiment may be as large
as 1%. Smaller errors of 0.3% may arise from sample positioning errors, and
errors as large as 2% can arise from the effects of nonhydrostatic stress.

2.3.1 Introduction

Although an x-ray diffraction experiment may have very high precision, there are many
factors that need to be controlled if an accurate result is to be obtained. This paper
discusses the potential systematic errors arising from peak asymmetry that are due to
axial divergence, sample positioning errors, and nonhydrostatic stress distributions.

2.3.2 Axial Divergence

Axial divergence, the effect that causes peak asymmetry for diffraction maxima near
$0°$ or $180°$ 2θ in powder-diffraction patterns, arises from the elliptical shape of the
intersection of diffraction cones with the cylinder that describes the opening of the
detector as it is swept through 2θ. The description of this effect and a method for

H. Aoki et al. (eds), *Physics Meets Mineralogy* © 2000 Cambridge University Press.

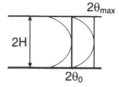

Figure 2.3.1 Origin of axial-divergence effects in energy-dispersive diffraction. The opening of the aperture perpendicular to the plane of diffraction is $2H$, the nominal diffractometer angle is $2\theta_0$, and diffraction occurs to a maximum angle of $2\theta_{max}$. The curvature is greatly exaggerated for clarity.

calculating the resulting profile have been presented by Finger et al. [1]. This correction can be applied directly for data measured with monochromatic radiation and a diffractometer. Axial divergence should not affect monochromatic powder patterns measured with flat-plate detectors, such as imaging plates or CCD detectors. Such patterns are usually processed to convert the circular rings on the flat two-dimensional detector into a pseudo-one-dimensional pattern by dividing the pattern into a number of wedges and integrating each piece [2]. As long as the integration is performed along a curved-line segment, the resulting pattern is free of the effects of axial divergence. If the integration is along a straight-line segment, the resulting pattern will have asymmetric broadening, although the form will be different from that for a diffractometer, as the magnitude of the effect will increase with increasing 2θ.

One of the side effects associated with axial divergence is a displacement of the diffraction angle for the peak from the maximum in the intensity. It is necessary therefore to correct the data, even if we plan to determine only the peak position. Of course, for Rietveld refinement, the shape of the profile must also be determined. The displacement of the peak position can also affect data measured with energy-dispersive techniques. The size of the aperture parallel to 2θ will affect only the resolution, not the asymmetry or the position of the maximum. It is well known that increases in the size of the perpendicular aperture will decrease resolution, but it is not generally appreciated that the peaks are skewed and the effective 2θ differs from the indicated value. The mathematics for this case, which to my knowledge have not previously been presented, are derived here.

As shown previously in Ref. [1] and graphically in Fig. 2.3.1, off-axis diffraction will occur at a lower diffraction angle than that arising along the centerline of the instrument. For the monochromatic case, such intensity is measured at a lower angle; however, for a polychromatic experiment, the measured radiation will have originated from diffraction at a higher 2θ value. In other words, the value Ed will systematically decrease as we move away from the centerline of the detector. Because d is constant, E must decrease, leading to an asymmetric peak.

The equation for the elliptical segment as a function of the distance from the centerline is given by

$$h(2\theta) = L[(\cos^2 2\theta / \cos^2 2\theta_0) - 1]^{1/2}, \qquad (2.3.1)$$

where L is the sample-to-detector aperture distance and $2\theta_0$ is the nominal value of the diffraction angle. To provide the most useful information, it is necessary to transform Eq. (2.3.1) from θ to energy space by use of the well-known relationship $Ed = C/\sin\theta$, where C is a constant. Substituting for d in terms of θ_0 and E_0, we obtain

$$h(E) = L[(\cos^2[\sin^{-1}(E_0 \sin\theta_0/E)] - (E_0 \sin\theta_0/E)^2)^2/\cos^2 2\theta_0]^{1/2}. \quad (2.3.2)$$

If we assume a point sample, the maximum value for $h(E)$ is L, which is the half-opening size of the aperture. From Eq. (2.3.1) we calculate the maximum value of $2\theta_{max}$ that contributes to the diffraction as

$$2\theta_{max} = \cos^{-1}[\cos 2\theta_0/(H^2/L^2 + 1)]. \quad (2.3.3)$$

In energy coordinates, this maximum angle corresponds to a minimum energy of

$$E_{min} = E_0 \sin\theta_0/\sin\theta_{max}. \quad (2.3.4)$$

The profile function $D(E, E_0)$ is found by integration over the ellipsoid to obtain

$$D(E, E_0) = [L/Hh(E) \cos 2\theta_0] \int_0^H dz. \quad (2.3.5)$$

For the simple case treated here, the integral can be replaced with H to yield

$$D(E, E_0) = [L/h(E) \cos 2\theta_0]. \quad (2.3.6)$$

The observed profile will be obtained by the convolution $D * R$, where R is the intrinsic profile of the diffraction line. This convolution is written as

$$y(E) = \frac{\int_{E_{min}}^{E_0} D(\Delta, E_0)R(E - \Delta, E_0)\,d\Delta}{\int_{E_{min}}^{E_0} D(\Delta, E_0)\,d\Delta}, \quad (2.3.7)$$

where the integral in the denominator corrects the intensities for the varying length of the diffraction ring. Equation (2.3.7) cannot be integrated, even if R is a simple Gaussian; therefore numerical integration is used to approximate the integral. The resulting equation is

$$y(E) = \frac{\sum_{n=1}^{M_{max}} w_n D(\delta_n, E_0)R(E - \delta_n, E_0)}{\sum_{n=1}^{M_{max}} w_n D(\delta_n, E_0)}, \quad (2.3.8)$$

where $\delta_n = E_{min} + (E_0 - E_{min})x_n$, w_n and x_n are the Gauss–Legendre weights and abscissas associated with the nth point, and N_{max} is the number of integration points.

To investigate the magnitude of this effect on lattice constants measured with energy-dispersive methods, I calculated profiles for diffraction patterns measured with instrumental parameters commonly used at beamline X17C of the National Synchrotron Light Source (NSLS), where the sample-to-detector distance is 240 cm and the normal opening of the detector aperture ($2H$) is 6 cm; thus $H/L = 0.0125$. Figure 2.3.2 shows the peak that would occur with a nominal 2θ of $8°$, a peak energy of 15 keV, a detector resolution of 10^{-2}, and a purely Gaussian peak shape. The peak is not visibly asymmetrically skewed; however, the intensity maximum is displaced from the correct position by roughly 0.06 keV, corresponding to a relative error in d of 0.4%. As

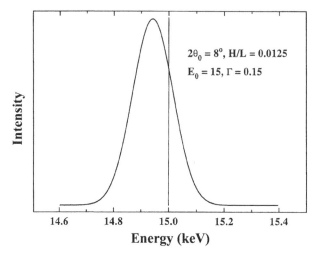

Figure 2.3.2 Effect of axial divergence in energy-dispersive diffraction. This diagram is calculated for conditions similar to those used at beamline X17C of the NSLS.

shown in Fig. 2.3.3, which shows the second-order harmonic of this peak, the offset is proportional to E; thus $\Delta d/d$ is constant. The increased seriousness of this effect at lower diffraction angles can be seen in Fig. 2.3.4, which is calculated for parameter values identical to those of Fig. 2.3.2, except that 2θ is reduced to 5°. In this case, the relative error in d is 1% and the loss of resolution is particularly obvious.

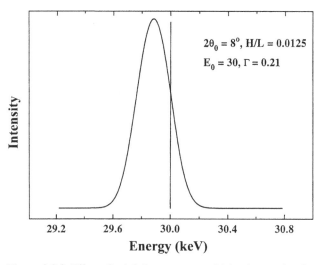

Figure 2.3.3 Effect of axial divergence on a higher harmonic of the peak shown in Fig. 2.3.2. The offset between the true peak position and the intensity maximum is proportional to energy as long as $2\theta_0$ and H/L are constant.

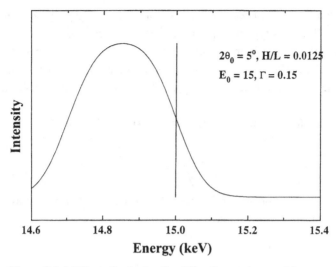

$2\theta_0 = 5°$, H/L = 0.0125

$E_0 = 15$, $\Gamma = 0.15$

Figure 2.3.4 Effect of reducing the diffraction angle on axial divergence. Except for $2\theta_0$, this profile is calculated with the same parameters as those of Fig. 2.3.2. The difference between the maximum in the intensity and the correct energy is now 1% and the peak is noticeably broadened.

The effect of this error will be minimized if we determine the value of 2θ by measuring the spectrum of a standard material, using exactly the same detector aperture. There will, however, be an offset error, and it will not be possible to change the detector angle without recalibrating. Alternatively, we could calibrate the angular position with a very small value of H/L and use Eq. (2.3.8) to determine the corrected value of E_0 for each peak. The worst case will occur when the aperture is closed for the standard, possibly to limit the intensity, and later opened for a weakly diffracting sample.

2.3.3 Sample Positioning Errors

When the sample is not placed at the correct position in the center of the instrument, errors in the measured values of d spacings will result. In this section the magnitudes of the resulting errors and techniques for eliminating or reducing them are presented.

In the following discussion, the coordinate system is defined to have x along the incident beam. For a diffractometer system, the z axis is perpendicular to the plane defined by the incident and the diffracted beams. For other systems, z is parallel to the sample rotation axis. In all cases, y is chosen to complete a right-handed system. For studies with a diamond-anvil cell, only the centering in the z direction can be reliably performed by optical means. The correct position along y is visible only when the sample is viewed exactly normal to the diamonds; otherwise, the high index of refraction leads to incorrect apparent positions. In the x direction, only the position of focus can be used, and then only if the thickness of the diamond is known precisely.

This problem has long been known and solved for single-crystal high-pressure diffraction [3, 4]. In this type of experiment, each diffraction maximum can be measured in eight different positions. From the observed angles, a number of quantities, including the offsets of the crystal from the center of the diffractometer, can be found. In addition, the values for the true angles, which have the various errors removed, can be estimated.

For diffractometer systems, King and Finger [4] list the errors in θ associated with offsets in x and y, namely,

$$\delta\theta = -\frac{\sin\theta}{2L}\delta x, \qquad \delta\theta = -\frac{\cos\theta}{2L}\delta y, \tag{2.3.9}$$

where L is the sample-to-detector distance. If we differentiate Bragg's law and apply a little algebra, it can be shown that

$$\frac{\delta d}{d} = -\frac{\sin\theta}{\cos\theta}\delta\theta. \tag{2.3.10}$$

Combining Eqs. (2.3.9) and (2.3.10), we obtain

$$\frac{\delta d}{d} = \frac{\cos\theta}{2L}\delta x, \qquad \frac{\delta d}{d} = \frac{\cos^2\theta}{2L\sin\theta}\delta y. \tag{2.3.11}$$

If we assume a 2θ value of $10°$, an error of 0.01 cm in the positions, and a 20-cm sample-to-detector distance, the associated values of $\delta d/d$ are 2×10^{-4} and 3×10^{-3} for the offsets in x and y, respectively. The former can be ignored for all but the most accurate diffraction experiments, but the second would lead to a 1% error, which is important in all cases. For those diffractometers capable of measuring both positive and negative 2θ values, it is possible to remove the y centering error as it will appear to be an error in the zero value for 2θ.

For a powder diffractometer, we can estimate errors in y positioning by measuring the reflection position at positive and negative 2θ. For such instruments, large errors in the z position will reduce the intensity and have a minor effect on the 2θ position through axial divergence. With a two-dimensional detector, errors in both y and z will lead to powder-diffraction rings that are off center relative to the x-ray beam. Uncorrected errors in the x position correspond to an unknown sample-to-crystal distance, particularly for the two-dimensional detectors. With both diffractometers and two-dimensional systems, the error in x cannot be estimated unless it is possible to spin the cell by $180°$. This condition is frequently not possible.

For experiments that are not hydrostatic, an additional error may result from off-center samples that is due to the variability of the pressure environment.

2.3.4   Nonhydrostatic Stress

If a powder sample is subjected to nonhydrostatic stress, here assumed to be uniaxial, a particular Bragg plane will be subjected to an effective pressure that depends on the angle between the normal to the plane and the unique stress axis. Because stress–strain relationships are not isotropic, the response will in general be anisotropic. The effect has been known for a long time [5]; however, this study concentrates on a few aspects of the consequences that are not generally appreciated.

As shown by Singh et al. [6], the uniaxial stress component is given by $t = \sigma_{33} - \sigma_{11}$. The equivalent hydrostatic pressure is given by $\sigma_P = (\sigma_{11} + \sigma_{22} + \sigma_{33})/3 = (\sigma_{11} + t/3)$. The strain produced by the deviation from hydrostatic conditions is given by

$$\epsilon_d(hkl) = [d_m(hkl) - d_P(hkl)]/d_P(hkl), \tag{2.3.12}$$

where $d_m(hkl)$ is the measured interplanar spacing and $d_P(hkl)$ is the spacing under hydrostatic pressure σ_P. The equations for lattice strains [7–12] can be transformed into

$$\epsilon_d(hkl) = (1 - 3\cos^2 \Psi)Q(hkl), \tag{2.3.13}$$

where

$$Q(hkl) = (t/3)\big[\alpha\{2G_R^X(hkl)\}^{-1} + (1 - \alpha)\{2G_V\}^{-1}\big], \tag{2.3.14}$$

where $G_R^X(hkl)$ and G_V are the shear moduli calculated under isostress and isostrain conditions, respectively, Ψ is the angle between the diffracting plane normal and the load axis, and α, which ranges between 0.5 and 1.0, is the relative importance of isostrain conditions. Under high-pressure loading, isostress dominates and $\alpha \rightarrow 1$. In this case,

$$d_m(hkl) = d_P(hkl)[1 + (1 - 3\cos^2 \Psi)]t\{6G_R^X(hkl)\}^{-1}. \tag{2.3.15}$$

The form of $G_R^X(hkl)$ for each of the seven crystal systems are available [6] in terms of the unique strain-tensor elements and the Miller indices hkl.

Estimation of the absolute error associated with ignoring the effects of nonhydrostatic stress are difficult to judge as they will depend on diffraction geometry and the yield strength of the material. If we consider transmission geometry with normal incidence to one of the diamonds, which is a common configuration, $\Psi = \pi/2 - \theta$. For monochromatic radiation, each reflection has a different value for Ψ and the systematics could be complicated. For energy-dispersive data, Ψ is constant; however, the differential strain depends on t and the shear moduli. It has been shown [6] that t/σ_P may be as high as 28% for pressures of the order of 5 GPa. It is clear that compressibility curves obtained without a hydrostatic medium and that are not corrected for this effect must be regarded with suspicion.

To correct for nonhydrostatic effects, it is necessary to measure the data at several values of Ψ. One means is to design a cell in which the normals of the diffracting planes may be oriented at any arbitrary angle with respect to the load axis (e.g. Ref. [6]). Equation (2.3.15) shows that d_m equals d_P when $3\cos^2 \Psi = 1$. For FeO at 8.3 GPa [6], the value of $d_m(111)$ ranged from 2.405 at $\Psi = 0°$ to 2.450 at $\Psi = 90°$. If the $\Psi = 90°$ data are used in a least-squares refinement, the resulting V_P is 1.7% smaller than V_m. In addition, the standard deviation is reduced by a factor of 6 by correcting for these effects.

2.3.5 Conclusions

Despite the simple geometry of many diamond-anvil-cell setups used for x-ray analysis, the experiment must be analyzed carefully; otherwise the systematic errors may bias the results by several percentage points. For highly compressible materials, it may be possible to obtain equations of state that are sufficiently precise. For hard materials, or experiments at very high pressures, the types of errors discussed here may render the results meaningless.

Acknowledgments

Thanks to the U.S. National Science Foundation for support and to D. E. Cox for forcing me to learn about axial divergence.

References

[1] L. W. Finger, D. E. Cox, and A. P. Jephcoat, J. Appl. Crystallogr. **27**, 892 (1994).

[2] A. P. Hammersley, ESRF Internal Rep. EXP/AH/95-01, FIT2D V5.18 Reference Manual V1.6 (European Synchrotron Radiation Facility, Grenoble, France, 1995).

[3] W. C. Hamilton, in *International Tables for X-Ray Crystallography* (Knyoch Press, Birmingham, U.K., 1974), Vol. 4, p. 273.

[4] H. E. King and L. W. Finger, J. Appl. Crystallogr. **12**, 374 (1979).

[5] J. Scott-Weaver, T. Takahashi, and W. A. Bassett, Eos Trans. Am. Geophys. Union **53**, 511 (1972).

[6] A. K. Singh, C. Balasingh, H. K. Mao, J. Shu, and R. J. Hemley, Phys. Rev. Lett. **80**, 2157 (1998).

[7] A. K. Singh, J. Appl. Phys. **73**, 4278 (1993).

[8] A. K. Singh, J. Appl. Phys. **74**, 5920 (1993).

[9] A. K. Singh, Philos. Mag. Lett. **67**, 379 (1993).

[10] A. K. Singh, in *High Pressure Science and Technology – 1993*, Vol. 309 of AIP Conference Proceedings Series (American Institute of Physics, New York, 1993), p. 1629.

[11] A. K. Singh and C. Balasingh, J. Appl. Phys. **75**, 4956 (1994).

[12] A. K. Singh and C. Balasingh, Bull. Mater. Sci. **19**, 601 (1996).

Chapter 2.4

Statistical Analysis of Phase-Boundary Observations

Abby Kavner, Terry Speed, and Raymond Jeanloz

A statistical method that uses a generalized linear model is presented to provide best-fit phase boundaries to experimentally determined phase-diagram data. The experimental determination of the exact locus of a phase boundary is inherently uncertain because of increasing difficulty in determining the presence of one phase and/or the absence of another as the phase boundary is approached from either side. We present a logistic model, which states that a phase transformation $\alpha \to \beta$ can be expressed in terms of the probability of observing β at a given set of measured state variables X and Y such that the phase boundary is defined as $P(\beta \mid X, Y) = 0.5$. As an example, high-pressure and high-temperature data for the melting curve of platinum are analyzed with this method.

2.4.1 Introduction

Phase diagrams are determined by recording the state of a material at a series of thermodynamic conditions, for example, whether the sample is molten or crystalline as a function of temperature, composition, and pressure. Although phase boundaries are thermodynamically well defined, the experimental observations that constrain phase diagrams often display significant scatter about the equilibrium boundaries. The scatter may be evident either within a single set of measurements or between distinct experiments from different laboratories.

In addition to random and systematic experimental errors (e.g., difficulty in controlling or characterizing the state of the system), there are intrinsic reasons to expect scatter in the observations of phase boundaries. As the phase transition is approached,

H. Aoki et al. (eds), *Physics Meets Mineralogy* © 2000 Cambridge University Press.

uncertainties arise because (1) more than one phase is present at the equilibrium condition of the phase transition, (2) a finite thermodynamic driving force (undercooling) is required in practice for inducing the transition either in the forward or the backward direction, and (3) kinetic hindrances in solid-state transformations can further complicate phase-boundary determinations. Cause (1) arises from considerations of equilibrium thermodynamics, so is always present in principle, whereas the extent of kinetic hindrances [causes (2) and (3)] will in general be extremely experiment dependent. The result is a phase diagram that displays a "region of indifference" in which observations do not provide a straightforward bracketing of the phase boundary (see, for example, Ref. [1]).

Historically, phase-diagram boundaries are typically drawn freehand based on a visual inspection of the experimental phase-equilibrium data (see, for example, Ref. [2]). Inconsistent results are either redetermined or the boundary is interpolated based on unquantified criteria. In short, there is no attempt to use the information contained in the scatter or inconsistencies among the experimental observations. An early application of statistical techniques [3] determines a best-fit phase-boundary curve by calculating the trace of the discriminant function, which is a method to determine a best-fit separation of a data set into two distinct classes (for example phase A and phase B). However, this method is purely empirical and can be tested only by comparison with freehand drawings of phase boundaries by the original investigator. Also, this method does not generate statistical confidence limits. More rigorous statistical methods were developed for which a cumulative error distribution was used [4]. This method, which combines the probability distributions for the location of each data point to determine the probability distribution for the location of the phase boundary, generates an error bar, but ignores data that are inconsistent with an exact bound or that lie in a region of indifference about the actual phase boundary.

Here we present a method for quantifying some of the uncertainties inherent in the experimental determination of a phase boundary. Our approach uses all of the measurements that are obtained, including scattered and (apparently) mutually inconsistent observations lying in a region of indifference about the phase boundary. The point is that because of the thermodynamic and kinetic effects noted above, as well as random sources of experimental error, scatter and inconsistencies among the measurements are not only expected to be present, but also directly provide additional information about the true uncertainties in the experimental results.

We illustrate these points by referring to recent work on the melting curve of platinum to ultrahigh pressures (Ref. [5]). This is a good example because the system consists of one component that is not expected to undergo phase transformations, so only the solid–liquid phase transformation is considered [6]. The melting curve of platinum is thermodynamically well defined and is unlikely to be influenced by kinetic hindrances.

Taken together, the data set (Fig. 2.4.1), showing both consistent and inconsistent bounds for different experiments at the same pressure, provides a measure of the uncertainty in the placement of the phase boundary, as we show below.

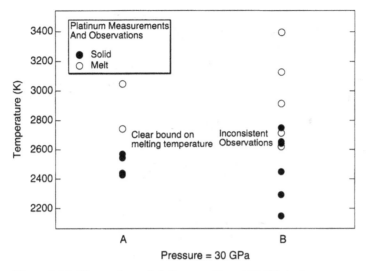

Figure 2.4.1 Observations of platinum melting at 30 GPa in the laser-heated diamond cell [5]. In one set of experiments (A), the peak temperatures at which melting is and is not observed (open and filled symbols, respectively) on increasing a decreasing temperature are mutually consistent and provide a bound on the melting temperature of platinum, between 2569 and 2743 K. In the second set of observations (B), the scattered data are mutually inconsistent but do bracket the same temperature range as that of A. The peak temperatures are determined with a precision comparable with the size of the symbols ($\leq \pm 10$–30 K), but absolute accuracy is approximately ± 100–300 K because of uncertainties in the optical properties of platinum at high pressures and temperatures [5].

For simplicity, we consider only a one-component system in our present analysis. Although the method is applicable to any type of phase boundary, we continue to use the melting curve of platinum as our example. Special emphasis is placed on establishing an objective statistical method to determine the melting curve because the presence of uncertain observations is inherent in measurements of phase boundaries and the presence of scattered or inconsistent observations may contain additional information about the proper placement of the boundary.

2.4.2 Generalized Linear Model

In the classical linear model it is assumed that a random variable y characterizing a response is related to a vector $\mathbf{x}$ of explanatory variables by the relation $y = \beta^T \mathbf{x} + \varepsilon$, where the expectation $E[y] = \mu = \beta^T \mathbf{x}$ is the linear model, the error ε has mean zero, and the solution is determining the model coefficients β^T [7]. Frequently the additional assumption that ε is normally distributed is made. Such linear models are not appropriate for responses that have values that are bounded: for example, responses that take only nonnegative values or only binary values (e.g., phase assemblage A

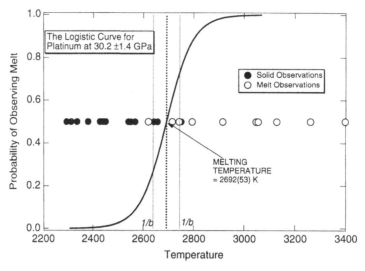

Figure 2.4.2 Logistic curve for platinum melting data at 30.2 (±1.4) GPa
(two out of the four data sets are shown in Fig. 2.4.1) depicts the
maximum-likelihood estimate of the probability of observing melt as a
function of temperature. The melting temperature T_m at this pressure is
interpreted to be at $p = 0.5$, with uncertainty given by $1/b$.

versus assemblage B, or molten and solid in the present instance). For bounded responses, a theory of generalized linear models (GLMs) has been developed in which
a transformation is chosen so that the expected value of the response is linearized [7].
With binary responses taking the values of 0 and 1, one natural transformation of the
expectation – the probability p of being 1 – is the logistic transformation

$$\text{logit}\,(p) = \ln\{p/(1-p)\}. \tag{2.4.1}$$

For the present discussion, we assume that pressure is fixed and that all sources of
systematic error have been accounted for. Then the logit of the probability of observing
melt as a function of temperature T is given by $\text{logit}\,\{p(T)\} = b(T - T_m)$, where T_m
is the melting temperature and b is a positive scale parameter. The probability of
observing melt at a given temperature and at a fixed pressure is therefore

$$p(T \mid T_m, b) = \frac{e^{b(T-T_m)}}{1 + e^{b(T-T_m)}}. \tag{2.4.2}$$

Thus the probability of observing melt lies in the interval [0, 1], increasing monoton-
ically from below 0.5 to above 0.5 as T increases from below T_m to above T_m, and
the rate of this increase depends on the parameter b (Fig. 2.4.2). In fact, $p(T_m) = 1/2$
and $\partial p/\partial T = b/4$ at $T = T_m$, and the value of b^{-1} can be taken as a measure of
uncertainty in the determination of T_m.

The most direct approach is to analyze the complete set of data by assuming that
a particular functional relationship connects melting temperature and pressure. To
remain within the GLM framework, this function must be expressible as a linear
combination of specified functions, for example powers of P (pressure), with possibly

unknown coefficients. When this is done, the unknowns can be estimated together
with the parameter b by the maximum-likelihood method, the standard approach to
fitting GLMs. Pressure can be reparameterized in terms of other variables, such as
compression, in order to examine other functional dependencies [5].

The likelihood function generated by the data when we write T_m as a quadratic
function $T_m = a + cP + dP^2$ is given by

$$L(a, b, c, d) = \prod_{i=1}^{n} p(T_i, P_i)^{y_i} [1 - p(T_i, P_i)]^{(1-y_i)}, \tag{2.4.3}$$

where $p(T_i, P_i)$ is the probability of observing melt at the measured temperature T_i and
pressure P_i, parameterized by Eq. (2.4.2) and the assumed relation between T_m and
P. The corresponding observation is $y_i = 1$ if melt and $= 0$ if solid. Depending on the
value of y_i, either the first or the second of the ith term in the product is unity, so each
pair of temperature and pressure measurements and corresponding visual observation
contribute a single term to the likelihood function.

To obtain the maximum-likelihood estimates of the coefficients in the relation between melting temperature and pressure, the logarithm of the likelihood function is
maximized in the variables a, b, c, and d. This could be done directly, but we take
advantage of the GLM function in the statistical package SPlus [8, 9].

The error in each temperature measurement, 15–60 K, is not propagated through the
GLM. The result of this analysis is a set of estimates of the parameters a, c, and d that
yield a best-fitting quadratic relation between T_m and P, an associated b parameter,
and the value of the maximum likelihood that is used to calculate what are termed
deviances in GLMs. In this example, the b parameter has units of inverse temperature,
and so its reciprocal is taken to be a measure of the temperature uncertainty in the
melting curve, assumed to be constant (i.e., independent of pressure) for the data set.
This is illustrated graphically in Fig. 2.4.2, which represents the logistic function (at
a single pressure) calculated for the data set shown in Fig. 2.4.1.

In theory, the presence of highly nonuniform sampling (for example, a large number
of observations of solids at 300 K and 30 GPa) should not affect the placement of the
logistic curve. This assumes, however, that the model being fitted for is correct. The
effect of nonuniform sampling may be to expose inadequacies in the model or the error
assumptions. A second issue can arise if there are many data points in one section of the
curve, but fewer in another (for example, many more observations at lower pressures
than at higher pressures). In this case, issues of invalid extrapolation can arise – the
computed model may not be valid in the less densely populated area. However, these
are generic problems with statistical analysis, not unique to GLMs.

2.4.3   Results: Analysis of Platinum Data

The complete set of observations on the melting of platinum as a function of pressure
is shown in Fig. 2.4.3, along with two of the fits obtained from the GLM (Table 2.4.1).

Table 2.4.1. *GLM fits to the melting-curve data of Fig. 2.4.3**

Melting relation		b^{-1} (K)	Residual deviance	Degrees of freedom
Constrained at room pressure to 2045 K				
Constant:	$T_m = 2913$	322	89	88
Linear:	$T_m = 2169 + 17.58 \times P$	109	58	87
Quadratic:	$T_m = 2076 + 26.26 \times P$ $-0.139 \times P^2$	97	54	86
Cubic:		92	54	85
Unconstrained				
Constant:	$T_m = 2871$	179	61	80
Linear:	$T_m = 2339 + 12.85 \times P$	86	40	79
Quadratic:	$T_m = 2451 + 6.44 \times P$ $+0.0750 \times P^2$	83	40	78

*The functional form is shown, along with the best-fit melting equation, the b^{-1} parameter, and the residual. Fits are shown for the data both constrained and unconstrained by the room-pressure melting temperature (2045 ± 50 K).

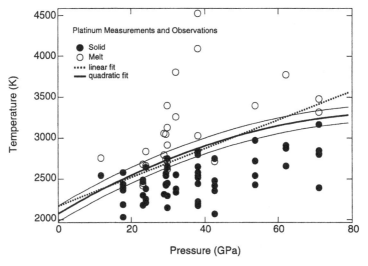

Figure 2.4.3 Complete set of experimental observations that determine the melting curve of platinum between 10 and 70 GPa. Peak temperatures are shown at a given pressure, with observations of the sample being molten or solid indicated by open or filled circles, respectively. The linear and the quadratic fits from the GLM are shown, corresponding to linear and quadratic functions of melting temperature with pressure. The fits, which increase the room-pressure melting point of platinum (2045 ± 10), are shown in black. The error envelope for the quadratic fit is shown as a solid thin curve.

Best-fit melting curves were calculated for both for the high-pressure data alone and for the high-pressure data anchored by the room-pressure melting temperature of 2045 (± 10) K. It is evident from Table 2.4.1 that the data require an increase in T_m with pressure, with b^{-1} (interpreted as a measure of the error in the determination of the melting curve), and the residual deviance computed by SPlus for the data set decreasing significantly as we go from a constant melting temperature to a linear fit. From our high-pressure data alone (unconstrained melting relation in Table 2.4.1), we cannot statistically distinguish between the linear and the quadratic fits to the data set because the residual deviance does not decrease further. However, when we include the room-temperature melting point for platinum in the data set, the quadratic fits yields a modest improvement in the fit, as shown by the decrease in the residual deviance, yet a further higher-order polynomial is unjustified.

The fit of the complete melting curve to a single logistic function depends on the assumption that b is constant throughout the pressure range of the measurements. To test this, T_m and b were determined separately for each set of platinum melting data at a given pressure. In the case in which the observations provide a consistent bracketing (no scatter points), the GLM fitting does not converge, so the T_m was chosen to be the exact midpoint between the lowest temperature melt and the highest temperature solid and b is undetermined. For the three data sets with inconsistent scatter among the observations, the b parameter can be determined with the GLM and is equal to 0.0116, 0.0107, and 0.0122 at 23, 29, and 43 GPa, respectively. These values are virtually constant ($b = 0.0115 \pm 0.0008$) over the pressure range examined, thus justifying the original assumption.

An interpretation of this constancy of b throughout the pressure range is that the bounding of the melting temperature at a given pressure is equally good, statistically speaking, if the data appear by eye to be well constrained (as in case A in Fig. 2.4.1) as well as in the case in which the data display some inconsistency (as in case B).

2.4.4   Results: Analysis of Previous Statistical Methods

As the last step, we applied this statistical analysis to two sets of data that were analyzed using the two previous statistical methods outlined above [3, 4]. The results of the re-analysis are shown in Figs. 2.4.4 and 2.4.5, along with the original curves drawn through the data and our new statistical analyses of the data sets. In both data sets, the b^{-1} error is small, approximately equivalent to the width of the maximum-likelihood GLM solution shown as a solid curve in each plot. In his analysis of the anorthite + vapor; vapor + liquid phase boundary, Chayes [3] determined a best fit by using temperature as a quadratic function of pressure, and also the inverse, pressure as a quadratic function of temperature. The phase boundaries predicted by Chayes are different when temperature or pressure is used as the dependent variable, a problem avoided when the GLM approach is used. (The large offset in the curve determined by $T_m = a + cP + dP^2$ may be due to a misprint in the Chayes paper.) The phase boundary determined with

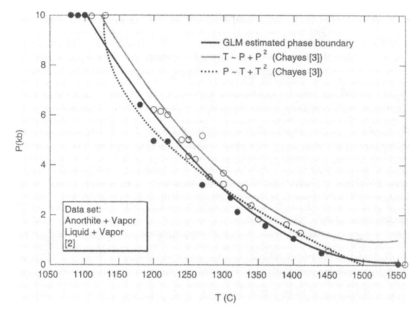

Figure 2.4.4 Phase-boundary data of anorthite + vapor; liquid + vapor [2] used by Chayes for his statistical analysis [3]. The dotted curve shows Chayes's discriminant function solution for the best-fit line $T_m = a + cT + dT^2$ through the phase-boundary data, and the gray curve shows his discriminant solution for the inverse, $P_m = e + fP + gP^2$ for the same data. The solid bold curve shows the fitted logistic model through the same data. The error parameter b^{-1} is equal to 0.046 K, which is not resolvable from the best-fit phase boundary drawn through the data.

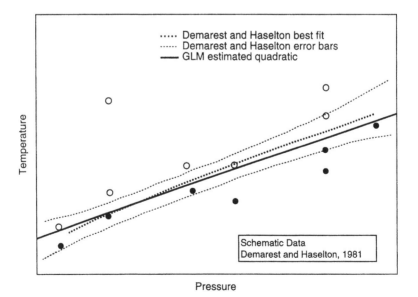

Figure 2.4.5 Data set (synthetic) used by Demarest and Haselton [4] to test methods to determine a best-fit phase boundary. The P and the T axes both range from 0 to 1. Their best fit is shown as a dotted curve, with associated error bars. The logistic model fitted to this data is shown as a bold curve. The error parameter b^{-1} is 0.0023 K, not resolvable on this plot.

$P_m = a + cT + dT^2$ does not provide a satisfying fit, especially at the extreme ends of the temperature range.

On the other hand, the GLM fit to the schematic data of Demarest and Haselton [4] (Fig. 2.4.5) closely agrees with their best fit; however, the GLM fit avoids a misclassified open circle that is not avoided in the previous best fit. In addition, the GLM-fitted phase boundary yields a significantly smaller error than the error of Demarest and Haselton. This suggests that the error bar calculated by Demarest and Haselton may be overestimated, particularly at the pressure extrema.

Although the GLM is applied here to the specific case of a melting curve of a single-component material measured as a function of pressure, the GLM is a powerful method to determine any type of phase boundary, even in the presence of complex phase assemblages [10]. Besides its simplicity and flexibility, a major advantage of the GLM is that it removes the possibility of experimenter prejudice by using a quantitative method to place a best-fit estimate of a boundary and its associated uncertainty through a real data set.

References

[1] P. W. Bridgman, *The Physics of High Pressure* (Bell, London, 1952).

[2] H. S. Yoder, Jr., in *Origins of Anorthosites and Related Rocks*, Y. W. Isachsen, ed. Memoir 18, New York State Museum and Science Service pp. 13–22. The University of the State of New York, Albany, NY 1969.

[3] F. Chayes, Am. Mineral. **53**, 359 (1968).

[4] H. H. Demarest, Jr., and H. T. Haselton, Jr., Geochim. Cosmochim. Acta **45**, 217 (1981).

[5] A. Kavner and R. Jeanloz, J. Appl. Phys., **83** 7553 (1998).

[6] N. C. Holmes, J. Moriarty, G. Gathers, and W. J. Nellis, J. Appl. Phys. **66**, 2962 (1989).

[7] T. J. Hastie and D. Pregibon, in *Statistical Models in S*, J. M. Chambers and T. J. Hastie, eds. (Wadsworth-Brooks/Cole Advanced Books+Software C. 1992, Pacific Grove, CA).

[8] P. Spector, *An Introduction to S and SPlus* (Duxbury, Belmont, CA, 1994).

[9] W. N. Venables and B. D. Ripley, *Modern Applied Statistics with S-Plus* (Springer-Verlag, New York, 1994).

[10] A. Kavner and R. Jeanloz, Geophys. Res. Lett. **25**, 4161 (1998).

Part III

New Findings in Oxides and Silicates

Chapter 3.1

Search for a Connection Among Bond Strength, Bond Length, and Electron-Density Distributions

G.V. Gibbs, M.B. Boisen, Jr., F.C. Hill, and Osamu Tamada

The Pauling electrostatic bond strength s was originally defined as a measure of the average strength of the bonds comprising a coordinated polyhedron in a complex ionic crystal, but well-developed correlations between bond strength and bond length observed for oxide crystals convinced Brown and Shannon that it is not only a direct measure of the strength of a bond, but that it is also a measure of bond type: the greater the bond strength, the more covalent the bond. Bond critical point properties of electron-density distributions calculated for molecules and observed for crystals, show that the bond strength parameter $p = s/r$ (r is the row number in the periodic table of the M cation) of an MO (oxygen) bond correlates with its bond length, the accumulation of the electron density at the bond critical point and the electronegativity of the M cation. Collectively, these trends lend support to the Brown and Shannon picture of bond strength as a measure of bond type.

3.1.1 Introduction

Only a handful of physical observables (bond length, force constant, dipole moment, polarizability) can be uniquely identified and associated with the individual bonds that bind atoms together into molecules and crystals. Over the years, each of these observables has played a pivotal role in advancing our understanding of the chemical bond. Of the four, bond length is not only the single most characteristic property of a bond, but also it is by far the most accessible and easy to measure. Ever since Bragg [1] undertook the first structural analysis of a crystal of rock salt, and continuing with

H. Aoki et al. (eds), *Physics Meets Mineralogy* © 2000 Cambridge University Press.

the thousands of structural analyses that have since been undertaken with diffraction and spectroscopic techniques, a wealth of bond-length data has been determined and is available for study. One of the most important observations afforded by these data is that the length of a bond is an intrinsic measure of bond strength: the shorter a particular MO bond, the greater its strength [2]. In advancing a set of principles that govern the structures of complex ionic crystals, Pauling [3] proposed that the average strength s of the MO electrostatic bonds comprising each of the MO_ν coordinated polyhedron in an oxide crystal should depend on the electric charge ze and the coordination number ν of the M^{+z} metal cation such that $s = z/\nu$ valence units (v.u.). With this simple yet powerful definition, he found that the sum of the strengths of each of the t bonds in a number of oxide crystals reaching each oxide anion,

$$\zeta = \sum_{i=1}^{t} s_i,$$

often equals 2.0 v.u., the valence of the anion with its sign changed. Hence, for an ionic crystal such as periclase, MgO, where each Mg cation resides in a corner sharing the MgO_6 coordinated polyhedron, and each oxide ion is bonded to six Mg cations, the strength of the MgO bond is $1/3$ v.u. and the sum of the bond strengths ζ reaching each oxide ion is 2.0 v.u., matching exactly the formal charge of the oxide ion with its sign changed. However, numerous exceptions to the rule have been observed over the years. For example, the ζ values observed for the individual oxide anions in diopside, $CaMgSi_2O_6$, range between 1.6 and 2.5 v.u. Reexamining the crystal structure of melilite, Smith [4] observed a direct connection between the ζ values of the individual oxide anions and the lengths of the SiO bonds: the greater the value of ζ, the longer the bonds. This was followed by studies by Zachariasen [5, 6] and Zachariasen and Plettinger [7], who prepared empirical bond-strength–bond-length curves for the bonds in uranyl and borate structures. The curves show that the bond lengths in these structures also decrease as the strength of each bond increases. They also observed that the resulting ζ values did not depart by more than 0.1 v.u. from the ideal values. A cubic expression was subsequently formulated by Clark [8] in modeling the relationship between bond strength and the individual SiO bond lengths observed for diopside and several other chain silicates. He discovered that the bonded interactions and the accompanying distortions of the silicate chains in these structures can be rationalized in terms of the electrostatic bond strengths, a finding that led him to conclude that the bonding in the basic building blocks of these structures can be explained in terms of an ionic model without invoking any covalent models such as the Pauling–Cruickshank $(d - p)\pi$-bonding model [9]. The following year, Baur [10] prepared $R(MO)$ versus ζ scatter diagrams for 15 different types of MO bonds and found that each set of the observed bond lengths $R(MO)$, correlates linearly with ζ. Using the slope b of each regression line, he found that the bond lengths for an oxide crystal can be generated to within ~ 0.02 Å, on average, with the linear expression $R(MO) = \langle R(MO) \rangle + b\Delta\zeta$, where $\langle R(MO) \rangle$ is the average observed bond length for a given coordinated polyhedra and $\Delta\zeta = \langle \zeta \rangle - \zeta$ is the difference between the mean ζ value, $\langle \zeta \rangle$, for the coordinated polyhedron and the individual ζ value for each of its oxide anions. Donnay [11], Donnay and Allman [12], and more recently Brown and Shannon [13] found that they could model the bond-strength–bond-length relationships with the power-law expression $s = (R_o/R)^{-N}$. The constants (R_o, N) in this expression have been derived by Brown and Shannon [13], Brown [14], and Brown and Altermatt [15] for the bond lengths observed for a large number of oxide crystals with the constraint that the sum of the strengths of the bonds reaching each cation in a structure is equal to its valence. Constants were derived not only for individual power-law equations for a relatively large number of common cations, but they were also derived for three universal power-law equations for first-, second-, and third-row cations. The sum of the strengths of the bonds reaching each atom in the oxides, when calculated with these equations, agrees to within 0.05 v.u. with the valence of the atom, regardless of whether the bond is predominantly ionic or covalent. Despite the fact that the bond-strength model was originally proposed to determine the structures of complex ionic crystals [3], Brown and Shannon [13] discovered that a single set of parameters is capable of modeling MO bonded interactions ranging from ionic to predominantly covalent. With this result they concluded that the strength of a bond, as defined by Pauling [3], is a direct measure of bond type: the larger the strength, the more covalent the bond [16]. Support for this conclusion has since been provided by a study by Brown and Skowron [17] (1990), who found that the average bond strengths derived from observed structures increase quadratically with the spectroscopic electronegativities of the M cations involved in the bonds.

Using molecular-orbital methods, Gibbs [18] and later Gibbs et al. [19] found for molecules containing main-group and closed-shell transition-metal cations M that they could reproduce the power-law expressions of Brown and Shannon [13] for MO bonds involving first-row (Li, Be, B, ...) and second-row (Na, Mg, Al, ...) M cations. They found with the definition of the bond parameter $p = s/r$, where r is the row number of the M cation, that the bond-length data for the molecules can be modeled with the expression $R(MO) = 1.39 \times p^{-0.22}$. They also found that observed bond-length data for main-group M cations from the six rows of the periodic table can likewise be modeled. Since then, similar expressions have been obtained for the nitride $R(MN) = 1.49 \times p^{-0.22}$, fluoride $R(MF) = 1.37 \times p^{-0.22}$, and sulfide $R(MS) = 1.93 \times p^{-0.22}$ molecules and crystals [20–22]. This modeling of the experimental and theoretical bond-length data for nitride, oxide, fluoride, and sulfide molecules and crystals suggests that the bond parameter p is directly related to the value of the electron density along the bonds: the greater the value of p, the greater the accumulation of the electron density [23]. Graph-theoretic calculations of resonance bond numbers n (the average number of electron pairs in a bond) for the bonds in representative fragments isolated from 10 silicate crystals were found to mimic Pauling bond strengths despite the assumption in the calculations of a purely covalent model. When n was divided by the row number of the cation, not only did the observed bond lengths plot as a single trend when plotted against n/r [as observed between $R(MO)$ and s/r], but also the data can be modeled by the expression $R(MO) = 1.39 \times (n/r)^{-0.22}$. This result lends support to the argument that the Pauling bond strength, although originally defined in terms of a purely ionic model, actually provides a rough measure of the average number of electron pairs that resonate among the bonds of a coordinated polyhedron. It likewise conforms with Pauling's conjecture that "If the bonds resonate among the alternative positions, the valence of the metal atom will tend to be divided equally among the coordinated atoms, and a rule equivalent to the electrostatic valence rule would express the satisfaction of the valences of the nonmetal atoms" [2]. In this chapter, it is demonstrated that a close connection exists between the value of the electron density along a MO bond and the bond parameter p for both molecules and crystals, with p tending to increase linearly with the accumulation of the electron density along the bond and the in situ electronegativity of the M cation bonded to the oxide anion.

3.1.2   Power-Law Relationships

The relationship between bond length and the bond parameter p is displayed in Fig. 3.1.1, in which $\sim$140 average MO bond lengths, $\langle R(MO) \rangle$, observed for main-group and transition-metal cations with closed-shell configurations for cations from all six rows of the periodic table [24] are plotted against p. As observed above for the main-group cations, the bond length versus p data scatter roughly along a single trend. Superimposed on the plot is the curve defined by the equation $R = 1.39 \times p^{-0.22}$ derived

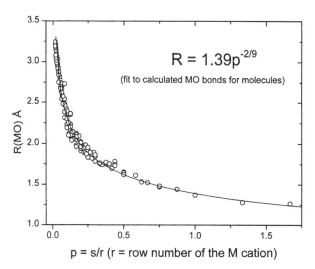

Figure 3.1.1 Scatter diagram of the average MO bond lengths $R(MO)$ observed for oxide crystals plotted against the bond-strength parameter $p = s/r$, where s is the Pauling bond strength and r is the row number of main-group and closed-shell transition-metal M cations. The curve was generated with the expression $R = 1.39 \times p^{-0.22}$ derived from MO bond-length data calculated for hydroxyacid molecules by molecular-orbtial methods [19].

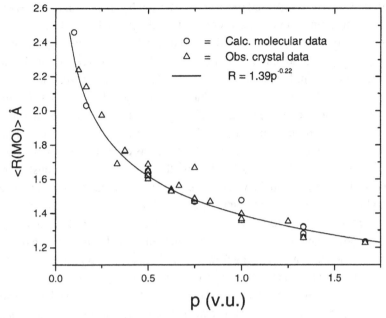

Figure 3.1.2 Scatter diagram of the average bond lengths, $R\langle(MO)\rangle$ (Å), plotted as open circles against the bond-strength parameter p, (v.u.) of the coordinated polyhedra of geometry optimized hydroxide and oxide molecules [25]. The open triangles represent experimental $R\langle(MO)\rangle$ (Å) vs. p data for the coordinated polyhedra of crystals described below. The bond-length data for the crystals were restricted to the structures [bromellite, BeO; danburite, $CaB_2Si_2O_8$; coesite, SiO_2; L-alanine; citrinin and bis(tetramethylammonium) hexanitrocobaltate (III)] for which electron-density distributions and critical-point properties have been experimentally determined [26–31].

from the results obtained in molecular-orbital calculations [18]. With a few exceptions, the experimental bond-length data scatter about the theoretical line with an average deviation of ~0.05 Å. Recently, Hill et al. [25] calculated the optimized geometries and the electron-density distributions for a number of hydroxyacid molecules also containing first- and second-row M cations. The average bond lengths, $\langle R(MO)\rangle$, calculated for the coordinated polyhedra of these molecules, are plotted as open circles in Fig. 3.1.2 against the p values of the polyhedra. Average bond lengths were plotted rather than individual ones because the Pauling bond strength and accordingly p are both measures of the average strength of the bonds comprising a coordinated polyhedron. For purposes of comparison, the average MO bond lengths observed for individual coordinated polyhedra in six crystals whose electron-density distributions have been determined by x-ray-diffraction methods are also plotted in the figure (open triangles) against p. These include the p values and the average BeO bond length observed for bromellite (BeO), [26], the average CaO, BO, and SiO bond lengths observed for danburite $(CaB_2Si_2O_8)$ [27], the average SiO bond lengths observed for coesite (SiO_2) [28], and the nonequivalent CO bond lengths observed for crystalline L-alanine $(C_3H_7NO_2)$

[29], citrinin ($C_{13}H_{14}O_5$) [30], and the average LiO and NO bond lengths observed for Li bis(tetramethylammonium) hexanitrocobaltate (III) $Li[N(CH_3)_4]_2[Co(NO_2)_6]$ [31]. As is apparent from the plot, the theoretical bond lengths calculated for the molecules and the experimental bond lengths determined for the crystals scatter together along basically the same trend. In fact, the trend is modeled very nicely with the power-law expression $R = 1.39 \times p^{-0.22}$, which is drawn in the figure for purposes of comparison. As observed above, it is clear from these results that the average bond lengths calculated for the coordinated polyhedra of the molecules and those observed for the crystals exhibit a similar dependence on p. On the basis of the power-law relationship that exists between bond length and p, it follows that a comparable relationship should exist between bond length and the value of electron density along the bond, underscoring the analysis that the strength (length) of a bond should increase (decrease) with the accumulation of electron density in the binding region of a bond [32]. In other words, the universal power-law connection between bond length and bond strength displayed in Fig. 3.1.1 is indicated as stemming from a more fundamental power-law connection between bond length and the accumulation of electron density between bonded atoms.

In a comparison of the bonded radii $r_b(O)$ for the oxide anions involved in MO bonds, Gibbs et al. [23] and more recently Hill et al. [25] observed that $r_b(O)$ tends to increase from ∼0.65 Å when bonded to highly electronegative N to ∼0.95 Å when bonded to intermediate electronegativity Si to ∼1.35 Å when bonded to highly electropositive Na. In other words, the bonded radius of the oxide anion, measured from theoretical electron-density distributions, is not constant but increases in a regular way as the ionic character of the MO bond increases. The bonded radii obtained from experimental electron-density maps have been found not only to be in close agreement with the theoretical values but also to increase with increasing bond length and ionic character. The trend between $\langle R(MO)\rangle$ and $\langle r_b(O)\rangle$ obtained for the hydroxyacid molecules [25] is displayed in Fig. 3.1.3, in which it is seen that average bonded radii of the oxide anions $\langle r_b(O)\rangle$ of the coordinated polyhedra increase nonlinearly with $\langle R(MO)\rangle$. As reported above, the bonded radii of the oxide anions reported for crystals are similar to those calculated for molecules [33]. This report is borne out by Fig. 3.1.3, in which the observed $\langle r_b(O)\rangle$ values for the coordinated polyhedra in the six crystals are plotted against $\langle R(MO)\rangle$, and it is seen that both the molecular and crystal data plot together in a single trend. As observed for the molecules, $\langle r_b(O)\rangle$ in the crystals increases with increasing bond length.

The calculation of the bond critical point (bcp) properties [34] for the electron-density distributions for the hydroxyacid molecules shows that the optimized MO bond lengths decrease with increasing $\rho(\mathbf{r}_c)$, the value of the electron density at the bcp's along each bond [25]. The average value of $\rho(\mathbf{r}_c)$, $\langle\rho(\mathbf{r}_c)\rangle$, for each of the MO_n coordinated polyhedra for the molecules is plotted as open circles against the average $R(MO)$ bond length $\langle R(MO)\rangle$ in Fig. 3.1.4. As above, the average MO bond lengths observed for the polyhedra in the six crystals are plotted as open triangles in the figure against $\langle\rho(\mathbf{r}_c)\rangle$ obtained in x-ray-diffraction experiments. As is apparent from the diagram, the theoretical data for the molecules and the experimental data

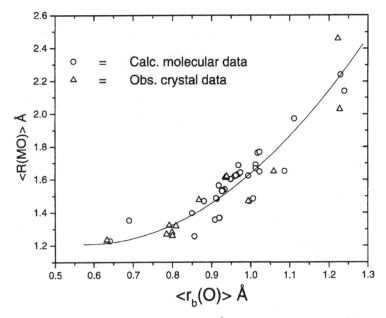

Figure 3.1.3 Scatter diagram of $R\langle(MO)\rangle$ (Å) vs. the average bonded radius of the oxide anions $\langle r_b(O)\rangle$ comprising the coordinated polyhedra of the molecules and crystals described in the legend of Fig. 3.1.2. As in Fig. 3.1.2, the molecular data are plotted as open circles and the crystal data are plotted as open triangles.

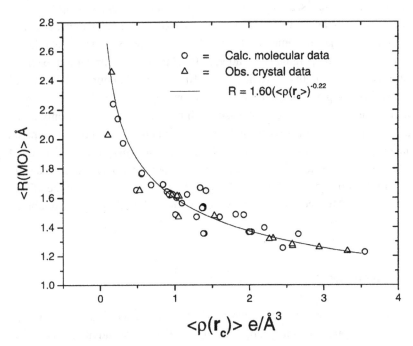

Figure 3.1.4 Scatter diagram of $R\langle(MO)\rangle$ (Å) vs. the average value of the electron density at the critical points $\langle \rho(r_c)\rangle$ for the bonds comprising the coordinated polyhedra of the molecules and crystals described in the legend of Fig. 3.1.2.

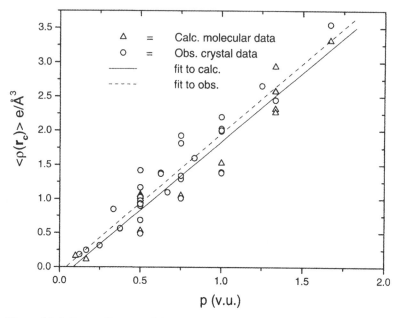

Figure 3.1.5 Scatter diagram of the average value of the electron density at the critical points $\langle \rho(\mathbf{r}_c) \rangle$ for the bonds comprising the coordinated polyhedra of the molecules and crystals described in the legend of Fig. 3.1.2 vs. the bond-strength parameter p.

recorded for the crystals scatter together basically along the same trend. The power-law expression fit to the combined theoretical and experimental data sets, $\langle R(MO) \rangle = 1.60 \times \langle \rho(\mathbf{r}_c) \rangle^{-0.22}$, has the same exponent as the power-law expression that relates MO bond length to its bond parameter p and n/r. In addition, it is similar to the empirical bond-strength–bond-length curves derived by Brown and Shannon [13] for the oxides for the first-row and particularly for the second-row cations ($R = 1.620 \times s^{-0.23}$). The upshot of these results is that the bond-strength parameter p must not only correlate with the average value of the electron density at the bcp's of the bonds comprising a coordinated polyhedron, but it must increase linearly with $\langle \rho(\mathbf{r}_c) \rangle$, as displayed in Fig. 3.1.5. Regression lines fit to the two data sets are almost identical; each has a slope of 2.0 and an intercept close to zero (see Fig. 3.1.4). Hence a rough estimate of $\langle \rho(\mathbf{r}_c) \rangle$ for the bonds of a MO_n coordinated polyhedron can be found by simply doubling the bond-strength parameter p.

3.1.3   Discussion

As observed above, Brown and Skowron [17] found that the average bond strength for a given MO bond is quadratically related to the electronegativity of the M cation involved in a MO bond. As discovered by Boyd and Edgecombe [35], the in situ electronegativities χ_M of the M atoms in number of diatomic MH hydride molecules can be estimated from the bcp properties of the electron-density distribution calculated for the molecules. Hill et al. [25] have since modified the Boyd–Edgecombe [35] relationship such that χ_M for an M cation in an oxide can be estimated with the expression

$$\chi_M = 1.31 \times F_M^{-0.23}, \tag{3.1.1}$$

where $F_M = \langle r_b(O)\rangle / [N \times \langle \rho(\mathbf{r}_c)\rangle]$, $\langle r_b(O)\rangle$ is the average bonded radius of the oxide anion bonded to an M cation and N is the number of valence electrons on M (recall that N is proportional to Z effective and that Allred and Rochow [36] found that the Pauling's [2] electronegativities are proportional to Z_{eff}/r^2, where r is the covalent radius of the atom). Hill et al. [25] recalculated the χ_M values for the M atoms for the hydride molecules (see Table 1 of Ref. [35]; note that the values in the table were converted to angstroms) with Eq. (3.1.1), assuming that $F_M = r_b(H)/[N \times \rho(\mathbf{r}_c)]$, where $r_b(H)$ is the bonded radius of the H atom and $\rho(\mathbf{r}_c)$ is the value of the electron density at the bcp's of the hydride molecules. The resulting χ_M values are plotted in Fig. 3.1.6 against Pauling's [2] empirical electronegativities in which the data fall along the identity line with a coefficient of determination r^2 of 0.99. In addition, the χ_M values calculated for hydroxyacid and nitride molecules are likewise highly correlated with Pauling's electronegativities [25, 37]. As observed by Boyd and Edgecombe [35], one of the main advantages of modeling χ_M with the bcp properties of an electron-density distribution is that it provides a straightforward procedure for estimating experimental χ_M values for the atoms in a crystal. The scatter diagrams presented in this study show

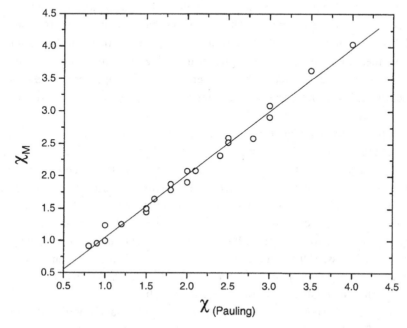

Figure 3.1.6 Electronegativity χ_M calculated with the bcp properties for the MX diatomic molecules given in Table 1 of Ref. [35] vs. Pauling's [2] thermochemical electronegativities, χ (Pauling).

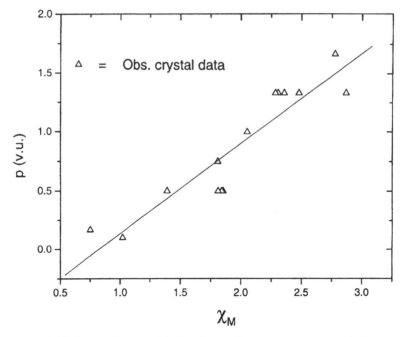

Figure 3.1.7 Scatter diagram of the bond-strength parameter p vs. the relative electronegativities of M cations estimated with Eq. (3.1.1) for the crystals described in the legend of Fig. 3.1.2.

that the theoretical bcp properties calculated for the hydroxyacid molecules are similar to those observed for the six crystals. They also show that the bond-strength parameter p is highly correlated with both $\langle R(MO) \rangle$ and $\langle \rho(r_c) \rangle$, suggesting that the parameter is a direct measure of bond type: the greater the value of p, the more covalent the bond. The χ_M values calculated with the experimental bcp properties for the crystals increase (Fig. 3.1.7) with increasing p, in support of the arguments that bond strength is a direct measure of bond type.

It is important to note that Hill et al. [25] in their study were unable to obtain an estimate of the in situ electronegativity χ_O for the oxide anion (no more than we were able in this study to obtain an estimate of the electronegativity of the oxide ions in the six crystals). This complicates the situation because, as χ_M changes for each M cation, so then should χ_O change. Inasmuch as the character of the bond is believed to be related to the electronegativity difference, $|\chi_M - \chi_O|$ [2], it then follows that our assumption that the covalency of a MO bond increases with increasing χ_M may not be valid. Nonetheless, as the trends reported here and by Hill et al. [25] and Feth et al. [37] between χ_M and the bcp properties seem to conform with chemical intuition, we believe that they are qualitatively correct and meaningful.

In a comprehensive review of electronegativity and molecular properties, Bergmann and Hinze [38] consider electronegativity to be a derived atomic property that depends on the valence state and the bonding of an atom in a molecule rather than a property that is independent of the environment of the atom. For example, the electronegativity of carbon χ_C is believed to increase in the hydrocarbons with decreasing bond length and coordination number and increasing s character of its hybrid orbitals, primarily because s electrons are lower in energy than p electrons [39]. In addition, a calculation of the average χ_C values for the C atoms in the molecules H_4CO_4 (2.15), H_2CO_3 (2.28), and CO_2 (2.52) [25] shows that χ_C also increases with decreasing bond length from 1.37 to 1.11 Å, with decreasing coordination number from 4 to 2, with increasing s character from sp^3 to sp hybridization, and with increasing bond strength from 1.0 to 2.0 v.u. Similar trends have been reported for the electron-density distributions for SiO and GeO bonds in which χ_{Si} and χ_{Ge} both increase with decreasing bond length and coordination number and increasing s character and bond strength p [40]. Thus, each of these examples and each of the trends reported in this chapter lends support to the argument of Brown and Shannon [13] that bond strength is a direct measure of bond type: the larger the strength, the shorter and the more covalent the bond.

Acknowledgments

The U.S. National Science Foundation is thanked for generously supporting this study with grant EAR-9627458. David Brown of the Brockhouse Institute for Materials Research at McMaster University is thanked for clarifying several questions that we had about the Brown–Shannon paper and for a number of important observations that improved the manuscript.

References

[1] W. L. Bragg, Proc. R. Soc. A **89**, 468 (1914).

[2] L. Pauling, *The Nature of the Chemical Bond*, 3rd ed. (Cornell U. Press, Ithaca, NY, 1960).

[3] L. Pauling, J. Am. Chem. Soc. **51**, 1010 (1929).

[4] J. V. Smith, Am. Mineral. **38**, 643 (1953).

[5] W. H. Zachariasen, Acta Crystallogr. **7**, 795 (1954).

[6] W. H. Zachariasen, Acta Crystallogr. **16**, 355 (1963).

[7] W. H. Zachariasen and H. A. Plettinger, Acta Crystallogr. **12**, 526 (1959).

[8] J. R. Clark, in *Pyroxenes and Amphiboles: Crystal Chemistry and Phase Petrology*, J. J. Papike, ed. (Mineralogical Society of America, Washington, D.C., 1969) p. 31.

[9] J. J. Papike, in *Pyroxenes and Amphiboles: Crystal Chemistry and Phase Petrology*, J. J. Papike, ed. (Mineralogical Society of America, Washington, D.C., 1969) p. 117.

[10] W. Baur, Trans. Am. Crystallogr. Assoc. **6**, 129 (1970).

[11] G. Donnay, Carnegie Inst. Washington Yearb. **68**, 292 (1969).

[12] G. Donnay and R. Allman, Am. Mineral. **55**, 1003 (1970).

[13] I. D. Brown and R. D. Shannon, Acta Crystallogr. Sect. A **29**, 266 (1973).

[14] I. D. Brown, *Structure and Bonding in Crystals*, M. O'Keeffe and

A. Navrotsky, eds. (Academic, New York, 1981), Vol. 2, p. 1.

[15] I. D. Brown and D. Altermatt, Acta Crystallogr. Sect. B **41**, 244 (1985).

[16] I. D. Brown, Acta Crystallogr. Sect. B **48**, 553 (1992).

[17] I. D. Brown and A. Skowron, J. Am. Chem. Soc. **112**, 3401 (1990).

[18] G. V. Gibbs, Am. Mineral. **67**, 421 (1982).

[19] G. V. Gibbs, L. W. Finger, and M. B. Boisen, Phys. Chem. Mineral, **14**, 327 (1987).

[20] L. A. Buterakos, G. V. Gibbs, and M. B. Boisen, Phys. Chem. Mineral. **19**, 127 (1992).

[21] J. S. Nicoll, G. V. Gibbs, M. B. Boisen, R. T. Downs, and K. L. Bartelmehs, Phys. Chem. Mineral. **20**, 617 (1994).

[22] K. L. Bartelmehs, G. V. Gibbs, and M. B. Boisen, Jr., Am. Mineral. **74**, 620 (1989).

[23] G. V. Gibbs, O. Tamada, and M. B. Boisen, Jr., Phys. Chem. Mineral. **24**, 432 (1997).

[24] R. D. Shannon, Acta Crystallogr. Sect. A **32**, 751 (1976).

[25] F. C. Hill, G. V. Gibbs, and M. B. Boisen, Phys. Chem. Mineral. **24**, 580 (1997).

[26] J. W. Downs, in *Diffusion, Atomic Ordering, and Mass Transport*, Vol. 8 of Advances in Physical Geochemistry Series, J. Ganguly, ed. (Springer–Verlag, Berlin, 1991), p. 91.

[27] J. W. Downs and R. J. Swope, J. Phys. Chem. **96**, 4834 (1992).

[28] J. W. Downs, J. Phys. Chem. **99**, 6849 (1995).

[29] C. Gatti, R. Bianchi, R. Destro, and F. Merati, J. Mol. Struct. **225**, 409 (1992).

[30] P. Roversi, M. Barzaghi, F. Merati, and R. Destro, Can. J. Chem. **74**, 1145 (1996).

[31] R. Bianchi, C. Gatti, V. Adovasio, and M. Nardelli, Acta Crystallogr. Sect. B **52**, 471 (1996).

[32] T. Berlin, J. Chem. Phys. **19**, 201 (1952).

[33] G. V. Gibbs, M. A Spackman, and M. B. Boisen, Jr., Am. Mineral. **77**, 741 (1992).

[34] R. F. W. Bader, *Atoms in Molecules* (Oxford Science, Oxford, U.K., 1990).

[35] R. J. Boyd and K. E. Edgecombe, J. Am. Chem. Soc. **110**, 4182 (1988).

[36] A. L. Allred and E. G. Rochow, J. Inorg. Nucl. Chem. **5**, 269 (1958).

[37] S. Feth, G. V. Gibbs, M. B. Boisen, and F. C. Hill, Phys. Chem. Mineral. **25**, 234 (1998).

[38] D. Bergmann and J. Hinze, Angew. Chem. Int. Ed. Engl. **108**, 162 (1996).

[39] L. C. Allen, J. Am. Chem. Soc. **111**, 9003 (1989).

[40] G. V. Gibbs, M. B. Boisen, F. C. Hill, O. Tamada, and R. T. Downs, Phys. Chem. Minerals. **25**, 574 (1998).

Chapter 3.2

MgO – The Simplest Oxide

R. E. Cohen

Periclase, or MgO, is a simple ionic mineral, but one in which many-body interactions are important. The thermal equation of state, elasticity, melting, thermal conductivity, and diffusivity have been studied by use of first-principles methods, i.e., from fundamental physics. There is generally good agreement with experiment. Studies of MgO provide fundamental information on the high-pressure behavior of minerals in the deep Earth.

MgO, periclase, is the simplest oxide and has been a subject of intense experimental and theoretical study. Oxides and silicates make up the bulk of the Earth's mantle and crust, and thus it is important to understand and predict their behavior. The behavior of MgO, as the prototypical oxide, is the key to understanding mineral and rock behavior in the bulk of the Earth. An important feature of MgO is the non-rigid behavior of the oxygen O^{2-} ion, which makes the interactions not describable by pairwise interactions. All other more complex oxides share this feature and add additional complications as well. MgO is simple in that it is an ionic material with no solid-state phase transitions until over 500 GPa according to the best computations. If we understand MgO we will not immediately understand all other oxides and silicates, but if we cannot understand MgO, we cannot understand any other oxide or silicate.

There is an enormous literature on MgO, and a comprehensive review is impossible here. Instead the main thrust is on the underlying fundamental physics of MgO. First-principles methods, as opposed to empirical methods, are emphasized. Properties that are addressed range from the thermal equation of state and elasticity to properties of defects, surfaces, and impurities.

H. Aoki et al. (eds), *Physics Meets Mineralogy* © 2000 Cambridge University Press.

The various theoretical methods give energies as functions of atomic or nuclear coordinates. In quasi-harmonic lattice dynamics (QHLD), analytic second derivatives allow determination of the dynamical matrix, which, when diagonalized, gives the vibrational frequencies and then allows the determination of free energies and thus thermodynamic properties. QHLD breaks down increasingly with increasing temperatures. Classical molecular dynamics (MD) is appropriate at high temperatures, above the Debye temperature. MD has analytic forces and integrates Newton's equation, $F = ma$, forward in time. In MD, the temperature is given by the average kinetic energy, and pressure can be obtained from the virial theorem. MD allows determination of free-energy differences by means of free-energy integration techniques and observation of dynamical processes.

3.2.1   Electronic Structure of MgO

MgO has been studied with a variety of methods ranging from nonempirical ionic models to self-consistent electronic-structure methods. The self-consistent calculations make no assumptions about bonding, shape of the charge density, form of the interactions,etc. Models are many orders of magnitude faster than self-consistent computations and can be justified by comparisons with the self-consistent electronic structure results.

In first-principles methods, properties of materials are derived without any experimental data. Most of the methods discussed here are based on density functional theory (DFT) [1], which states that the ground-state properties of a system depend on only its charge density and give a prescription for computing the charge density, total energy, etc. [2]. The total energy includes the electronic kinetic energy, the electrostatic energy between electrons and between electrons and nuclei, and the quantum-mechanical exchange and correlation energy E_{xc}. The charge density is found by the solution of effective one-electron Schrödinger equations with an effective potential that includes, in addition to the self-consistent field electrostatic potential, an effective exchange-correlation potential, $V_{xc} = \delta E_{xc}/\delta\rho$, where ρ is the charge density. The exchange-correlation potential V_{xc} and energy E_{xc} are given by the interactions in a uniform electron gas in the local-density approximation (LDA). There are different version of the LDA, depending on the parameterization of the interactions in the uniform electron gas. In most cases, these interactions are obtained from Monte Carlo simulations for the electron gas [3].

3.2.1.1   Self-Consistent Methods

There are many different self-consistent methods for solving the Kohn–Sham (KS) equations [2] of DFT. The main differences are in the basis sets that are used. Some basis sets are more efficient (faster) and others are more accurate; some are straightforward to converge and others are more complicated. Another major difference among

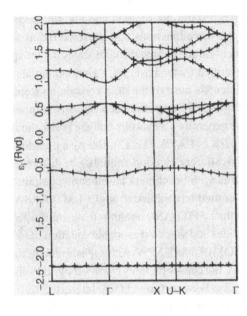

Figure 3.2.1 Band structure for MgO. Solid curves are LAPW bands at zero pressure ($V = 18.1$ Å^3). Symbols are generated from the potential-induced-breathing potential, overlapping spherical ions. Clearly the model of overlapping ions is an excellent approximation to the crystal potential and the charge density [8].

different methods is whether all electrons are included or whether a pseudopotential approximation is used to account for the deeper states that do not interact significantly among different atoms.

Important early first-principles studies of MgO at high pressures are the augmented-plane-wave (APW) computations of Bukowinski [4, 5]. The APW method uses a muffin-tin potential (spherical inside spheres around each atom and flat between the atoms, in the interstitial region). In the linearized APW (LAPW) method [6, 7] a basis set is used that is optimal for describing the space between atoms where the wave functions are slowly varying as well as the regions closer to the nuclei where the wave functions vary rapidly. There are no shape approximations for the charge density or potential and no pseudopotential approximation; all electrons including core electrons are included. It is very computationally intensive but solves the KS equations accurately. Figure 3.2.1 shows the band structure for MgO [8]. The band structure is fairly simple. The valence bands consist of O $2p$ states. The O $2s$ states show some dispersion and need to be considered as fat-core or semicore states; in other words the O $2s$ states are not noninteracting. The lowest conduction bands are Mg $3s$ states. The width of the valence bands indicates the interactions of the O^{2-} ion with its environment. An isolated O^{2-} ion would have a single eigenvalue energy (actually split by a very small spin-orbit splitting). The O $2p$ bandwidth is due primarily to O–O $pp\sigma$ interactions according to the tight-binding parameters of Kohan and Ceder [9]. With increasing pressure the valence bandwidth increases from 5.0 eV at 0 GPa to 7.1 eV at 350 GPa, and the DFT bandgap increases from a direct gap of 4.9 eV to an indirect gap (Γ–X) of 8.4 eV (the LDA gives bandgaps that are too small by a large factor; this is well understood. Qualitative behavior of the bandgap should be correct, however). Similar results were obtained more recently by linear muffin-tin orbital–atomic sphere approximation (LMTO–ASA) [10].

In pseudopotential methods, the core states are omitted and are replaced with an effective potential designed so that the wave functions outside some radius from the nuclei agree with the exact wave functions and so that the core states do not appear in the Hamiltonian and need not be described by the basis set. Pseudopotentials allow a plane-wave basis, which is very efficient as most of the time-consuming steps can be performed as fast Fourier transforms, which are very efficient on modern computers and scale well. The best pseudopotentials generally give results that are in excellent agreement with the all-electron methods such as LAPW [11, 12]. Because pseudopotential generation is not a unique process and different pseudopotentials give different results, it is important for them to be compared with benchmark all-electron computations.

There have been a number of other methods applied to MgO. LMTO-ASA is very fast but less accurate than LAPW, as the LMTO-ASA assumes that a muffin-tin potential is applied to the alkaline-earth oxides and gives a reasonable equation of state but a much lower B1–B2 transition for MgO [10] than LAPW. Although the energetics differ from the more accurate LAPW results, the band structures are in very good agreement.

Hartree–Fock computations have also been performed [13]. Interestingly, they give too stiff an equation of state even when correlation corrections are added. Local correlation corrections [14] to Hartree–Fock computations give cohesive energies comparable with those obtained with the LDA, which overbinds crystals in general, so there is clearly room for improvement.

3.2.1.2 Nonempirical Models

Self-consistent electronic-structure computations give reliable energetic properties, but are computationally intensive. Whereas self-consistent computations at static zero-temperature conditions are now relatively straightforward, finite-temperature properties are much more difficult. MgO is an ideal test bed for studying finite-temperature properties of oxides and silicates as accurate ab initio models have been developed. Two types of models have been developed: ionic models and tight-binding models. Because most work on MgO has evolved around the ionic model, the tight-binding method is not reviewed here, but see Ref. [9] for a promising model.

Ionic Models

The lack of covalency seen in the electronic structure of MgO suggests the use of overlapping-ion-type, or Gordon–Kim-type, models. These models have been very successful, and numerous studies of dynamical properties of MgO have been computed with only this assumption and the LDA. These are reviewed below. Experimental studies of the charge density support the overlapping-ion model for MgO [15] and cluster CI computations strongly support the fully ionic model for MgO and other alkaline-earth oxides [16]. The ionic nature of MgO was pointed out in the early LDA calculations of Bukowinski [4].

Application of the overlapping-ion model for MgO is complicated by the fact that O^{2-} is unstable in the free state. In a crystal the O^{2-} ion is stabilized by the crystal potential. Thus, in an overlapping-ion model, the anions must be stabilized somehow. Most studies have used a 2+ charged sphere surrounding the nucleus with a radius chosen to give a potential inside the sphere that equals the Madelung (electrostatic) potential at the site in the crystal [17]. This leads to the potential-induced-breathing (PIB) model in which the O^{2-} ion breathes in response to atomic motions [18, 19]. In the PIB model, the charge density is modeled as overlapping O^{2-} and Mg^{2+} ions, and the total energy is computed with the LDA. The total energy is then a sum of the Madelung energy, the self-energy of each ion, and the overlap energies. The overlap energy is a sum of the short-range electrostatic electron–electron and electron–nuclear energies, the exchange and correlation interactions, and the overlap kinetic energy. Only the kinetic energy is different from that of KS self-consistent calculations; the Thomas–Fermi kinetic energy is used for the overlap part. In current applications, the self-energy is the KS total energy computed for each ion in the Watson sphere (with the Watson sphere interactions subtracted). In some earlier papers, the ions were computed normally but the energy was computed with the Thomas–Fermi functional for both the self-energy and the overlap energy. Later work showed that using the more accurate KS kinetic energy is more reliable.

Initially it was thought that the PIB model improved LO-TO splitting from rigid-ion models [20], but it was found later that actually the PIB model incorrectly gave divergent LO-mode frequencies because of the dependence on the absolute value of the potential, so that uniform potential shifts, as occur locally in a LO mode, would cause changes in the energy [21]. This behavior was rectified when a correction was included that removed the divergence in lattice dynamics [19]. The resulting model proved to be robust and predictive for MgO, Al_2O_3, and other ionic materials [22–28], although for most materials PIB results are not as dependable as self-consistent computations. However, they are many orders of magnitude faster, so for some problems they compete not with self-consistent methods, but with empirical potential models.

The success of the PIB model is due to the fact that PIB charge densities are very close to the self-consistent charge densities for ionic compounds. The eigenvalues generated from the overlapping charge density are in excellent agreement with the self-consistent LAPW band structure (Fig. 3.2.1) [8]. The PIB charge density also agrees very well with the self-consistent LAPW charge density. (Fig. 3.2.2).

As mentioned above, the main problems with the PIB model are the behavior of LO modes. In the related variational-induced-breathing (VIB) model, the Watson sphere radii are chosen variationally rather than by use of the Madelung potential; that is, the total energy is minimized with respect to all of the Watson sphere radii [29]. The LO-TO splitting then comes out as in the rigid-ion model. The VIB model, although computationally more demanding than PIB because the Watson sphere radii must be varied, is quite reliable for ionic materials such as MgO, and MD has been quite successful is predicting high-temperature properties accurately [30].

We obtain further understanding by generating a series of charge densities with different Watson sphere potentials ($P_{wat} = Z_{wat}/R_{wat}$) and comparing the band structure resulting from the potential generated by each charge density with the self-consistent LAPW charge densities. We can then find the P_{wat} that gives the best-fit charge density

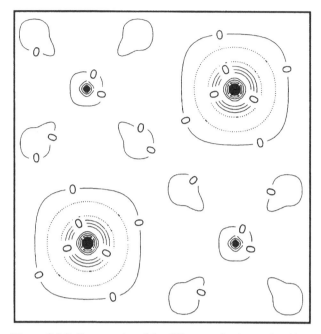

Figure 3.2.2 Contour plot of the differences in charge densities
between PIB and LAPW for MgO at zero pressure. The contour
interval is 5 me/bohr^3. Most of the contours shown are at
zero difference. The only significant differences are around the
nuclei; the PIB charge density does not make contact with the or-
thogonalization spikes around the core of the neighboring atoms.
This appears to be unimportant energetically, at least in MgO [8].

as a function of volume. Results of this exercise are shown in Fig. 3.2.3 and are compared with the PIB and the VIB models for all of the B1 alkaline-earth oxides and B2
CaO. Figure 3.2.3 shows that the PIB model underestimates the change in charge density with pressure (i.e., the compression of the O^{2-} ion) but is very close to giving the
correct charge density at zero pressure (vertical lines). The VIB model [Figs. 3.2.3(b)
and 3.2.3(c)] does much better at reproducing the change in charge density with pressure, but there is a systematic offset (except for BaO) to lower potentials (which may
be an artifact of using the Thomas–Fermi approximation for the kinetic energy in the
early VIB computations from which this figure was derived).

There have been some efforts at modeling the stabilizing potential for an oxygen
ion more accurately than the Watson sphere model by use of the crystal potential [32].
Zhang and Bukowinski [33] developed a similar model but neglected the effective
potential from the kinetic energy, which resulted in a destabilizing rather than a stabilizing potential that had to be augmented with a Watson-sphere-like potential. More
recently, a self-consistent effective stabilizing potential model was developed [34, 35]
that gives results similar to those of VIB and PIB for MgO but has promise for more-
complicated materials, especially with advances in treating nonspherical densities [36].
Another self-consistent ionic model was developed and applied to alkaline-earth oxides [37] and gives promising results, although it appears to be less accurate and more

computationally intensive than the VIB model. Because fewer approximations are made in the self-consistent ionic models, the decrease in accuracy must be due to either numerical problems in the more elaborate methods or to compensating errors in the VIB and the PIB models, which commonly occur in Gordon–Kim models, leading to their high level of accuracy with very low computational demands.

A wave-function-oriented, rather than density-oriented, ionic model was developed by Pyper [38, 39]. However, this model, even when parameterized by a fitting of self-consistent results, gives incorrect optic-mode behavior, with TO frequencies too high by a factor of ∼2, as well as incorrect behavior of the LO modes [40]. It appears that overlapping ionic densities give a more accurate model for ionic materials than localized wave functions.

3.2.2   **Equation of State**

Table 3.2.1 shows the equation-of-state parameters from a number of theoretical and experimental studies compared with experiments. Note that the LDA gives a density that is slightly too high and that Hartree–Fock with correlation corrections (P91) gives a bulk modulus that is too large. Ab initio model computations such as VIB are in excellent agreement with the static equation of state and are as accurate as more sophisticated self-consistent computations. Figure 3.2.4 shows the equation of state computed with the VIB model and MD compared with the experimental equation of state, and Fig. 3.2.5 shows the thermal expansivity of MgO [30] compared with experiment [49]. At 300 K the MD VIB thermal expansivity α is high (Table 3.2.1) because of neglect of quantum-vibrational statistics in classical MD, but it agrees well with experiment at high temperatures [Fig. 3.2.5(a)]. The thermal expansivity is in excellent agreement with experiment up to 200 GPa, the highest pressure measurements [Fig. 3.2.5(b)] [48]. Figure 3.2.6 shows the computed bulk modulus compared with zero-pressure experiments [50]. Again there is excellent agreement with the temperature dependence of the bulk modulus, showing that the VIB model in conjunction with MD gives excellent properties for MgO and indicating again that MgO is highly ionic, with properties obtainable from the energetics of overlapping ions computed with the LDA. See also Ref. [51], in which the thermal equation of state of MgO is compared with that of Fe and the qualitative differences in behavior of the thermal parameters are discussed. Table 3.2.2 gives the raw PVT results from the MD simulations of Inbar and Cohen [30] so future comparisons can be made by use of other types of equations of state. Table 3.2.3 shows the high-temperature equation-of-state parameters derived from the MD study. Of particular interest are the higher-order parameters γ, δ, and q. These are shown in Figs. 3.2.7–3.2.9 and are compared with other studies and models.

The decrease in $\delta_T = (\frac{\partial \ln \alpha}{\partial \ln V})_T$ with pressure is very important because it leads to an increase in the seismic parameter $\eta = d \ln V_s / d \ln V_p$ with pressure to values over 2 [54]. Such an increase has been observed in seismological studies. The observed increase may be partly due to partial melt in the lower mantle [55], but the intrinsic

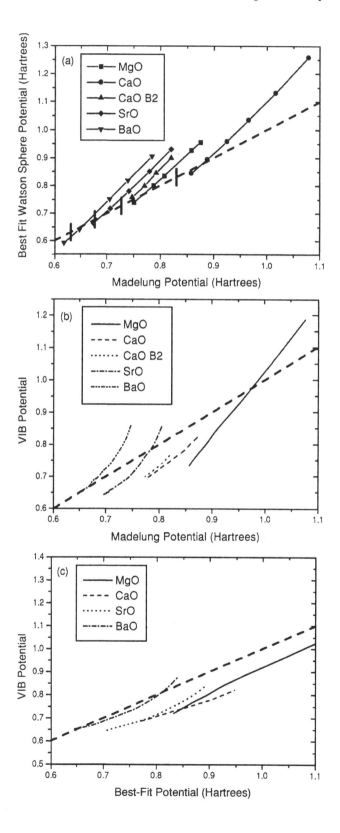

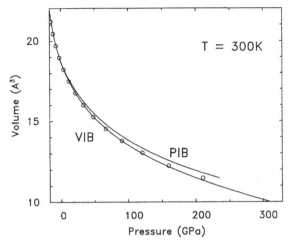

Figure 3.2.4 Equation-of-state isotherms at 300 K. The solid curves are a fourth-order fit to MD results with the VIB potential and a third-order fit to the PIB QHLD results [26]. The circles represent a third-order fit to static, shock and ultrasonic data [48].

pressure-induced rise in η in oxides, as shown theoretically in MgO, may also be important.

3.2.3 Elasticity

Elastic constants are sensitive to details of the electronic interactions in a crystal, as becomes evident when we attempt to compute them. Elastic constants are very sensitive to convergence parameters and k-point sampling in self-consistent computations [56].

Figure 3.2.3 Comparison of LAPW, VIB, and PIB charge densities for alkaline-earth oxides by comparing the best fit, VIB, and PIB Watson sphere potentials. (a) PIB Watson sphere potential that gives a charge density whose potential generates a band structure that best fits the LAPW band structure. The PIB model uses the Madelung potential as P_{wat}, so this is indicated as the bold dashed one-to-one line. The pressure increases to the left for each material, and the Madelung potential $P_{mad} = \alpha Z/r$, where $Z = 2$ for the alkaline-earth oxides and r is the near-neighbor distance in bohrs (1 bohr = 0.52917706 Å), P is in hartrees (1 Hartree = 27.21 eV), and $\alpha = 1.7476$ for the B1 structure and 1.7627 for B2. The bold vertical lines show the Madelung potential at zero pressure, so PIB does an excellent job of giving accurate zero-pressure charge densities. The slope with pressure is, however, underestimated by PIB; that is, it underestimates the compression of the O^{2-} ion as it includes only the electrostatic contribution to the compression of the ion, and not the short-range repulsion part. (b) VIB potential as a function of pressure (shown as Madelung potential). Comparison with (a) shows that VIB does a better job of reproducing the compression of the O^{2-} ion, as it includes the short-range repulsion in determining the Watson sphere radius. (c) Comparison of VIB with the best-fit potential. VIB gives the correct slope but there is a small offset from a perfect fit, except for BaO. The offset may be due to use of an early VIB model in this figure. These results are from Lu and Cohen [31].

Table 3.2.1. *Equation-of-state parameters for MgO*

Parameter	$T = 300$ K						Static					
	Exp.	VIB MD[a]	QH PIB[b]	QH VIB[c]	EM[d]	EM[e]	APW[f]	LAPW[g]	Pseudo[h]	HF[i]	HF[j]	HF + P91[j]
V (Å^3)	18.66[k]	18.70	18.66	18.69	18.73	—	18.80	18.09	19.20	18.54	18.59	17.29
K_T (GPa)	160[l]	153	180	177	232	161	155	172	160	158	180	212
K'_T	4.22[m]	4.68	4.15	—	—	—	4.16	4.09	4.26	3.53	—	—
K''_T	—	−0.05	−0.03	—	—	—	—	—	—	—	—	—
$\alpha \times 10^{-6}$ (K^{-1})	32[n]	39	24	—	19	31	—	—	—	—	—	—

[a] Inbar and Cohen [30]
[b] QH, quasi-harmonic; Isaak et al. [26]
[c] Wolf and Bukowinski [21]
[d] EM, empirical model; Reynard and Price [41]
[e] Agnon and Bukowinski [42]
[f] Bukowinski [5]
[g] Mehl et al. [8]
[h] Karki et al. [12]
[i] Causa et al. [43]
[j] McCarthy and Harrison [13]
[k] Mao and Bell [44]
[l] Sumino et al. [45]
[m] Chang and Barsch [46]
[n] Touloukian [47]

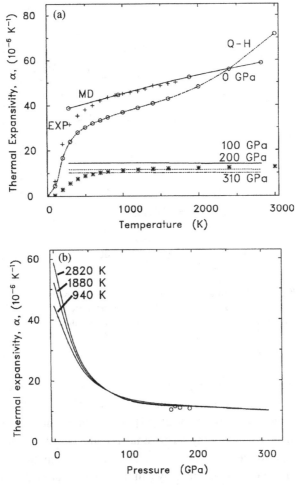

Figure 3.2.5 Thermal expansivity of MgO. (a) A comparison as a function of temperature of lattice dynamics with the PIB model [26], MD with the VIB model [30], and experiment [47]; (b) thermal expansivity as a function of pressure. Symbols are experimental measurements from Ref. [49].

The most comprehensive self-consistent study of elasticity in MgO as a function of pressure was performed by Karki et al., who used a plane-wave basis set and pseudopotentials [12]. The results are static (i.e., zero temperature) but show interesting behavior, such as the behavior of the violation of the Cauchy condition $c_{12} - c_{44} = 2P$, which indicates the importance of noncentral forces. Karki et al. found that the Cauchy violation given by the PIB model [26] is close to that obtained self-consistently, again indicating that the overlapping-ion model, with the PIB (or VIB) approximation for the stabilizing potential for the O^{2-} ions, contains the essential physics of MgO. The pseudopotential study also showed important changes in anisotropy with pressure, consistent with experiment [48], and present, but unremarked, in the earlier PIB study [26]. There

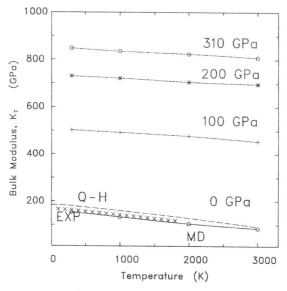

Figure 3.2.6 Bulk modulus for MgO computed with the
VIB model and MD as a function of temperature for
different pressures [30]. The results are in excellent
agreement with the zero-pressure measurements
(×'s) [50].

are no self-consistent studies of temperature dependencies on elasticity in MgO. PIB
lattice dynamics, however, gave reasonable estimates of temperature derivatives and
higher-order cross derivatives in MgO [26, 57]. Rather than using the normal quasi-
harmonic model, in which the energy versus strain is computed at a volume given by P
and T, Isaak et al. performed lattice dynamics free-energy computations as functions
of strain and found that only approximately half of the temperature dependence is due
to thermal expansivity, with the rest being an anharmonic temperature effect at constant
volume. The study by Isaak et al. remains one of the few studies of temperature effects
on elasticity of an oxide at high pressures. Karki et al. give much more information
and more reliable self-consistent results for elasticity at $T = 0$.

3.2.4   Thermal Conductivity

Heat flow from the Earth's core into the mantle is governed by the thermal conductivity of the phases present. Thermal conductivity at high pressures is extremely difficult
to measure experimentally. There are a couple of measurements made with conventional methods to ~5 GPa [58, 59], and the only higher-pressure estimates of thermal
conductivity were obtained when heat transfer was modeled in laser-heated diamond-
anvil experiments on Fe embedded in MgO and Al_2O_3. Manga and Jeanloz obtained
an average estimate of the relative increase in thermal conductivity with pressure of
4.3% GPa^{-1} from data at 58 and 125 GPa [60]. There have also been a number of

Table 3.2.2. *P–V–T results from MD calculations*

Temp (K)	Pressure (GPa)	Volume (Å^3)
300.1	0.14	18.677
305.2	1.06	18.66
304.4	46.4	15.38
297.6	91.8	13.74
288.0	177.9	11.86
295.4	310.6	10.01
1001.2	−0.3	19.29
1000.5	0.62	19.21
1000.9	56.1	15.16
999.6	96.5	13.75
1003.4	178.0	11.95
999.9	315.0	10.01
2000.9	1.08	19.98
2001.9	52.9	15.61
2000.2	94.0	14.03
2001.4	205.3	11.6
2001.1	321.1	10.0
3004.0	0.96	21.06
3003.4	95.7	14.21
3001.7	189.25	12.05
3000.6	234.9	11.28
3002.0	310.25	10.21

Table 3.2.3. *Calculated equation-of-state parameters for MgO for temperatures above the Debye temperature from MD simulations*

Parameter	1000 K	2000 K	3000 K
V (Å^3)	19.29	20.29	21.38
K_T (GPa)	132.09	104.15	83.93
K_T'	5.01	5.44	5.66
K_T''	−0.067	−0.07	−0.09
$\alpha \times 10^{-6}$ (K^{-1})	44.56	52.41	58.75

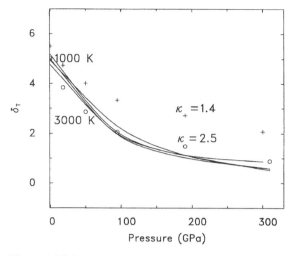

Figure 3.2.7 Pressure variation of δ_T at different
temperatures. The solid curves are MD results at 300, 1000,
2000, and 3000 K [30]. The + signs represent proposed fit
of Anderson and Isaak, $\delta_T = \delta_{T_0} \, \eta^{1.4}$ at 1000 K [52]. The
circles (o) are the same fit but with $\kappa = 2.5$ instead of
$\kappa = 1.4$, which fit the results much better. δ_T is temperature
independent and becomes pressure independent at pressures
higher then 200 GPa.

theoretical estimates of pressure effects [61] made with Debye theory [62], which
gives

$$k = \frac{1}{3} C \bar{v} l, \tag{3.2.1}$$

where C is the specific heat per volume, $\bar{v}$ is the average phonon velocity, and l is the
mean free path. Kieffer [63], for example, modeled the parameters in the Boltzmann–
Peierls model [64] as functions of temperature in terms of the Gruneisen parameter.
Such models, although useful, are not fundamental, relying on approximations, for
example, as all normal modes depend in the same way on volume.

A fundamental way of calculating thermal conductivities is to use Green–Kubo
theory, which relates the autocorrelation function of the energy current to the thermal
conductivity. The vibrational thermal conductivity of MgO was calculated by this
technique for MgO as a function of pressure from 0 to 275 GPa [65] with the VIB
potential described above. MD simulations were performed and the thermal conductivity evaluated from the autocorrelation of the energy current. The resulting calculated
thermal conductivity takes into account all anharmonicities in the potential and, unlike
the above expression, is an exact procedure for a given potential, with only statistical
uncertainties that are due to finite simulation times. The results show an increasing
thermal conductivity with increasing pressure and decreasing with temperature, as
expected. However, radiative effects were not included and will become important at
high temperatures in transparent materials.

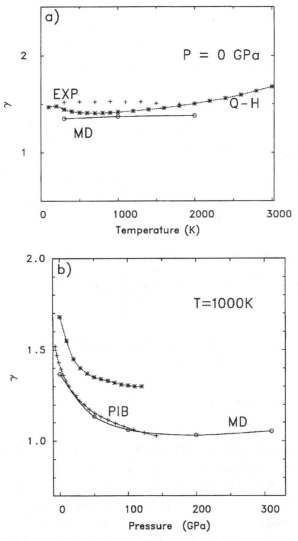

Figure 3.2.8 (a) Temperature dependence of the Grüneisen parameter at zero pressure. The solid line represents MD results [30], the dashed curve is from calculations by Isaak et al., who used the QHLD approximation [26], and the plus signs (+) are experimental results from Sumino et al. [45] and Isaak et al. [53] (b) Pressure variation of the Grüneisen parameter at 1000 K. Results from this study are shown by circles (○), plus signs (+) represent results from calculations by Isaak et al. [26] and the stars (*) are calculations from an empirical model by Agnon and Bukowinski [42].

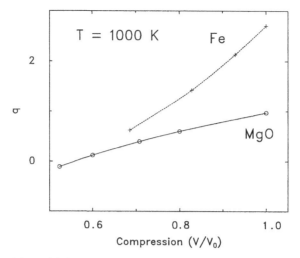

Figure 3.2.9 Parameter $q = (\partial \ln \gamma / \partial \ln V)_T$ as a function
of the compression $\eta = V/V_0$. The solid curve represents
MD results for MgO, and the dashed curve represents
results from a nonempirical study of Fe [51].

MD simulations were performed similar to those described above to find the equation of state [30]. The thermal conductivity is given by

$$k = \frac{V}{k_B T^2} \int_0^\infty dt \, \langle J(t) J(0) \rangle , \qquad (3.2.2)$$

where V is the volume, T is the absolute temperature, t is time, and J is the energy current [66, 67]. The angle brackets represent a statistical average over the system. It is necessary to define the energy current in terms of the microscopic model. This requires giving the location of the energy or the local energy density. The underlying philosophy is that energy is accounted for where it makes most physical sense and is tractable. Therefore the self-energy is positioned at each nucleus, and the interaction energy between two atoms is divided equally between the two atoms. Then the energy current is given by

$$\mathbf{J} = \sum_a S_a \mathbf{v}_a - \frac{1}{2} \sum_{b \neq a} (\mathbf{F}_{ab} \cdot \mathbf{v}) \mathbf{r}_{ab} + \sum_{b \neq a} \left(\frac{\partial \phi_{ab}}{\partial P_{wat}} \right) \left(\frac{\partial P_{wat}}{\partial t} \right) \mathbf{r}_{ab} , \qquad (3.2.3)$$

where S_a is the self-energy of ion a, $\mathbf{v}_a$ is the velocity of ion a, $\mathbf{F}_{ab}$ is the contribution of the pairwise force between atoms a and b, ϕ_{ab} is the pairwise short-range potential between atoms a and b, P_{wat} is the Watson sphere potential for atom a, and $\mathbf{r}$ is the vector between a and b. The second term is the normal contribution we obtain for a rigid-ion model, and the first and the third terms are due to the many-body nature of the VIB model. The first term is the energy transport from the self-energy of the ions, and the third term is from the effect of changes of Watson sphere potential on the pair potentials.

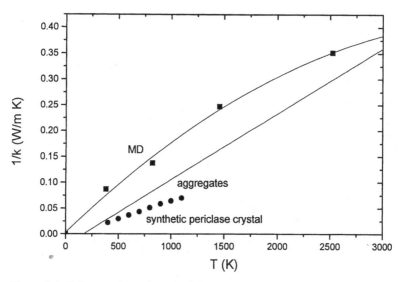

Figure 3.2.10 Inverse thermal conductivity computed with MD compared with experiments on single crystals and aggregates.

$J(t)$ was obtained from the MD simulations, the autocorrelation function was computed with the Fourier transform method [68], and the thermal conductivity was computed. Error bars were obtained from the statistics of the fluctuations. The system size was 64 atoms in periodic boundary conditions. A limited number of tests indicated that the thermal conductivity k converged at high temperatures for this system size. Most of the runs were done at an energy corresponding to an average temperature of 2500 K.

First, the behavior near zero pressure is considered with respect to temperature. At high temperatures, experimental data can usually be reduced as

$$1/k = a + bT, \tag{3.2.4}$$

where a is a sample-dependent constant primarily extrinsic in nature, depending on defects, impurities, and grain size, and b is related to anharmonicity or phonon–phonon scattering that leads to a finite thermal conductivity (a purely harmonic crystal would have infinite thermal conductivity). Actually, this simple relationship should be obeyed at constant volume, although much experimental data does fit this form at constant $P = 0$ as well. A set of simulations at a constant density of 3.35 g/cm^3 was performed at different temperatures (Fig. 3.2.10). With a linear fit, there is a small offset [$a \neq 0$ in Eq. (3.2.4)] and $1/k = 0.0419 + 1.26 \times 10^{-4}T$ is obtained. However, a should be zero for the perfect crystal. A quadratic fit goes through zero, however, and gives $1/k = 1.97 \times 10^{-4}T - 2.34 \times 10^{-8}T^2$ (see Fig. 3.2.10).

The high-temperature saturation observed in Fig. 3.2.10 is discussed further below. The aggregate data, interestingly, have the same slope as the linear fit to the MD results, with $1/k = -0.0215 + 1.27 \times 10^{-4}T$ [69, 70]. Single-crystal experiments [70, 71] give $1/k = -0.00531 + 6.944 \times 10^{-5}T$.

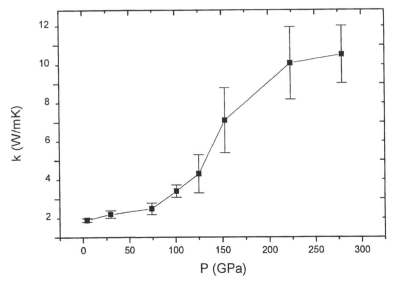

Figure 3.2.11 Thermal conductivity versus pressure at 2500 K for MgO predicted by the VIB model and MD.

Figure 3.2.11 shows the computed lattice thermal conductivity versus pressure at 2500 K. Only a very moderate increase is found to 130 GPa, with pressures corresponding to the base of the mantle. Katsura quotes a relative increase of 3.8% GPa^{-1} over the 5-GPa range of his experiment [58], but when reanalyzing his data, I obtained $5\pm3\%$ 2σ at 373 K and $6\pm3\%$ 2σ at 1473 K. The MD results give a much smaller initial slope of 0.3% GPa^{-1}at 2500 K (or $0.5/K_0$). Manga and Jeanloz [60] analyzed thermal conductivity in terms of power-law and linear models and obtained initial slopes of $(7\pm1)/K_0$ for the linear model and $(3.0\pm0.7)/K_0$ for the power-law model. An initial slope of $7/K_0$ is expected in the Debye model considered by Roufosse and Jeanloz [72], in which the thermal conductivity is given by Eq. (3.2.1) and the mean free path l is assumed to be given by

$$l = \frac{d}{\alpha\gamma T}, \tag{3.2.5}$$

where d is the interatomic distance, α is the thermal expansivity, γ is the Gruneisen parameter. The velocity $\bar{v}$ in Eq. (3.2.1) is estimated from elastic constants. Typical values then give $(\partial \ln k/\partial \ln \rho) = 7$. For comparison in percentage of change, the linear model gives $5\pm1\%$ 2σ GPa^{-1} and the power-law model gives $2\pm1\%$ 2σ GPa^{-1}.

These results are analyzed further in Fig. 3.2.12, in which $\ln k$ is plotted versus $\ln \rho$ and compared with various models. Manga and Jeanloz considered a power-law model $k/k_0 = (\rho/\rho_0)^7$ and a linear model $(k-k_0)/k_0 = 7(\rho-\rho_0)/\rho_0$, both consistent with their analysis of the data [60]. Although at high compressions the thermal conductivity begins to behave as expected in the power-law model, at low pressures it does not. The linear model also predicts too large an increase at low compressions compared with the MD results. At low pressures a simple proportionality is more consistent with the MD results.

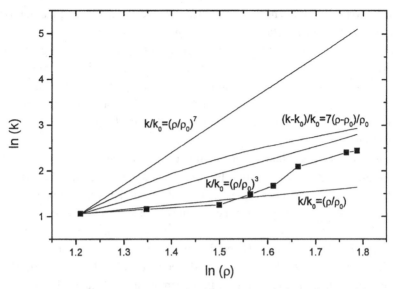

Figure 3.2.12 Predicted thermal conductivity compared with various models.

Considering the difficulty and preliminary nature of the simulations, the difficulties of the experiments, and sparsity of data, it is impossible to draw firm conclusions, but because thermal conductivity at the base of the mantle is an important geophysical parameter, further experimental and theoretical studies are justified. The present results suggest that anharmonicity and phonon–phonon scattering may behave quite differently than has been assumed with compression, with a much smaller initial pressure effect than expected from theory and limited experimental data. Perhaps the low initial slope is due to the high MD temperatures, over 1000 K higher than previous low-pressure data. One possibility is that the mean free path becomes so small at these temperatures that the thermal conductivity reaches the minimum thermal conductivity where the Debye theory breaks down and saturates. From Eq. (3.2.5) and values from Inbar and Cohen [30], $l = 5$ at 2500 K compared with $\sim$75 at 300 K, and Eq. (3.2.5) may underestimate l at very high temperatures. At higher pressures, the mean free path rises because of the decrease in thermal expansivity α. The saturation of the thermal conductivity would be consistent with a small pressure dependence at low pressures. The low-pressure data (Fig. 3.2.10) does indeed seem to indicate saturation at high T, and, as discussed above, a linear fit gives an unexpected residual term, further bolstering the quadratic behavior. Simulations at lower temperatures as functions of compression would clarify this.

3.2.5 **Melting**

It is important to know the melting temperatures, enthalpy of melting, and density of melts and solids as functions of composition to model both the chemical and the physical evolution of the Earth. Determining melting curves experimentally at extreme pressures is very difficult. There are significant discrepancies among laboratories on melting curves of the important geophysical materials Fe [73, 74] and $MgSiO_3$ perovskite [75–78]. It is also very difficult to accurately compute melting curves theoretically, because accurate potentials or electronic-structure methods are needed to obtain the forces among atoms, long simulations are needed to equilibrate and obtain thermodynamic properties of the liquid, and because free energies cannot be directly calculated, thermodynamic integrations or reversals must be performed to obtain the melting point.

Cohen and Gong [28] obtained the melting curve for MgO clusters to 300 GPa by reversing melting and crystallization by using MD simulations and the PIB model. Given the accuracy of the model for properties of the solid, it was surprising when the first experimental measurements for the melting curve of MgO showed a discrepancy of over a factor of 3 in the dT/dP slope, with the experiments showing a much shallower slope [79]. Because the MD and the lattice dynamics calculations showed that crystalline properties of MgO were well predicted by the models, such a large discrepancy could indicate a problem with the liquid simulations. Because the potentials do not use any information that is particular to the crystalline state and because they are based on fundamental physics, only one possibility seemed open – that there was a problem with the liquid simulations that was due to the use of finite clusters. Cohen and Gong used finite clusters of 64 to 1000 atoms and then extrapolated to the bulk by fitting the finite cluster results as functions of $1/L$ and then extrapolated as $1/L \to 0$, where L is the length (i.e., dimension) of the clusters.

I simulated crystalline and liquid MgO with periodic boundary conditions, i.e., with no surfaces, by using the VIB model to obtain the melting curve [80]. The changes in enthalpy and volume, $\Delta H = \Delta E + P\Delta V$ and ΔV, were obtained as functions of T and P between the solid and the liquid, which at the melting point T_m give the melting slope through the Clapeyron equation,

$$\frac{dT}{dP} = \frac{T_m \Delta V_m}{\Delta H_m}. \qquad (3.2.6)$$

Because the primary discrepancy with experiment is the melting slope, and both theory and experiment agree on the melting point at zero pressure, the zero-pressure melting point was fixed at 3200 K, and then the Clapeyron equation was integrated to give the melting curve.

Systems of 64 atoms were simulated. To check for system-size effects, some simulations were also carried out for 512-atom systems. Simulations were performed at $P = 0$, 12.5, 25, 50, and 100 GPa. At each pressure, MD runs were performed at various temperatures near the expected melting point in both solid and liquid. Initially the kinetic energies of the atoms were scaled to obtain approximately the desired temperature for ~2 ps. After equilibration, each simulation was run for an additional 6–15 ps, during which the system volume, enthalpy, and kinetic energy were monitored. The enthalpy was a constant of the motion. The temperature was calculated from the average kinetic energy, and the volume was averaged.

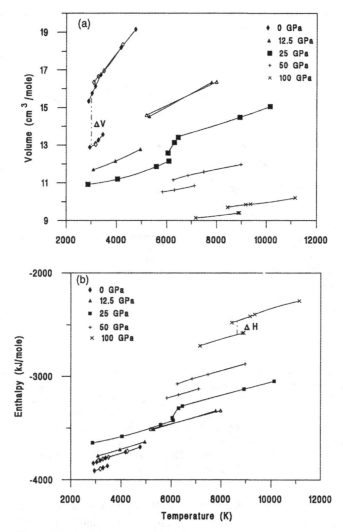

Figure 3.2.13 Volumes and enthalpies of crystalline and liquid
MgO as functions of temperature at different pressures. The open
diamonds and triangles indicate 512-atom supercells; the other
symbols indicate 64-atom supercells. The agreement for the
different sized periodic systems indicates that we are converged for
the volume and enthalpy. ΔV and ΔH are indicated at 0 GPa and
100 GPa respectively. (a) Volume, (b) enthalpy.

After these simulations were performed at several temperatures in the solid, the
solid melted, as shown from presence of diffusion, the loss in intensity of a simulated
Bragg reflection intensity, and a large drop in temperature that was due to the enthalpy
of melting. Simulations were the performed in the liquid state to obtain H and V of
the liquid phase as functions of temperature.

Results of the simulations showing the enthalpies and volumes of solid and liquid
at $P = 0$ are shown in Fig. 3.2.13. No system-size effects were seen in volume or

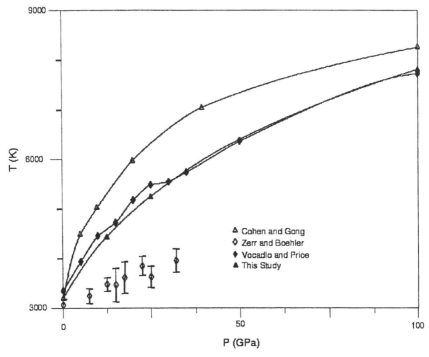

Figure 3.2.14 Melting curve for MgO. The curve for this study is computed from the MD VIB volume and enthalpy data by integration of the Clapeyron equation [Eq. (3.2.6)] starting at 3200 K at 0 GPa. The Cohen and Gong curve was obtained by reversal of the melting transition for clusters and extrapolation to infinite system size with the PIB model [28]. The Vocadlo and Price curve was obtained by the melting of periodic systems with an empirical potential [81]. There is almost perfect agreement between the current results and the results of Vocadlo and Price. There is a factor of 3 discrepancy in the slope with the experimental results of Zerr and Boehler [79].

enthalpy. For any temperature, we may obtain the difference in volume and enthalpy between solid and liquid either by interpolating or extrapolating. From the figure we can see that there is some temperature variation in both of these quantities. However, because both are increasing functions of temperature, their quotient is less sensitive.

To obtain the melting curve, the melting temperature at each pressure was estimated to be the temperature of the liquid just beyond the limit of superheating of the solid. ΔV and ΔH of melting were then calculated for those temperatures, and the Clapeyron equation was numerically integrated to obtain new estimates of the melting temperatures. This process was repeated until the melting temperature converged. The resulting melting curve is shown in Fig. 3.2.14.

The fractional change in volume on melting and ΔH_m versus pressure are shown in Fig. 3.2.15. Agreement is quite good for the volume of melting with the cluster results [28] and with the results of Vocadlo and Price [81]. The enthalpy of melting agrees within the precision of the earlier cluster results of Cohen and Gong, but there is an unexplained discrepancy with the results of Vocadlo and Price, which do not seem consistent with their melting curve.

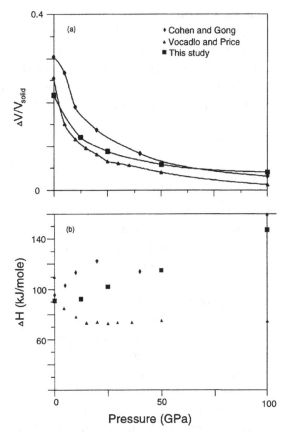

Figure 3.2.15 Fractional (a) volume of melting and (b) enthalpy of melting versus pressure.

There is a large discrepancy between the predicted melting curve and the experimental melting results of Zerr and Boehler [79], amounting to a factor of 3 in the slope dT/dP (114 K/GPa versus 35, respectively). This implies a discrepancy in ΔV_m and/or ΔH_m through the Clapeyron equation. Unfortunately, neither ΔH_m nor ΔV_m have been measured directly for any alkaline-earth oxide. Nevertheless, the value for ΔH_m is consistent with literature estimates [82]. There may be some systematic errors in the experimental melting slope. Further experiments are called for.

Cohen and Weitz [80] also studied premelting in MgO clusters and found a large increase in diffusivity before melting. This is also reflected in a low-frequency peak in $S(q, \omega)$ that develops before melting. It is not clear if this is a cluster effect or would also be observed in the bulk.

3.2.6 Defects and Diffusion

Two other properties of materials important to geophysics are chemical diffusivity and rheology or mechanical properties, and little is known about their high-pressure behavior. MgO is again an ideal test case for understanding these properties in minerals. Crystalline diffusion is too slow to be studied directly except at temperatures near melting, but diffusivity can be obtained if the free energies of vacancy formation and migration are known, as well as the attempt frequencies for hops. Ita and Cohen [83] performed free-energy integrations with MD and the VIB model to obtain diffusion as a function of pressure and temperature in MgO and derived likely rheological properties of MgO in the Earth [84].

The self-diffusion coefficient D is [85]

$$D = Z_f \frac{Z_m}{6} l^2 \nu \, \exp \left[\frac{(\Delta G_f / W) + \Delta G_m}{k_b T} \right], \tag{3.2.7}$$

where Z_f is the number of equivalent ways of forming a vacancy type, Z_m is the number of equivalent diffusion paths, l is the jump distance, ν is the attempt frequency, ΔG_f and ΔG_m are the energies of formation and migration, respectively, and W is a solubility factor that depends on defect associations. For Schottky defects (uncorrelated vacancies) $W = 2$, $Z_f = 1$, and $Z_m = 12$ for rock-salt-structured (B1) crystals such as MgO. For bound pairs (highly correlated defects) $W = 1$ and $Z_f = 6$. Symmetry and energy considerations determine the value of Z_m. In either case, $l^2 = a^2/2$, where a is the cubic cell parameter for rock salt. All of the above quantities can be determined from MD simulations [83]. The binding energy was computed from MD simulations, and it was found that the vacancies should be unbound Schottky pairs above 1000 K, so that $W = 2$, $Z_f = 1$, and $Z_m = 12$.

The free energies of formation ΔG_f and migration ΔG_m required in Eq. (3.2.7) require free-energy integration techniques [86], as the free energy is not directly accessible in an MD simulation. Ita and Cohen [83] used the adiabatic switching technique [87], in which one performs a single simulation, varying the effective potential or state of the system from one state to another along a progress variable λ that describes the state and varies from 0 to 1. Then the Helmholtz free-energy difference ΔF is given by

$$\Delta F = \int_0^1 \left\langle \frac{\partial H(\lambda)}{\partial \lambda} \right\rangle_\lambda d\lambda, \tag{3.2.8}$$

and ΔG is obtained when PV is added. The free energies of formation ΔG_f were obtained by a simulation for a perfect crystal and a defective crystal with one Mg and O vacancy in each periodic supercell and by use of λ to vary from the VIB potential to an Einstein crystal (each atom attached to its equilibrium site by a perfect spring), for which the free energy is known. Because the final state was the same per atom (atoms do not interact with each other in the Einstein crystal), ΔG_f could be obtained as the difference in the two free-energy differences. The migration free energy ΔG_m was obtained by use of λ to vary an atom position neighboring a vacancy from its equilibrium position up to the saddle point, as described by Milman et al. [88]. Details are given in Ref. [83].

Figure 3.2.16 shows the predicted diffusion constants as functions of P and T. Ionic conductivity measurements indicate that Mg diffusion (D_{Mg}) is controlled by impurities [89] whereas O diffusion (D_O) is intrinsic in nature and directly comparable with the simulations. The predicted D_O is in excellent agreement with experiment.

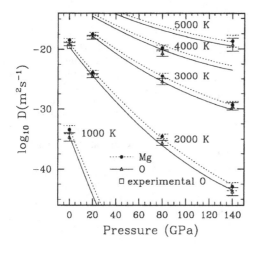

Figure 3.2.16 Predicted pressure and temperature dependence of the self-diffusion coefficients in MgO. Curves represent the best fit to the coefficients by use of the activation energy–volume relation given by Eq. (3.2.9).

The computed diffusion constants were fit to the relation

$$
\begin{aligned}
\ln D = \ln(a^2 v) + S_0^* + P S_0^{*'} \\
- (E_0^* + P V_0^* + P^2 V_0^{*'})/k_b T,
\end{aligned}
\tag{3.2.9}
$$

giving the zero-pressure activation entropy $S_0^* = 3-(4)k_b$, its pressure derivative $S_0^{*'} = 0.03-(0.02)k_b$, activation energy $E_0^* = 9.0-(9.4) \times 10^{-19}$ J, activation volume $V_0^* = 16.0-(16.7)$ Å^3, and its pressure derivative $V_o^{*'} = -0.031-(-0.038)$ Å^3/GPa for Mg–(O). The curves in Fig. 3.2.16 were from Eq. (3.2.9) with these parameters.

Estimates are often made with the assumption that diffusion scales with the temperature relative to the melting temperature at a given pressure, by use of the homologous temperature relation [90]:

$$
D = D_0 \exp(g T_m/T).
\tag{3.2.10}
$$

Using the effective diffusion coefficient, $D_{\text{eff}} = 2D_{\text{Mg}}D_O/(D_{\text{Mg}} + D_O)$ for D [91], Ita and Cohen tested the homologous temperature model by using the theoretical melting curve of Cohen and Weitz [80] discussed above (Fig. 3.2.14) and the extrapolated experimental melting curve of Zerr and Boehler [79], which has a lower dP/dT. The homologous relation worked well with the theoretical melting relation, but the experimental melting curve is not consistent with the present diffusion results within the homologous temperature model. Ita and Cohen also tried using only zero-pressure diffusion results in Eq. (3.2.10) and found that extrapolations to high pressure with the melting curve were reasonably reliable (Fig. 3.2.17).

3.2.7   Summary and Conclusions

This chapter should make it clear that the ground-state electronic and energetic properties of MgO are well understood and can be predicted accurately with fundamental physics. MgO is an ionic material but has the added complication that the O^{2-} ion

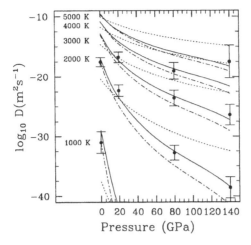

Figure 3.2.17 Test of homologous melting relation [Eq. (3.2.10)]. Solid curves: fits made with the theoretical melting curve [80]; short-dashed curves: fits made with the extrapolated experimental melting curve [79]; dash–dotted curves: fits made with the theoretical melting curve and only 0-GPa diffusion coefficients [83].

depends on its environment. The VIB model is a reliable technical for studying the properties of MgO, and self-consistent methods have also been applied. Thermal properties, elasticity, melting, and diffusivity have been studied for MgO as functions of pressure, and generally excellent agreement is found with experiment, the notable exception being the melting curve of MgO. MgO will remain a useful benchmark material for experimental and theoretical understanding of minerals at high pressures.

Acknowledgments

This paper is dedicated to Y. Matsui for paving the way and showing that theory could be used fruitfully to study mineral properties at high pressures. Three important papers of his that helped lead me into this field are Refs. [92–94]. This work was done in collaboration with Z. Gong, I. Inbar, D. G. Isaak, J. Ita, M. Kluge, H. Lu, and J. Weitz, without whom it would not have been possible. This work was supported by U.S. National Science Foundation grant EAR94-18934. MD computations were performed on the Cray J916/12-1024 at the Geophysical Laboratory, Carnegie Institution of Washington, purchased with support from U.S. National Science Foundation grant EAR 95-12627 and the Keck Foundation.

References

[1] P. Hohenberg and W. Kohn, Phys. Rev. **136**, 864 (1964).

[2] W. Kohn and L. J. Sham, Phys. Rev. A **140**, 1133 (1965).

[3] L. Hedin and B. I. Lundqvist, J. Phys. C **4**, 2064 (1971).

[4] M. S. T. Bukowinski, J. Geophys. Res. **85**, 285 (1980).

[5] M. S. T. Bukowinski, Geophys. Res. Lett. **12**, 536 (1985).

[6] S. H. Wei and H. Krakauer, Phys. Rev. Lett. **55**, 1200 (1985).

[7] D. J. Singh, *Planewaves, Pseudopotentials, and the LAPW Method* (Kluwer Academic, Boston, 1994).

[8] M. J. Mehl, R. E. Cohen, and H. Krakauer, J. Geophys. Res. **93**, 8009 (1988); **94**, 1997 (1989).

[9] A. F. Kohan and G. Ceder, Phys. Rev. B **54**, 805 (1996).

[10] G. Kalpana, B. Palanivel, and M. Rajagopalan, Phys. Rev. B **52**, 4 (1995).

[11] O. Schutt, P. Pavone, W. Windl, K. Karch, and D. Strauch, Phys. Rev. B **50**, 3746 (1994).

[12] B. B. Karki, L. Stixrude, S. J. Clark, M. C. Warren, G. J. Ackland, and J. Crain, Am. Mineral. **82**, 51 (1997).

[13] M. I. McCarthy and N. M. Harrison, Phys. Rev. B **49**, 8574 (1994).

[14] K. Doll, M. Dolg, and H. Stoll, Phys. Rev. B **54**, 13529 (1996).

[15] J. M. Zuo, M. O'Keeffe, P. Rez, and J. C. H. Spence, Phys. Rev. B **78**, 4777 (1997).

[16] F. Illas, A. Lorda, J. Rubio, J. B. Torrance, and P. S. Bagus, J. Chem. Phys. **99**, 389 (1993).

[17] C. Muhlhausen and R. G. Gordon, Phys. Rev. B **23**, 900 (1981).

[18] M. J. Mehl, R. J. Hemley, and L. L. Boyer, Phys. Rev. B **33**, 8685 (1986).

[19] R. E. Cohen, L. L. Boyer, and M. J. Mehl, Phys. Rev. B **35**, 5749 (1987).

[20] L. L. Boyer, M. J. Mehl, J. L. Feldman, J. R. Hardy, J. W. Flocken, and C.Y. Fong, Phys. Rev. Lett. **54**, 1940 (1985).

[21] R. E. Cohen, L. L. Boyer, and M. J. Mehl, Phys. Chem. Minerals **14**, 294 (1987).

[22] R. E. Cohen, Geophys. Res. Lett. **14**, 1053 (1987).

[23] A. P. Jephcoat, R. J. Hemley, and H. K. Mao, Physica B **150**, 115 (1988).

[24] A. P. Jephcoat, R. J. Hemley, H. K. Mao, R. E. Cohen, and M. J. Mehl, Phys. Rev. B **37**, 4727 (1988).

[25] R. E. Cohen, M. J. Mehl, and L. L. Boyer, Physica B **150**, 1 (1988).

[26] D. G. Isaak, R. E. Cohen, and M. J. Mehl, J. Geophys. Res. **95**, 7055 (1990).

[27] H. Cynn, D. G. Isaak, R. E. Cohen, M. F. Nicol, and O. L. Anderson, Am. Mineral. **75**, 439 (1990).

[28] R. E. Cohen and Z. Gong, Phys. Rev. B **50**, 12301 (1994).

[29] G. H. Wolf and M. S. T. Bukowinski, Phys. Chem. Minerals **15**, 209 (1988).

[30] I. Inbar and R. E. Cohen, Geophys. Res. Lett. **22**, 1533 (1995).

[31] H. Lu and R. E. Cohen, "Comparing charge densities of alkaline earth oxides generated from self-consistent calculations and a model" (unpublished, 1988).

[32] P. J. Edwardson, Phys. Rev. Lett. **63**, 55 (1989).

[33] H. Zhang and M. S. T. Bukowinski, Phys. Rev. B **44**, 2495 (1991).

[34] L. L. Boyer, H. T. Stokes, and M. J. Mehl, Ferroelectrics **164**, 177 (1995).

[35] H. T. Stokes, L. L. Boyer, and M. J. Mehl, Phys. Rev. B **54**, 7729 (1996).

[36] L. L. Boyer, H. T. Stokes, and M. J. Mehl, Ferroelectrics **194**, 173 (1997); Phys. Rev. Lett. **84**, 709 (2000).

[37] P. Cortona and A. V. Monteleone, J. Phys. Condens. Matter **8**, 8983 (1996).

[38] N. C. Pyper, Philos. Trans. R. Soc. **352**, 89 (1995).

[39] N. C. Pyper, J. Phys. Condens. Matter **8**, 5509 (1996).

[40] M. Wilson, P. A. Madden, N. C. Pyper, and J. H. Harding, J. Chem. Phys. **104**, 8068 (1996).

[41] B. Reynard and G. D. Price, Geophys. Res. Lett. **17**, 689 (1990).

[42] A. Agnon and M. S. T. Bukowinski, Geophys. Res. Lett. **41**, 7755 (1990).

[43] M. Causa, R. Dovesi, C. Pisani, and C. Roetti, Phys. Rev. B **33**, 1308 (1986).

[44] H. -K. Mao and P. M. Bell, J. Geophys. Res. **84**, 4533 (1979).

[45] Y. Sumino, O. L. Anderson, and I. Suzuki, Phys. Chem. Minerals **9**, 38 (1983).

[46] Z. P. Chang and G. R. Barsch, J. Geophys. Res. **74**, 3291 (1969).

[47] S. Touloukian, *Thermal Conductivity: Nonmetallic Solids* (IFI/Plenum, New York, 1970).

[48] T. Duffy, R. J. Hemley, and H.-K. Mao, Phys. Rev. Lett. **74**, 1371 (1995).

[49] T. S. Duffy and T. J. Ahrens, Geophys. Res. Lett. **20**, 1103 (1993).

[50] O. L. Anderson, H. Oda, and D. Isaak, Geophys. Res. Lett. **19**, 1987 (1992).

[51] E. Wasserman, L. Stixrude, and R. E. Cohen, Phys. Rev. B **53**, 8296 (1996).

[52] O. L. Anderson and D. G. Isaak, J. Phys. Chem. Solids **54**, 221 (1993).

[53] D. G. Isaak, O. L. Anderson, and T. Goto, Phys. Chem. Minerals **16**, 704 (1989).

[54] D. G. Isaak, O. L. Anderson, and R. E. Cohen, Geophys. Res. Lett. **19**, 741 (1992).

[55] T. Duffy and T. J. Ahrens, in *High-Pressure Research: Application to Earth and Planetary Sciences*, Y. Syono and M. H. Manghnani, eds. (American Geophysical Union, Washington, D.C., 1992), pp. 197–206.

[56] M. J. Mehl, B. M. Klein, and D. A. Papaconstantopoulos, in *Intermetallic Compounds, Principles and Practice*, J. H. Westbrook and R. L. Fleischer, eds. (Wiley, London, 1994), Vol. I.

[57] D. G. Isaak, Phys. Earth Planet. Inter. **80**, 37 (1993).

[58] T. Katsura, Phys. Earth Planet. Inter. **101**, 73 (1997).

[59] H. Yukutake and M. Shimada, Phys. Earth Planet. Inter. **17**, 193 (1978).

[60] M. Manga and R. Jeanloz, J. Geophys. Res. **102**, 2999 (1997).

[61] J. M. Brown, Geophys. Res. Lett. **13**, 1509 (1986).

[62] R. Berman, *Thermal Conduction in Solids* (Oxford U. Press, Oxford, U.K., 1976).

[63] S. W. Kieffer, J. Geophys. Res. **81**, 3025 (1976).

[64] R. E. Peierls, *Quantum Theory of Solids* (Clarendon, Oxford, U.K., 1955).

[65] R. E. Cohen, Reviews of High Press. Sc. and Technol. 7, 160–162 (1998).

[66] M. P. Allen and D. J. Tildesley, *Computer Simulation of Liquids* (Clarendon, Oxford, U. K., 1989).

[67] A. J. C. Ladd, B. Moran, and W. G. Hoover, Phys. Rev. B **34**, 5058 (1986).

[68] Y. H. Lee, R. Biswas, C. M. Soukoulis, C. Z. Wang, C. T. Chan, and K. M. Ho, Phys. Rev. B **43**, 6573 (1991).

[69] S. P. Clark, Jr., in *The Earth's Crust and Upper Mantle*, P. J. Hart, ed. (American Geophysical Union, Washington, D.C., 1969), pp. 622–626.

[70] C. Clauser and E. Huenges, in *Rock Physics and Phase Relations: A Handbook of Physical Constants*, T. J. Ahrens, ed. (American Geophysical Union, Washington, D.C., 1995), pp. 105–126.

[71] H. Kanamori, N. Fujii, and H. Mizutani, J. Geophys. Res. **73**, 595 (1968).

[72] M. C. Roufosse and R. Jeanloz, J. Geophys. Res. **88**, 7399 (1983).

[73] W. W. Anderson and T. J. Ahrens, J. Geophys. Res. **101**, 5627 (1996).

[74] R. Boehler, Geochim. Cosmochim. Acta **60**, 1109 (1996).

[75] D. L. Heinz, E. Knittle, J. S. Sweeney, Q. Williams, and R. Jeanloz, Science **264**, 279 (1994).

[76] J. S. Sweeney and D. L. Heinz, EOS Trans. Am. Geophys. Union **76**, F553 (1995).

[77] J. S. Sweeney and D. L. Heinz, Geophys. Res. Lett. **20**, 855 (1993).

[78] A. Zerr and R. Boehler, Science **262**, 553 (1993).

[79] A. Zerr and R. Boehler, Nature (London) **371**, 506 (1994).

[80] R. E. Cohen and J. Weitz, in *High Pressure–Temperature Research: Properties of Earth and Planetary Materials*, M. H. Manghnani and T. Yagi, eds. (American Geophysical Union, Washington, D.C., 1998), pp. 185–196.

[81] L. Vocadlo and G. D. Price, Phys. Chem. Minerals **23**, 42 (1996).

[82] M. W. Chase, Jr., et al., *"JANAF thermochemical tables,"* J. Phys. Chem. Ref. Data Suppl. **14**, (1985).

[83] J. Ita and R. E. Cohen, Phys. Rev. Lett. **79**, 3198 (1997).

[84] J. Ita and R. E. Cohen, Geophys. Res. Lett. **25**, 1095 (1998).

[85] R. J. D. Tilley, *Defect Crystal Chemistry and its Applications* (Blackie, London, 1987).

[86] D. Frenkel, in *Molecular-Dynamics Simulation of Statistical-Mechanical Systems*, G. Ciccotti and W. G. Hoover, eds., Vol. XCVII of International School of Physics Enrico Fermi Proceedings Series (North-Holland, New York, 1986), pp. 151–188.

[87] J. E. Hunter, W. P. Reinhardt, and T. F. Davis, J. Chem. Phys. **99**, 6856 (1993).

[88] V. Milman, M. C. Payne, V. Heine, R. J. Needs, J. S. Lin, and M. H. Lee, Phys. Rev. Lett. **70**, 2928 (1993).

[89] D. R. Sempolinski and W. D. Kingery, J. Am. Ceram. Soc. **63**, 664 (1980).

[90] C. G. Sammis, J. C. Smith, and G. Schubert, J. Geophys. Res. **86**, 10707 (1981).

[91] J. P. Poirier, *Creep of Crystals* (Cambridge U. Press, Cambridge, U.K., 1985).

[92] Y. Matsui and K. Kawamura, Nature (London) **285**, 648 (1980).

[93] Y. Matsui K. Kawamura and Y. Syono, in *High Pressure Research in Geophysics*, S. Akimoto and M. H. Manghnani, eds. (Center for Academic Publications, Tokyo, Japan, 1982), pp. 511–524.

[94] Y. Matsui and K. Kawamura, in *Materials Science of the Earth's Interior*, I. Sunagawa, ed. (Terra Scientific, Tokyo, Japan, 1984), pp. 3–23.

Chapter 3.3

First-Principles Theoretical Study
of the High-Pressure Phases of MnO and FeO:
Normal and Inverse NiAs Structures

Z. Fang, K. Terakura, H. Sawada, I. Solovyev, and T. Miyazaki

The phase stability of transition-metal monoxides MnO and FeO under ultrahigh
pressure, which reaches the range in the Earth's lower mantle, was studied with
the first-principles calculations based on density function theory. The plane-wave
basis pseudopotential method was used to perform the structure optimization
efficiently, and the electron–electron interaction was treated by the generalized
gradient approximation (GGA) supplemented by the LDA + U method
(LDA is local-density approximation). Two related structures, normal NiAs
(nB8) and inverse NiAs (iB8) types, are emphasized. Our results predict that the
high-pressure phase of MnO should take the nB8 structure rather than the CsCl
(B2) structure and that a metastable nonmagnetic B1 structure can be realized
for MnO in the intermediate pressure range. A very unique iB8 structure rather
than the nB8 structure is predicted as the high-pressure phase of FeO, although no
materials have ever been known to take the iB8 structure. The novel feature of the
iB8 FeO is that the system should be a band insulator in the antiferromagnetic state
and that the existence of a bandgap leads to special stability of the phase. The larger
c/a ratios for both nB8 MnO and iB8 FeO were explained based on our analysis
of the cation radius–anion radius ratios versus c/a for series of similar materials.

3.3.1 Introduction

The high-pressure phases of metal monoxides [including alkaline-earth-metal monoxides, transition-metal monoxides (TMMOs), etc.], with the rock-salt (B1) structure
at normal pressure and room temperature, are important for both condensed-matter

H. Aoki et al. (eds), *Physics Meets Mineralogy* © 2000 Cambridge University Press.

physics and earth science because their crystal structure is simple and moreover MgO and FeO are considered to be important constituents of the Earth's deep mantle. TMMOs have been occupying a special position in condensed-matter physics for decades as prototypical examples of the Mott insulator [1]. They serve as typical systems for studying of the basic nature of the insulating state, superexchange, and orbital magnetism. The basic properties of TMMOs are specified by the electron correlation, whose strength is measured by U/W, where U is the effective Coulomb interaction integral between d electrons and W is the d-band width. As the d-band width is directly controlled by pressure, studies of the high-pressure behaviors of TMMO have been regarded as a useful way to understand their basic properties. Insulator–metal transition, magnetic-moment collapse, and so on, are possible transitions that may occur under high pressure. In addition, structural phase transition induced by pressure is also an attractive subject; particularly for FeO, its stable high-pressure phase may have important implications in earth science.

It is found experimentally that almost all of the alkaline-earth-metal monoxides undergo a structure phase transition from the B1 to the CsCl type (B2) structure or distorted B2 structure under high pressure [2–5], except MgO, for which no phase transition has been observed up to 230 GPa [6]. However, TMMOs behave rather differently from alkaline-earth-metal monoxides.

For both MnO or FeO, an antiferromagnetic (AF) rhombohedrally distorted B1 (rB1) phase (compressing for MnO [7] and stretching for FeO [8, 9] along a body diagonal of the rock-salt cell) is observed experimentally under normal pressure, when temperature is decreased below the Néel temperature. It is also known experimentally that, with increasing pressure, both distortions of MnO and FeO in rB1 phase are enhanced [7, 9].

A recent shock-compression experiment on MnO showed the existence of a possible high-pressure phase above 90 GPa [10], although some earlier static high-pressure experiments on MnO indicated absence of phase transition up to a pressure of 90 GPa at room temperature [11, 12]. However, the crystal structure of the high-pressure phase has not been determined yet. In the plane of the critical pressure versus cation ionic radius, alkaline-earth monoxides (except MgO) are connected by a smooth line; MnO is located at a smoothly extrapolated part of the line whereas FeO is clearly off the line. It was therefore speculated that MnO may take the B2 structure under high pressure. More recently, a high-pressure experiment with static compression at room temperature was performed for MnO, and x-ray-diffraction patterns were obtained for various pressures [7]. Indeed, this static-compression experiment reproduced the phase transition, but the x-ray diffraction cannot give us an unambiguous answer about the crystal structure because of the limited range of diffraction angle and, moreover, multiphase coexistence.

In the case of $Fe_{1-x}O$, shock-compression experiments indicated the existence of a pressure-induced phase transition at ~ 70 GPa [13, 14], although the room-temperature static-compression experiments in the diamond-anvil cell did not detect this phase transition up to 120 GPa [9]. Metallic behavior of $Fe_{1-x}O$ has been observed at

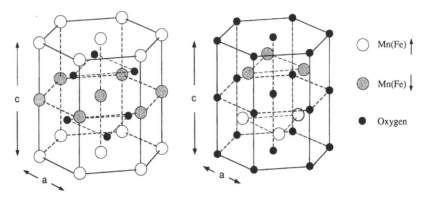

Figure 3.3.1 Comparison of AF normal B8 (nB8) and inverse B8 (iB8) structures. Oxygen is denoted by a small black filled circle, Mn (or Fe) with up-spin by a large open circle, and Mn (or Fe) with down-spin by a large gray filled circle. These two structures are definitely different, and there is no inversion symmetry for the AF iB8 structure.

static pressure and elevated temperatures [15] and also in shock-resistivity measurements [16] around the pressure range at which the phase change was observed by shock compression. This high-pressure phase of $Fe_{1-x}O$ was recently assigned to be the NiAs (B8) type structure by the analysis of x-ray-diffraction peak positions [17,18]. On the analogy of all the known examples of the transition-metal compounds with the B8 structure, a natural idea for the B8 (NiAs) FeO may be such that Fe occupies the Ni site and O the As site. This structure is given as nB8 for short hereafter. However, another structure, which is called inverse B8 (iB8), is possible if the Fe and the O positions are exchanged, although to our knowledge no transition-metal compounds with the iB8 structure exist. A comparison of nB8 with iB8 structure is shown in Fig. 3.3.1.

The first-principles calculations based on density function theory (DFT) [19–22] enable us to obtain valuable hints about the stable high-pressure phases of MnO and FeO. The pioneer work on silica [23] shows us a good example about how the first-principles density function calculations can be used in mineralogy as a powerful tool for studying the possible structural phase transition. Recent progresses on this method makes it possible for us to study very large systems with high accuracy. In the present work, the electronic structures and the phase stability of MnO and FeO under pressure were studied with the first-principles plane-wave basis pseudopotential calculations. It was predicted that the high-pressure phase of MnO should take the nB8 structure rather than the B2 structure. A unique AF iB8 structure rather than the nB8 structure was predicted as the high-pressure phase of FeO. The novel feature of this AF iB8 FeO is that the system should be a band insulator in the ordered AF state and the existence of a bandgap leads to special stability of this phase. In Sections 3.3.2 and 3.3.3 brief introductions to the first-principles calculations based on DFT and the plane-wave basis pseudopotential method are given, respectively. The results and discussion are given in Section 3.3.4. Finally, Section 3.3.5 is the summary of our results.

3.3.2 First-Principles Calculations Based on DFT

The development of DFT [19] has reduced a complicated many-electron problem to an effective single-electron problem without loss of rigor, so that the problem has become computationally solvable from first principles. This scheme is definitely different from some other similar methods such as the Hartree–Fock (HF) method in the sense that electron density $\rho(\mathbf{r})$ has been selected as the basic variable, instead of the many-electron wave function $\psi(\mathbf{r}_1, s_1, \mathbf{r}_1, s_2, \ldots, \mathbf{r}_n, s_n)$, which is very difficult to deal with.

DFT is based on the following two theorems [20]:

1. The nondegenerate ground state of a given many-electron system is uniquely determined by the one-electron density $\rho(\mathbf{r})$. For example, with the Kohn–Sham prescription, the total energy of the system in the ground state can be expressed as

$$E[\rho(\mathbf{r})] = T_s[\rho(\mathbf{r})] + \frac{1}{2}\int d^3r d^3r' \frac{\rho(\mathbf{r})\rho(\mathbf{r}')}{|\mathbf{r}-\mathbf{r}'|} + \int d^3r v(\mathbf{r})\rho(\mathbf{r})$$
$$+ E_{xc}[\rho(\mathbf{r})], \tag{3.3.1}$$

where the first term is the kinetic energy of a fictitious noninteracting electron system with the given density $\rho(\mathbf{r})$, the second term is the classical electrostatic interaction energy, the third term is the interaction of electrons with the potential from the nuclei, and the last term is the exchange-correlation energy that contains all other complicated contributions of the real interaction. Hartree atomic units ($m = |e| = \hbar = 1$) are used in this chapter.

2. The energy functional $E[\rho(\mathbf{r})]$ attains its minimum when $\rho(\mathbf{r})$ is the ground-state density.

The one-electron density $\rho(\mathbf{r})$ can be obtained self-consistently by the solution of the following effective one-electron problem [Eqs. (3.3.2)–(3.3.4)] [21], with the effective one-electron potential $v_{\text{eff}}(\mathbf{r})$ depending on the one-electron density $\rho(\mathbf{r})$ in reverse:

$$\left[-\frac{1}{2}\Delta + v_{\text{eff}}(\mathbf{r})\right]\psi_i(\mathbf{r}) = \epsilon_i \psi_i(\mathbf{r}), \tag{3.3.2}$$

$$v_{\text{eff}}(\mathbf{r}) = v(\mathbf{r}) + \int d^3r' \frac{\rho(\mathbf{r}')}{|\mathbf{r}-\mathbf{r}'|} + \frac{\delta E_{xc}[\rho(\mathbf{r})]}{\delta\rho(\mathbf{r})}, \tag{3.3.3}$$

$$\rho(\mathbf{r}) = \sum_{i:occ} |\psi_i(\mathbf{r})|^2. \tag{3.3.4}$$

Once we obtain the self-consistent one-electron density $\rho(\mathbf{r})$ from the above equations, we can obtain the total energy of the system in the ground state by using Eq. (3.3.1). However, to make the DFT computationally solvable, we must know the functional form of the exchange-correlation energy E_{xc} as a function of the electron density

$\rho(\mathbf{r})$. As the true functional form is not known, some approximations for the exchange-correlation energy have to be used.

Most of the first-principles electronic-structure calculations with DFT have been based on the local-density approximation (LDA) [19], in which the exchange-correlation energy is approximated as

$$E_{xc}^{LDA} = \int d^3\mathbf{r}\varepsilon_{xc}(\rho(\mathbf{r}))\rho(\mathbf{r}). \tag{3.3.5}$$

As for the exchange-correlation energy density $\varepsilon_{xc}(\rho)$, the results for uniform electron gas with a given electron density ρ is used [22].

Despite its dramatic simplification, LDA has made significant contributions to condensed-matter physics. However, several pieces of evidence have accumulated that indicate some systematic errors of LDA, because LDA underestimates the exchange energy significantly [24]. Proposals have been made to go beyond LDA [25, 26]. Recent studies [27, 28] suggest that the gradient correction to LDA, which is usually called a generalized gradient approximation (GGA) [29], can improve most, but not necessarily all, of the quantitative problems of LDA, especially for $3d$ transition metals and compounds. In the GGA, the exchange-correlation energy per particle is allowed to depend on not only the electron density at point $\mathbf{r}$, but also on the electron-density gradients. This generalizes Eq. (3.3.5) to the form

$$E_{xc}^{GGA} = \int d^3\mathbf{r}\epsilon_{xc}[\rho(\mathbf{r}), \nabla\rho(\mathbf{r})], \tag{3.3.6}$$

where the function ϵ_{xc} is chosen by some set of criteria and a variety of different forms for the function ϵ_{xc} have been proposed and applied in the literature. Here we used the form given by Perdew in 1991 [29].

Moreover, for the system with the strong electron correlation, another method, called LDA $+ U$, is usually used [30]. In the LDA $+ U$ method, the strong correlation between localized d electrons is explicitly taken into account through the screened effective Coulomb interaction parameter U_{eff}, so that the underestimation of correlation effects by GGA can be compensated.

In the present work, first-principles calculations based on DFT were performed to study the high-pressure phases of MnO and FeO. The electron–electron interaction is treated by the GGA as used in similar calculations [31–33]. However, the GGA is still not powerful enough to treat a strong correlation system such as that of Mott insulators. Therefore the GGA calculation is supplemented by the LDA $+ U$ method, particularly in the low-pressure regime in which MnO and FeO are regarded as typical Mott insulators.

3.3.3 Plane-Wave Basis Pseudopotential Method

In Section 3.3.2 we explained how we convert a many-electron problem into a computationally solvable effective single-electron problem by using DFT based on LDA or GGA. However, even for a single-electron problem, it is not simple to solve the Schrödinger equation with a general form of a potential. Most of the first-principles calculations based on DFT adopt the plane waves as the basis functions to solve the single-electron problem. There are several reasons for this. First, the calculation of the atomic force is simple [especially for first-principles molecular dynamics (FPMD)] because the plane waves do not depend on the atomic position. Second, the use of the fast Fourier transformation (FFT) reduces the number of operations. Therefore, despite the use of a large number of basis functions, the work load can be rather modest. Third, the numerical accuracy can be systematically controlled by changing the number of plane waves.

Nevertheless, the plane-wave expansions of eigenfunctions can be practical only with the use of pseudopotentials. The efficiency and the accuracy of a calculation depend crucially on the quality of pseudopotentials. Practically, two kinds of pseudopotential schemes have been used.

1. Norm-conserving pseudopotential [34]: The pseudo wave function must coincide with the all-electron wave function for $r > r_{cl}$, where r_{cl} is the core radius for a given angular momentum l. For $r < r_{cl}$, the pseudo wave function does not have any node but its norm in the core region must be the same as that of the all-electron wave function. This norm conservation is crucial to the self-consistent calculations.

2. Ultrasoft pseudopotential [35]: The basic requirement for using a pseudopotential is to reduce the number of plane waves needed to express a pseudo wave function, while keeping the eigenenergies to be correctly reproduced within a finite energy range of interest. We call this kind of pseudopotential soft pseudopotential. With the requirement of norm conservation, the pseudopotential will not become dramatically soft. A completely new idea of removing the norm-conservation requirement was introduced by Vanderbilt. Vanderbilt's ultrasoft pseudopotential can reduce significantly the number of plane waves to express the pseudo wave functions, though additional calculation on the charge deficit has to be introduced and the whole algorithm becomes rather cumbersome.

In the present study, the plane-wave basis pseudopotential method is used to perform the structural optimization efficiently. The $2p$ states of oxygen and $3d$ states of Mn and Fe are treated by the Vanderbilt ultrasoft pseudopotential. For other states, we use the optimized norm-conserving pseudopotentials. The cutoff energies for the wave functions and charge density expansions are 36 and 200 Ry, respectively, for all calculations. A tetrahedron method is adopted for the k-space integration. For different structures, we choose the sampling of k-points to give the absolute total energy convergence to better than 1 mRy/fomular unit. Moreover, note that the convergence in the relative energies between different structures is generally an order of magnitude better than the convergence in the absolute total energies. High credibility of the present approach has been well confirmed. However, we emphasize that all of our calculations are for stoichiometry materials at zero temperature.

3.3.4 **Results and Discussion**

With the approximations mentioned above, highly converged total energy calculations were performed for different crystal structures (B1, B2, nB8, iB8) and spin structures [AF, antiferromagnetic, ferromagnetic (FM), and nonmagnetic (NM)]. For all the systems, the structures are fully optimized; i.e., with each fixed volume we should optimize the c/a ratio. Forces acting on atoms, stress tensors on unit cells, and the total energies are used as guides in the structural optimization process.

The most important results obtained from our calculations are shown in Figs. 3.3.2 (a) and 3.3.2 (b), which give the calculated total energies versus volume for MnO and FeO, respectively, with different structures. To avoid confusion, only the important phases are presented here, although more phases have been treated. With the calculated total energies, it is possible to compare the relative stability of different phases.

Before going into the discussion about the high-pressure phases, first let us pay attention to the low-pressure region for a while, in which both MnO and FeO should take the AF rB1 structure. By fitting the calculated total energies versus volume to Murnaghan's equation of state [36], we can determine the zero-pressure lattice parameter a_0, the bulk modulus B_0, and the pressure derivative of the bulk modulus B' for the AF rB1 phases. Table 3.3.1 shows these fitted equation-of-state parameters, together with the calculated magnetic moments of the AF rB1 phase of MnO and FeO within a sphere radius (2.4 Å for MnO and 2.2 Å for FeO) and comparison with some experimental results. We can see good agreement between the theoretical and experimental results, although the experimental data scatter a bit. The calculations based on LDA + U are explained below.

Moreover, our calculations clearly show that for both MnO and FeO the rhombohedrally distorted B1 structures are more stable than the B1 structures without distortion (see Ref. [37] for details). The sign of the rhombohedral distortion (compressing for MnO and stretching for FeO) can be predicted correctly by our calculations. It is also shown that two distortions will both be enhanced with increasing pressure, being consistent with the experimental results. All of these seem to suggest that the present calculations based on GGA can account well for the properties of MnO and FeO. However, we emphasize that, as is already well known [32, 38], band-structure calculations based on LDA or GGA even with lattice optimization incorrectly predict a metallic ground state of FeO. For MnO, although an insulating ground state can be obtained, a much smaller bandgap is predicted compared with that of the experimental data [39] (~40% by our GGA calculation). These reflect the fact that the GGA is not yet powerful enough to treat strong correlation systems such as those of Mott insulators, although GGA gives much better results than the simpler LDA. However, for the high-pressure region, because of the reduction of electron correlation that is due to the increase of electron screening, we can expect that the GGA will give a better description than in the low-pressure region.

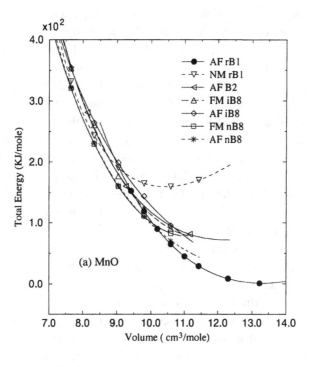

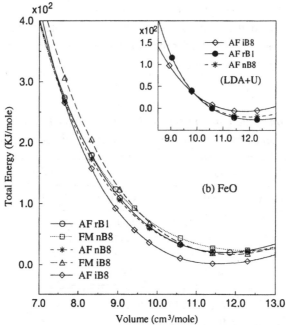

Figure 3.3.2 Total energies of (a) MnO and (b) FeO as functions of volume for different phases. The least-squares-fitted curves to Murnaghan's equation of state are shown. For FeO, the results based on the LDA $+ U$ calculations with $U_{\mathrm{eff}} = 4$ eV are shown in the inset.

Now we move into the discussion about the high-pressure phases. From Fig. 3.3.2(a), we can see clearly that the most stable phase of MnO at high pressure is the nB8 structure rather than the B2 structure, which definitely has a higher total energy than the nB8 phase. Within the whole volume range, the FM and NM B2 phases have higher energies than the AF B2 phase, so only the result for AF B2 phase is shown in Fig. 3.3.2(a). A detailed comparison of the total energies for the FM and AF nB8 structures predicts that the FM nB8 structure of MnO has the lowest energy under high pressure rather than the AF nB8 structure, although the energy difference is rather marginal. A first-order phase transition is expected to occur from the insulating AF rB1 structure to the metallic FM nB8 structure. This can be seen more clearly from Fig. 3.3.3, which shows the calculated enthalpy as a function of pressure for some different phases. The cross points of these enthalpy–pressure curves give the phase-transition critical pressures.

This theoretical result can explain the recent experiment [7] on MnO in a high-pressure range (>120 GPa) very well. For example, the consistency of the assignment of the experimental x-ray-diffraction peaks at 137 GPa by the nB8 structure is shown in Table 3.3.2. Almost an exact fit of peak positions and good agreement of the intensity profile can be obtained, except one peak for $d_{\text{exp}} = 1.844$ Å, which is analyzed as originating from the metastable NM rB1 phase (as discussed below). The intensity of this peak is actually reduced after annealing. The volume ($= 7.89$ cm^3/mole) and the c/a ($= 2.08$) estimated by fitting the experimental peak positions to the nB8 structure are in good agreement with the present calculated results, which give volume $= 7.92$ cm^3/mole, $c/a = 2.2$ for the FM state and volume $= 7.94$ cm^3/mole, $c/a = 2.1$ for the AF state at 137 GPa.

Table 3.3.1. *Equation-of-state parameters and magnetic moments of* AF rB1 *phases of* MnO *and* FeO *obtained by present calculations and comparison with literature*

Parameter	MnO		FeO		
	GGA	Exp.	GGA	LDA + U	Exp.
B_0 (GPa)	157	151[a], 162[b]	180	196	142[e], 180[f]
B'	3.23	3.6[a], 4.8[b]	3.55	4.9	4.9[f]
a_0 (Å)	4.46	4.435[c]	4.28	4.33	4.334[e]
Moment (μ_B)	4.47	4.45[d]	3.43		3.32[g]

[a] Noguchi et al., Geophys. Res. Lett. **23**, 1469 (1996).
[b] Jeanloz et al., J. Geophys. Res. **92**, 11433 (1987).
[c] L. F. Mattheiss et al., Phys. Rev. B **5**, 290 (1972).
[d] P. Dugek et al., Phys. Rev. B **49**, 10170 (1994).
[e] C. A. McCammon et al., Phys. Chem. Mineral. **10**, 106 (1984).
[f] I. Jackson et al., J. Geophys. Res. **95**, 21671 (1990).
[g] W. L. Roth et al., Phys. Rev. B **110**, 1333 (1958).

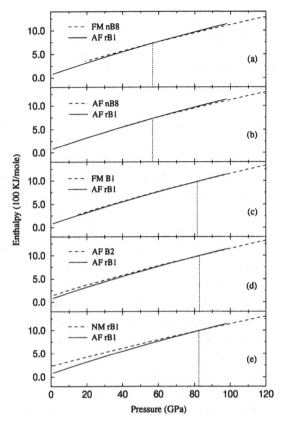

Figure 3.3.3 Calculated enthalpy against pressure for different phases of MnO: (a) transition from AF rB1 to FM nB8, (b) transition from AF rB1 to AF nB8, (c) transition from AF rB1 to FM B1, (d) transition from AF rB1 to AF B2, (e) transition from AF rB1 to NM rB1.

However, the critical pressure for the first-order transition from the AF rB1 structure to the nB8 structure (either FM or AF state) is estimated to be ~60 GPa [see Figs. 3.3.3(a) and 3.3.3(b)], which is much lower than the experimental transition pressure. We suggest two possibilities to interpret this discrepancy. First, as discussed above, transition-metal monoxides under lower pressure are typical Mott insulators with strong electron correlation, and even the GGA cannot describe their electronic structures properly. This implies that the real energy–volume curve of the insulating AF rB1 structure should be lower under low pressure than the one calculated with the GGA (this kind of situation can be seen more clearly in the case of FeO), leading to a higher critical pressure. At the same time, this effect will make the transition from AF rB1 to AF B2 structure, which has been predicted by our calculations to occur at ~85 GPa [see Fig. 3.3.3(d)], more difficult. We have checked this possibility by performing the LDA + U calculations with pseudopotentials for one point in the low-pressure range. The calculated results indeed show that the relative difference between the insulating AF rB1 structure and the FM metallic nB8 structure increases, depending on the selected value of U_{eff}. The second possibility is that the transition barrier from rB1 to nB8 structure may be fairly high because of the significant rearrangement of the atom positions. The existence of this transition barrier will generally make the actual transition pressure higher than the calculated one.

Experimentally, the high-pressure nB8 phase of MnO seems not to be realized directly from the low-pressure AF rB1 phase. There may exist some intermediate-pressure phases [7]. The experimental x-ray-diffraction pattern within the intermediate-pressure range (90 $\sim$ 120 GPa) cannot be explained simply by a single phase. Even in the high-pressure region (>120 GPa), there is an additional x-ray peak not coming from the nB8 structure. This situation suggests that MnO may undergo some complicated phase-transition course within the intermediate-pressure region between the normal-pressure insulating AF rB1 phase and the high-pressure metallic nB8 phase. Our calculations suggest that one possible intermediate phase is the NM rB1 structure [37]. From Figs. 3.3.2(a) and 3.3.3(e), we can find that a phase transition from the AF rB1 to the NM rB1 structure (so-called magnetic collapse) may occur around 85 GPa, which is close to the experimental transition pressure. Compared with the transition from AF rB1 to FM nB8 or AF B2, this transition occurs more easily with the simple change in the c/a ratio. We performed a simulation of the x-ray-diffraction pattern corresponding to the NM rB1 structure at 137 GPa. Interestingly, the sharp peak corresponding to $d_{exp} = 1.844$ Å can be assigned as the strongest reflection, i.e., the (102) peak, of the NM rB1 structure, and most of the other peaks of the NM rB1 structure nearly overlap with the peaks coming from the AF or FM nB8 structure. The observed broad peaks may be partly caused by the superposition of the

Table 3.3.2. *Observed and fitted x-ray-diffraction patterns of* MnO *at 137 GPa after laser annealing*

$d_{exp}{}^a$	$I_{exp}{}^b$	$I_{fit}{}^b$	$d_{fit}{}^a$ (nB8)	h	k	l	$d_{exp} - d_{fit}$
2.534	w	m	2.538	0	0	2	−0.004
2.110	s	s	2.114	1	0	0	−0.004
1.955	w	w	1.952	1	0	1	0.003
1.844	s		not coming from nB8				
1.628	s	s	1.624	1	0	2	0.004
1.218	m	m	1.220	1	1	0	−0.002
1.099	m	w	1.100	1	1	2	−0.001

$^a d_{exp}$ and d_{fit} are experimental and fitted d spacings, respectively, in units of Å. Experimental data come from Ref. [7]. The fitted hexagonal unit cell has $a = 2.441$ Å and $c = 5.076$ Å.

^{b} The relative intensities of the peaks are described as strong (s), medium (m), and weak (w).

peaks of two structures. However, as mentioned above, the peaks originating from the NM rB1 structure diminish by laser annealing, implying that the NM rB1 structure is metastable under pressure as high as 137 GPa. Some of the x-ray peaks in the intermediate-pressure range (90–120 GPa) can be assigned as originating from the NM rB1 structure.

Figure 3.3.4(a) shows the pressure–volume relation for MnO obtained by the present calculation and also by experiments. Clearly, in the low-pressure range, the sample is in the AF rB1 phase, and the calculated pressure–volume curve is in good agreement with the experimental results. The experimental data points around 120 GPa by shock compression are located just in between the curves corresponding to the AF rB1 and FM nB8 structures, suggesting that the sample is in a mixed phase. On the other hand, the highest pressure data point obtained by the static compression followed by laser annealing is just on the line of the pressure–volume curve for the nB8 structure.

The calculations for FeO is similar, and the results are shown in Figs. 3.3.2(b) and Fig. 3.3.4(b). Surprisingly, we found that the iB8 structure with the AF ordering is the most stable for FeO among several structures (including the rB1) in the whole volume range. Deferring the discussion about the relative stability between the distorted rB1 and iB8 structures at normal pressure for a while, we pay attention to the relative stability between the iB8 and nB8 structures in the compressed volume range, in which the GGA calculation will be reliable because of the reduced U/W. Clearly for FeO, the iB8 structure is more stable than that of nB8.

At this stage, two fundamental questions are to be answered:

1. Why is the nB8 structure realized rather than iB8 for most of the transition-metal compounds with the B8 structure (including the high-pressure phase of MnO)?

2. What is special about FeO, leading to such a strong stability of the iB8 structure?

As already mentioned, one of the important differences between the nB8 and the iB8 structures is that neither the oxygen atom nor the transition-metal atom have inversion symmetry in the AF iB8 structure and the transition-metal atom is located at the inversion center in the AF nB8 structure. Because of this difference, the strength of the p–d hybridization is weaker in the AF iB8 structure than in the AF nB8 structure. Such a reduction in the p–d hybridization in the iB8 structure compared with that in the nB8 structure was confirmed in the present calculation by estimation of the energy separation between the centers of the p and the d bands. As the p–d hybridization contributes to the stability of the structure, this aspect favors the nB8 structure and explains the general feature that the nB8 structure is actually realized in most cases. To understand any reason for the special situation of FeO, we calculated the density of states for both nB8 and iB8 with different magnetic structures. Among eight possible combinations of magnetic orderings (FM, AF, and additional two possible 120° spin arrangements within the ab plane) and atom positions (nB8 and iB8), the combination

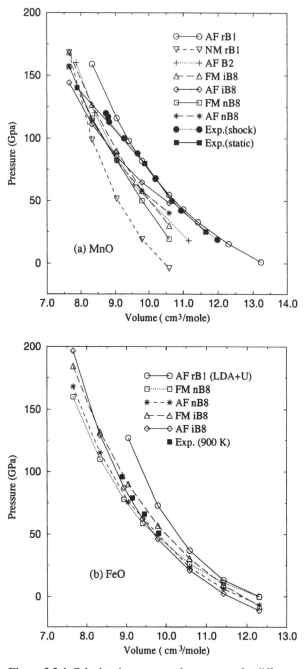

Figure 3.3.4 Calculated pressure–volume curves for different phases of (a) MnO and (b) FeO and comparison with experimental results. For MnO, two experimental results with shock compression [10] and static compression [7] (at room temperature) are shown. For FeO, the data from the experiment of Fei and Mao [17] on the isothermally (900 K) decompressed high-pressure phase of FeO are shown.

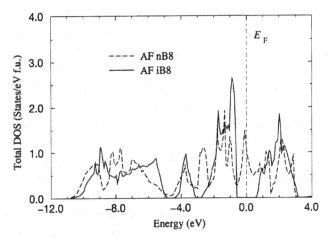

Figure 3.3.5 Calculated electronic density of states for AF iB8 and AF nB8 FeO with the use of experimentally determined lattice parameters at 96 GPa and 800 K [17].

of AF and iB8 is very unique in the sense that the electronic structure is insulating. Figure 3.3.5 shows our calculated electronic density of states (DOS) for AF iB8 and AF nB8 FeO at a compressed volume corresponding to the experimental lattice parameters for the high-pressure phase of FeO. We can see clearly that a well-defined bandgap exists just at the Fermi level for the AF iB8 structure. For this AF iB8 structure, the majority-spin band is completely occupied and only one subband is occupied in the minority-spin state to accommodate additional one electron because Fe^{2+} has a $d_\uparrow^5 d_\downarrow^1$ electron configuration. This occupied subband of minority-spin state has a mixed character of $3z^2 - r^2$, $x^2 - y^2$, and xy orbitals, all of which hybridize with the oxygen $2p$ orbitals only weakly. Appearance of the bandgap at the Fermi level will contribute to special stability of the AF iB8 structure. It should be noted that the AF iB8 FeO is a band insulator rather than a Mott insulator because it is insulating even in the high-pressure range where the Mott insulating condition breaks down [31, 40]. The calculated bandgap even increases slightly with pressure: 0.7 eV at normal pressure and 1.0 eV at 96 GPa. Such a behavior will not be expected for Mott insulators.

A puzzling feature of Fig. 3.3.2(b) is that the iB8 structure is predicted to be significantly more stable than the rB1 structure, even at zero pressure. This reflects again the fact that the GGA cannot describe the electronic structure of Mott insulators properly: the GGA incorrectly predicted that FeO is metallic at normal pressure. To reproduce the correct ground state of FeO, we have to take into account the local electron correlation and the spin-orbit interaction (SOI) as well. We adopt the LDA + U method with the linear muffin-tin orbital (LMTO) basis [41] to estimate the corrections caused by these ingredients. As the LMTO method in the atomic sphere approximation cannot predict the structural energy difference accurately, this method is used only to estimate the energy change $\Delta E_A(U_{\text{eff}}, \text{SOI})$ caused by the effective Coulomb interaction U_{eff} and SOI for a given system A. Then, $\Delta E_A(U_{\text{eff}}, \text{SOI})$ is added to the total energy $E_A^{\text{pp}}(\text{LSDA})$

obtained by the pseudopotential method within the local-spin-density approximation (LSDA). This method, with a reasonable value of 4 eV as the effective Coulomb interaction parameter U_{eff}, can reproduce a bandgap of AF rB1 FeO in good agreement with the result of an elaborate theoretical analysis [42] and at the same time predict that the AF rB1 structure is unambiguously more stable than the AF iB8 and AF nB8 structures at normal pressure. For a compressed volume with experimentally determined lattice parameters [17], the same calculation still predicts that the AF iB8 structure is most stable. After the U_{eff} correction is included, the total energies of the AF rB1 and AF nB8 phases relative to those of the AF iB8 phase are shown as an inset of Fig. 3.3.2(b). We can see that the puzzling features of Fig. 3.3.2(b) are explained by the electron correlation effect.

According to the present calculation, the high-pressure phase of FeO with the iB8 structure should be insulating in contrast to the experimental observation of metallic behavior [15, 16]. There are two possible origins of this disagreement. Our calculation assumes stoichiometric FeO, whereas real samples contain ∼5% Fe deficiency. As the AF iB8 FeO is a band insulator, itinerant carriers will be introduced by means of Fe deficiency. Another possibility is the mixture of other metallic phases in the temperature range of the observed metallic behavior (above 1000 K). According to the present calculation, even if the crystal structure is the iB8 type, FM order makes the system metallic and, moreover, the AF and FM nB8 phases are also metallic. The possible mixture of phases of the NiAs type will not affect the measured x-ray-diffraction peak positions. Nevertheless, the more than 19-kJ/mole enthalpy gain of the AF iB8 phase relative to other phases at 100 GPa is large enough to make the AF iB8 phase dominant, even at 1000 K. This statement is based on the following analysis of the entropy contribution to the Gibbs free energy. As the AF and the FM states are energetically very close in the nB8 structure [Fig. 3.3.1(a)], a magnetic order of ∼1000 K may be totally random, whereas in the iB8 structure the large energy difference of ∼50 kJ/mole between the AF and the FM states suggests the existence of at least strong short-range AF order. Therefore as an extreme case the magnetic entropy is taken into account for only the nB8 structure. This contributes approximately −9 kJ/mole to the Gibbs free energy at 1000 K. Another contribution of −5 kJ/mole at 1000 K from the electronic entropy through the metallic behavior of the AF and the FM nB8 phases has to be added. Note that the above estimation of the entropy contribution to the Gibbs free energy was made so as to be favorable to the nB8 structure, and yet the estimated enthalpy gains still support the stability of the AF iB8 phase.

We point out more strong evidence for the iB8 structure as the high-pressure phase of FeO. We have found that the intensity profile of the observed x-ray-diffraction pattern can be reproduced by only the iB8 type but not by the nB8 type. To be more concrete, we consider the relative intensity between (100) and (101) peaks as an example. Experimentally, the latter is stronger than the former [17]. The nB8 structure gives the reverse trend, whereas the iB8 structure reproduces the observed trend correctly.

The calculated pressure–volume relation is shown in Fig. 3.3.4(b) for FeO together with some experimental points corresponding to the high-pressure phase. Even if we

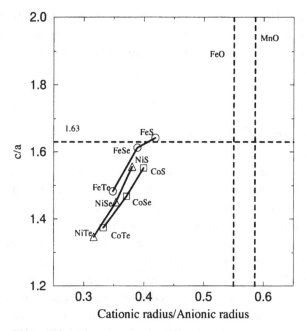

Figure 3.3.6 The c/a value for different materials, which
have the nB8 structure, at normal pressure and room
temperature as functions of (cation radius/anion radius)
ratio. Here 1.63 is the ideal c/a value for close-packed
materials.

admit some ambiguity caused by the difference in the temperature between the present
calculation (0 K) and the experiment (800 K), the theoretical pressure–volume curve
for the AF iB8 structure seems to account well for the experimental data.

To the best of our knowledge, no materials have ever been known to take the iB8
structure. The uniqueness of FeO in this sense may become even clearer when FeO is
compared with FeS, which, at normal pressure, takes a structure whose basic building
block is the nB8 structure rather than the iB8 structure. The present calculation correctly
predicts stronger stability of nB8 compared with iB8 for FeS. This is because the AF
iB8 phase of FeS has a bandgap of only 0.18 eV, which is too small to stabilize the
iB8 structure.

Finally, we make a brief comment on the c/a value of the nB8 structure of MnO and
the iB8 structure of FeO. In the high-pressure phases, the c/a value of both systems
exceeds 2.0, being unusually large compared with the values for most other related systems, which generally have a c/a value close to the ideal value [$c/a = (8/3)^{1/2} \approx 1.63$]
for the close-packed structure. However, we have found that c/a is an increasing function of r_c/r_a, where r_c (r_a) denotes the cation (anion) ionic radius. This can be seen
clearly from Fig. 3.3.6, which gives the c/a values of different materials at normal
pressure and room temperature as functions of r_c/r_a. All of these materials have the
nB8 structure at normal pressure and room temperature. Our calculations show that
both the nB8 structure for MnO and the iB8 structure for FeO have c/a values of $\sim$1.9

at normal pressure. This c/a value is not so unusual, because it is well on the extrapolated line of the general trend. With increase of pressure, our calculations show that the r_c/r_a ratio will increase because of a decrease in charge transfer. Consequently, the c/a value for both nB8 MnO and iB8 FeO will increase from ~ 1.9 to ~ 2.0 at high pressure.

3.3.5   Summary of Results

In the present work, the phase stability of transition-metal monoxides MnO and FeO under pressure were studied with the first-principles plane-wave basis pseudopotential calculations based on the GGA supplemented by the $LDA + U$ method. The present calculations predict that the high-pressure phase of MnO should take the nB8 structure rather than the B2 structure and that the metastable NM B1 structure can be realized for MnO in the intermediate pressure range. A very unique AF inverse B8 (iB8) structure rather than the nB8 structure is predicted as the high-pressure phase of FeO. To our knowledge, there are no other transition-metal compounds with the iB8 structure. The stability of the AF iB8 structure for FeO is caused by the existence of a bandgap at the Fermi level, implying that the AF iB8 FeO is a band insulator at low temperature even at high pressure. Analysis of x-ray-diffraction experiments provides further support to the present theoretical predictions for both MnO and FeO.

After submitting this article, the following related papers have appeared: (1) I. I. Mazin, Y. Fei, R. Downs, and R. Cohen, Am. Mineralogist, **83**, 451 (1998); (2) Z. Fang, K. Terakura, H. Sawada, T. Miyazaki, and I. V. Solovyev, Phys. Rev. Lett. **81**, 1027 (1998); and (3) Z. Fang, I. V. Solovyev, and K. Terakura, Phys. Rev. B **59**, 762 (1999).

Acknowledgments

The authors thank Y. Syono, Y. Noguchi, T. Yagi, and T. Kondo for many valuable comments and for providing us with their experimental data before publication. Thanks are also given to Y. Morikawa for much help in numerical calculations. The present work is partly supported by the New Energy and Industrial Technology Development Organization and also by the Grant-in-Aid for Scientific Research from the Ministry of Education, Science and Culture of Japan.

References

[1] N. F. Mott, *Metal-Insulator Transitions* (Taylor & Francis, London, 1974).

[2] R. Jeanloz, T. J. Ahrens, H. K. Mao, and P. M. Bell, Science **206**, 829 (1979).

[3] Y. Sato and R. Jeanloz, J. Geophys.

Res. **86**, 11773 (1981).

[4] L. Liu and W. A. Bassctt, J. Geophys. Res. **77**, 4934 (1972).

[5] S. T. Weir, Y. K. Vohra, and A. L. Ruoff, Phys. Rev. B **33**, 4221 (1986).

[6] T. S. Duffy, R. J. Hemley, and H. K. Mao, Phys. Rev. Lett. **74**, 1371 (1995).

[7] T. Kondo, T. Yagi, Y. Syono, T. Kikegawa, and O. Shimomura, Rev. High Pressure Sci. & Tech. **7**, 148 (1998) (Proc. Joint AIRAPT-16 & HPCJ-38, Kyoto, 1997).

[8] B. T. M. Willis and H. P. Rooksby, Acta Crystallogr. **6**, 827 (1953).

[9] T. Yagi, T. Suzuki, and S. Akimoto, J. Geophys. Res. **90**, 8784 (1985).

[10] Y. Noguchi, K. Kusaba, K. Fukuoka, and Y. Syono, Geophys. Res. Lett. **23**, 1469 (1996).

[11] R. Jeanloz and A. Rudy, J. Geophys. Res. **92**, 11433 (1987).

[12] S. L. Webb, I. Jackson, and J. D. F. Gerald, Phys. Earth Planet. Inter. **52**, 117 (1988).

[13] R. Jeanloz and T. J. Ahrens, Geophys. J. R. Astron. Soc. **62**, 505 (1980).

[14] T. Yagi, K. Fukuoka, H. Takei, and Y. Syono, Geophys. Res. Lett. **15**, 816 (1988).

[15] E. Knittle and R. Jeanloz, Geophys. Res. Lett. **13**, 1541 (1986).

[16] E. Knittle, R. Jeanloz, A. C. Mitchell, and W. J. Nellis, Solid State Commun. **59**, 513 (1986).

[17] Y. W. Fei and H. K. Mao, Science **266**, 1678 (1994).

[18] H. K. Mao, J. F. Shu, Y. W. Fei, J. Z. Hu, and R. J. Hemley, Phys. Earth Planet. Inter. **96**, 135 (1996).

[19] See, for example, *Density-Functional Theory of Atoms and Molecules*, R. G. Parr and W. Yang, eds. (Oxford, New York, 1989).

[20] P. Hohenberg and W. Kohn, Phys. Rev. B **136**, 864 (1964).

[21] W. Kohn and L. J. Sham, Phys. Rev. A **140**, 1133 (1965).

[22] J. P. Perdew and Y. Wang, Phys. Rev. B **45**, 13244 (1992).

[23] K. T. Park, K. Terakura, and Y. Matsui, Nature (London) **336**, 670 (1988).

[24] See, for example, J. P. Perdew, J. A. Chevary, S. H. Vosko, K. A. Jackson, M. P. Pederson, D. J. Singh, and C. Fiolhais, Phys. Rev. B **46**, 6671 (1992).

[25] A. D. Becke, Phys. Rev. A **38**, 3098 (1988).

[26] J. P. Perdew, Physica B **172**, 1 (1991).

[27] Y. M. Juan and E. Kaxiras, Phys. Rev. B **48**, 14944 (1993).

[28] P. Dufek, P. Blaha, V. Sliwko, and K. Schwartz, Phys. Rev. B **49**, 10170 (1994).

[29] J. P. Perdew, *Electronic Structure of Solids*, P. Ziesche and H. Eschrig, eds. (Akademie-Verlag, Berlin, 1991).

[30] I. V. Solovyev, P. H. Dederichs, and V. I. Anisimov, Phys. Rev. B **50**, 16861 (1994).

[31] R. E. Cohen, I. I. Mazin, and D. G. Isaak, Science **275**, 654 (1997).

[32] D. G. Isaak, R. E. Cohen, M. J. Mehl, and D. J. Singh, Phys. Rev. B **47**, 7720 (1993).

[33] D. M. Sherman and H. J. F. Jansen, Geophys. Res. Lett. **22**, 1001 (1995).

[34] D. R. Hamann, M. Schlüter, and C. Chiang, Phys. Rev. Lett. **43**, 1494 (179).

[35] D. Vanderbilt, Phys. Rev. B **41**, 7892 (1990).

[36] F. D. Murnaghan, Proc. Nat. Acad. Sci. USA **30**, 244 (1944).

[37] See our detailed paper, Z. Fang, I. V. Solovyev, H. Sawada, and K. Terakura, Phys. Rev. B **59**, 762 (1999).

[38] K. Terakura, T. Oguchi, A. R. Williams, and J. Kübler, Phys. Rev. B **30**, 4734 (1984).

[39] J. van Elp, R. H. Potze, H. Eskes, R. Berger, and G. A. Sawatzky, Phys. Rev. B **44**, 1530 (1991).

[40] O. Gunnarsson, E. Koch, and R. M. Martin, Phys. Rev. B **54**, R11026 (1996).

[41] O. K. Andersen, Phys. Rev. B **12**, 3060 (1975).

[42] M. Takahashi and J. Igarashi, Phys. Rev. B **54**, 13566 (1996).

Chapter 3.4

Computer-Simulation Approach to the Thermoelastic, Transport, and Melting Properties of Lower-Mantle Phases

Atul Patel, Lidunka Vočadlo, and G. David Price

This chapter is in honour of Y. Matsui and his significant contributions to mineral physics. We review some of our recent computer-simulation studies of the lower-mantle phases of MgSiO$_3$ perovskite and MgO with respect to the predicted thermoelastic properties and diffusion behaviour. The geophysical significance of these calculations is outlined. Then we outline the theory behind the lattice dynamics of perfect and defective systems and molecular dynamics techniques. We also discuss the atomistic diffusion theories in relation to our computer simulation approaches. Finally, we present the results of our studies.

3.4.1 Introduction

We discuss work that we have carried out on geophysically important phenomena, namely (1) the equations of state of MgSiO$_3$ perovskite and (2) diffusion in MgO (see Fig. 3.4.1).

1. To model the composition of the lower mantle it is necessary to obtain accurate thermoelastic parameters that are used in equations of state of the component minerals. Experiments at these extreme pressure and temperature conditions are difficult and can lead to large uncertainties in some of the thermoelastic constants. Recent reports reveal that there are discrepancies in the data obtained by various experimental studies of silicate perovskite, the most abundant mineral in the Earth. Both the x-ray-diffraction diamond-anvil-cell measurements of Mao et al. [1] and the multianvil high-pressure experiments of Wang et al. [2] are in agreement in their measurement of compressibility and the associated Birch–Murnaghan equation of state, but they differ for the

H. Aoki et al. (eds), *Physics Meets Mineralogy* © 2000 Cambridge University Press.

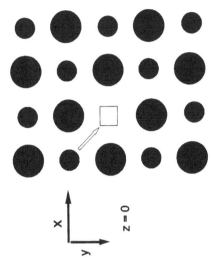

Figure 3.4.1 Migration
direction for ionic diffusion
in MgO.

values of the Grüneisen and the Anderson–Grüneisen parameters inferred.
Moreover, the thermodynamic analyses of Anderson and Masuda [3] and the
spectroscopic measurements of Chopelas [4] have yielded thermoelastic data
that are in good agreement with the measurements of Wang et al. [2] but not
with those of Mao et al. [1]. Theory provides a way of accessing a wide range
of P/T space at no expense of precision. Hence, in an attempt to resolve this
dichotomy, we have performed atomistic molecular dynamics (MD) and lattice
dynamics (LD) calculations on the pure end-member $MgSiO_3$ perovskite to
obtain estimates of the controversial thermoelastic data.

2. To gain insight into the rheological properties of the lower mantle, an
   understanding of the mechanisms underlying solid-state transport in such
   mantle materials must be obtained. Therefore a study of the theoretical basis
   underlying the diffusion equation and its application to the relatively simple
   structure of MgO at conditions within the Earth's interior will further our
   knowledge of mass transport and energy flows in the mantle and thus will
   eventually provide us with further information concerning the dynamical
   properties of the Earth's deep interior. Direct measurements of diffusion
   properties are extremely difficult to obtain, especially at the high temperatures
   and pressures required for simulating the Earth's interior; therefore our
   complementary computational approach is an asset to the study of mantle
   rheology.

3.4.2 Computer-Simulation Techniques and Diffusion Models

3.4.2.1 The Atomic Model

We may establish the physical properties of a solid by solving the Schrödinger equation explicitly, thereby obtaining precisely the energy surfaces associated with the

interactions of electrons and nuclei within a given system. This ab initio approach is not universally practicable because of size and complexity of the calculations that are often required. A simpler and more approximate treatment is provided through atomistic modelling, which describes the interactions between atoms or ions within a crystal structure, as opposed to the finer details of each electronic interaction. Such an approximation is usually based on the model of Born of solids [5], whereby only forces that are due to electron clouds are included.

To model such interatomic interactions, it is first necessary to understand the potential-energy functions that describe them. This is achieved initially through consideration of a two-body system; many-body systems are generally prohibitively complex, although simple three-body corrections may be included. When no net forces are acting on the constituent atoms, the sum of the attractive and the repulsive potential energies between each pair of atoms in a crystalline solid at 0 K is termed the static lattice energy:

$$U_L(\mathbf{r}_{ij}) = \sum_{ij} \frac{q_i q_j}{\mathbf{r}_{ij}} + \sum_{ij} \varphi_{ij} + \sum_{ijk} \theta_{ijk}. \tag{3.4.1}$$

The first term on the right-hand side is the contribution to the static lattice energy from the long-range Coulombic attraction for an infinite array of atoms. The summation of the long-range Coulombic interactions is very slowly convergent and therefore is computationally time consuming and expensive. Hence we use the Ewald method to overcome these convergency difficulties [6–8]. The second term in Eq. (3.4.1) accounts for the diffuse nature of the electron clouds surrounding the nucleus; it includes the short-range interactions associated with Pauli repulsion between neighbouring charge clouds and the short- and the long-range components of van der Waals attraction. The third term represents three-body interactions that, for highly ionic solids with dominant pairwise interactions, may be negligible.

In the rigid-ion model, the short-range interactions predominantly affect nearest-neighbour ions. Short-range potential functions may be represented by pairwise potentials such as the Buckingham potential, which takes the form

$$\varphi_{ij} = A_{ij} e^{-\frac{\mathbf{r}_{ij}}{B_{ij}}} - \frac{C_{ij}}{\mathbf{r}_{ij}^6}, \tag{3.4.2}$$

where A_{ij}, B_{ij}, and C_{ij} are constants and $\mathbf{r}_{ij}$ is the interatomic separation. The first term in φ_{ij} is that due to short-range repulsion, and the second is due to van der Waals-induced dipole–dipole attraction.

In addition to the above, models may also include the shell description of ionic polarisability [9]. In this, the ion (usually oxygen) is represented by a massless charged shell (representing the outer valence electron cloud) attached to a massive core by a harmonic spring, so that

$$U_s = \sum_i k_i^s \mathbf{r}_i^2, \tag{3.4.3}$$

where k_i^s is the spring constant and $\mathbf{r}_i$ is the core–shell separation. This gives a simple mechanical description of ionic polarisability, necessary for the calculation of defect

energies and high-frequency dielectric constants. The resulting free-ion polarisability
is given by

$$\alpha_i = \sum_i \frac{(Y_i e)^2}{k_i^s}, \tag{3.4.4}$$

where Y_i is the shell charge and e is the charge on the electron.

The choice, suitability, and accuracy of the potentials used in studies such as ours
are discussed in, for example, Refs. [10] and [11].

3.4.2.2 Static Simulations

Lattice Dynamics

The LD method is a semiclassical approach that uses the quasi-harmonic approximation to describe a cell in terms of independent quantised harmonic oscillators whose
frequencies vary with cell volume, thus allowing for thermal expansion (e.g., Ref. [5]).
The motions of the individual particles are treated collectively as lattice vibrations or
phonons. We have used the computer code PARAPOCS [12], in which the phonon
frequencies $\omega(\mathbf{q})$ of ions of mass m are obtained by the solution of

$$m\omega^2(\mathbf{q})\mathbf{e}_i(\mathbf{q}) = \mathbf{D}(\mathbf{q})\mathbf{e}_j(\mathbf{q}), \tag{3.4.5}$$

and the dynamical matrix $\mathbf{D}(\mathbf{q})$ is defined by

$$\mathbf{D}(\mathbf{q}) = \sum_{ij} \left(\frac{\partial^2 U}{\partial \mathbf{u}_i \partial \mathbf{u}_j} \right) \exp(i\mathbf{q} \cdot \mathbf{R}_{ij}), \tag{3.4.6}$$

where $\mathbf{R}_{ij}$ is the interatomic separation, $\mathbf{u}_i$ and $\mathbf{u}_j$ are the atomic displacements
from their equilibrium position, and U is the net interatomic potential (see Sub-
section 3.4.2.1). For a unit cell containing N atoms, there are $3N$ eigenvalue solutions $[\omega^2(\mathbf{q})]$ for a given wave vector $\mathbf{q}$. There are also $3N$ sets of eigenvectors
$[\mathbf{e}_x(\mathbf{q}), \mathbf{e}_y(\mathbf{q}), \mathbf{e}_z(\mathbf{q})]$ that describe the pattern of atomic displacements for each normal mode. For ions interacting within the shell model, the cores and the shells may
be treated separately. Once these predicted vibrational frequencies are calculated, a
number of thermodynamic properties may be calculated, such as free energy and heat
capacity, which are direct functions of these vibrational frequencies [12].

Phonon Sampling Techniques

When performing a LD calculation, ideally we would like to sample all points within
the first Brillouin zone of the crystal under investigation. However, such sampling
is unfeasible as it would require infinite computer time and memory. The sampling
points may be greatly reduced by symmetry, but this still remains time consuming and
expensive, so a number of averaging techniques have been developed [13–15]. The
degree of occupancy of phonon modes is highly temperature dependant. Above the
Debye temperature, all modes are occupied and the sampling scheme is less important;
however, such high-temperature computer studies suffer because of the breakdown

of the quasi-harmonic approximation as intrinsic anharmonicity becomes important. Below the Debye temperature, many more low-frequency modes are favoured and therefore the sampling technique used becomes very relevant. The free energy of the crystal is a logarithmic function of phonon frequency and is therefore fairly robust to sampling; however, the volume may well be affected. We refer the reader to Refs. [10] and [11] for details of the precise sampling schemes used in this type of study.

The Defective Lattice

Modelling point defects in a crystal structure can be approached in one of two ways: a single defect, such as an ion vacancy, may be modelled in an infinitely extending lattice, or a defect may be modelled within an enlarged unit cell, a supercell, with repeating boundary conditions. A defect embedded in an infinitely extending lattice structure can be described by the Mott–Littleton approach [16]. The lattice surrounding the defect is divided into two regions with a transition region between them. Region I is that in the immediate vicinity of the defect, containing maybe 100 ions, for which the interaction energies are calculated explicitly for the relaxed, heavily distorted lattice; region II extends indefinitely away from region I and the energies associated with region II are calculated by more approximate continuum methods.

In the LD supercell method, the defect is embedded within a grown unit cell, containing several hundred ions, with periodic boundary conditions. The energies are calculated by means of LD, taking into account the defect–defect interactions between supercells. However, if the supercell is sufficiently large, there will be little distortion at the cell boundaries and the defect–defect interaction will be small enough so as to be negligible. The advantage this method has over the preceding one is that the constant-pressure variable-volume Gibbs free energy of the defective lattice may be calculated directly with the code PARAPOCS, permitting not only the calculation of defect enthalpies, but also the constant-pressure entropic term. It is also possible to calculate the jump frequency of a particular defect by means of Vineyard theory (see below). In the diffusion study reported in Subsection 3.4.3.2 we use the LD supercell method.

The Diffusion Equations

At the microscopic level, diffusion in a crystal lattice may be described by particle jumps over potential-energy barriers within the minimum energy configuration of the lattice. To diffuse, an atom has to leave its position within the minimum energy configuration of the lattice, pass through a less stable higher-energy position (not usually occupied by atoms), and then fall into a minimum energy position again either at another lattice site or between lattice sites. The probability of an atom's overcoming such potential-energy barriers is governed by statistical mechanics; as the atom vibrates about its equilibrium position, there is some probability that it will vibrate with a sufficiently large amplitude to jump out of its equilibrium position, after which it may either fall into a neighbouring vacant site or become trapped in a potential well between atoms at neighbouring sites (Fig. 1). By simulating the motion of atoms in a crystal lattice and considering the pairwise forces between them, we may perform calculations to evaluate the free-energy changes associated with such migrating ions. For face-centred cubic crystals, such as MgO, the self-diffusion coefficient may be written as

$$D_{sd} = D_V N_V, \tag{3.4.7}$$

where N_V is the atomic fraction of vacancies and D_V is the coefficient for vacancy diffusion given by

$$D_V = \frac{Z}{6}\left(\frac{a}{\sqrt{2}}\right)^2 \upsilon \exp\left(\frac{\Delta S_m}{k}\right) \exp\left(-\frac{\Delta H_m}{kT}\right), \tag{3.4.8}$$

where Z is the coordination environment of the diffusing species, a is the cell parameter, $a/\sqrt{2}$ is the jump distance for MgO, υ is the attempt frequency, ΔS_m and ΔH_m are the activation entropy and enthalpy of migration, respectively, k is Boltzmann's constant, and T is the temperature in degrees Kelvin (e.g., Ref. [17]).

For lower-temperature extrinsic diffusion, the atomic fraction of vacancies N_V is determined by trace-element concentration (e.g., $3Mg^{2+} \leftrightharpoons 2Al^{3+} + \square$) or quenching effects; i.e., there are vacancies already present and the diffusion process involves only the migration of ions. For higher-temperature intrinsic diffusion, defects are thermally generated. The equilibrium atomic fraction of intrinsically generated vacancies in a crystal structure N_V at a given temperature has an exponential dependence on the vacancy formation energy and is given by

$$N_V = \frac{n_V}{N} = \exp\left(\frac{\Delta S_f}{k}\right) \exp\left(-\frac{\Delta H_f}{kT}\right), \tag{3.4.9}$$

where n_V is the number of vacant lattice sites, N is the total number of lattice sites, ΔS_f is the formation entropy, and ΔH_f is the formation enthalpy.

Therefore, substituting Eqs. (3.4.8) and (3.4.9) into Eq. (3.4.7) gives the intrinsic self-diffusion coefficient as

$$D_{sd} = \frac{Z}{6}\left(\frac{a}{\sqrt{2}}\right)^2 \upsilon \exp\left(\frac{\Delta S_f + \Delta S_m}{k}\right) \exp\left(\frac{-\Delta H_f - \Delta H_m}{kT}\right). \tag{3.4.10}$$

The problem we have therefore is to model these defect parameters and attempt frequencies. In both theory and experiment, the least well-constrained parameters in Eq. (3.4.10) are the entropy terms $\Delta S_{sd} (=\Delta S_f + \Delta S_m)$ and the attempt frequency υ (which is equivalent to the vibrational frequency of the migrating species along the migration path). We use Vineyard theory as an alternative expression for

$$\upsilon^* = \upsilon e^{\frac{\Delta S_m}{k}}, \tag{3.4.11}$$

where υ^* is obtained by the product of lattice vibrational frequencies (see below), which we calculate for our interatomic potential model by means of quasi-harmonic LD analysis, the basis of which is outlined above.

Vineyard Theory

There are several approaches to predicting the atomic-jump frequencies of diffusing species. They can simply be estimated as being typical of lattice modes (i.e., $\sim 10^{13}\,\text{s}^{-1}$) or they can be calculated from reaction-rate theory, the Green's functions, or dynamical theory (see, for example, Ref. [18]). In this study we have chosen to use Vineyard theory, as it has proved successful in the past and is one of the more efficient methods for determining the preexponential factor in the diffusion equation. Vineyard theory [19] is based on absolute rate theory and enables us to obtain an estimate for the attempt frequency v^*, which is defined by

$$v^* = \frac{\prod_{j=1}^{N} \omega_j}{\prod_{j=1}^{N-1} \omega_j'}, \tag{3.4.12}$$

where ω_j are the lattice frequencies with the defect in its stable state (i.e., one vacancy) and ω_j' are those of the defective lattice with the defect at the saddle point (i.e., two vacancies plus a migrating ion). This relation is valid only within the harmonic approximation [19], as motions near the saddle point are treated by the theory of small oscillations, and therefore the anharmonicity that may arise at the saddle surface is not adequately treated and may require an anharmonic correction.

3.4.2.3 Molecular Dynamics

MD is another technique that enables us to use the potential model description of interatomic interactions to predict the physical properties of matter. The essence of MD calculations (e.g., Ref. [20]) is the solution of Newton's laws of motion over a finite time period for a simulation box containing N ions, calculating the dynamic properties iteratively as the system evolves. Normally, periodic boundary conditions applied to the ensemble generate the required infinite system. The ions are initially assigned positions and velocities within the simulation box; their initial coordinates are usually chosen to be at the crystallographically determined sites, and their velocities v_i are chosen such that they concur with the required system temperature, such that both energy and momentum are conserved:

$$\sum_i m_i v_i(0) = 0, \tag{3.4.13}$$

$$\sum_i m_i [v_i(0)]^2 = 3Nk_B T, \tag{3.4.14}$$

where m_i are the atomic masses, k_B is Boltzmann's constant, and T is the initial chosen simulation temperature.

To calculate subsequent positions and velocities, the forces acting on any individual ion must be calculated from the first derivative of the potential function, and the new position and velocity of each ion may be calculated at each time step by the solution of Newton's equations of motion:

$$\mathbf{F}_i = -\frac{\partial U_L}{\partial \mathbf{r}_i} = m_i a_i(t) = m_i \frac{\mathrm{d}^2 \mathbf{r}_i}{\mathrm{d} t^2}, \tag{3.4.15}$$

where the terms take their usual meaning. This may be numerically integrated to generate a set of positions $\mathbf{r}_i(t + \Delta t)$ and velocities $v_i(t + \Delta t)$ as the system evolves. For an infinitesimally small time step the updating equations are

$$\mathbf{r}_i(t + \Delta t) = \mathbf{r}_i(t) + v_i(t)\Delta t, \tag{3.4.16}$$

$$v_i(t + \Delta t) = v_i(t) + \frac{\mathbf{F}_i(t)}{m_i}\Delta t. \tag{3.4.17}$$

For a finite time step, the above equations lose their accuracy and higher powers of Δt are required; in our calculations we solved the equations of motion by using the Gear predictor–corrector algorithm [21] whereby a fourth-order Taylor series expansion of the displacements with respect to time generates predicted evolving positions that are then corrected for iteratively until a convergent solution to the trajectories and velocities is obtained.

The instantaneous temperatures are calculated at each time step for a system of N particles from the equation

$$T = \frac{2}{3NK_B} E_{\text{kin}}, \tag{3.4.18}$$

where

$$E_{\text{kin}} = \frac{1}{2} \sum_{i=1}^{N} m_i v_i^2. \tag{3.4.19}$$

The MD code uses the constant-temperature algorithm devised by Nosé [22]. The pressure of the simulated system is calculated from the virial theorem as

$$P = \frac{NK_B T}{V} - \frac{NK_B T}{6V} \left(\sum_{i=1}^{N} \sum_{j>i}^{N} r_{ij} \frac{\partial U_{ij}}{\partial r_{ij}} \right), \tag{3.4.20}$$

where N is the number of particles within the system, V is the volume, r_{ij} is the distance between ions i and j, $\partial U_{ij}/\partial r_{ij}$ is the first derivative of the potential energy of ion i with respect to ion j, and k_B is Boltzmann's constant.

The MD code [23] used in our studies use the Ewald method to calculate the Coulombic term in the potential function more efficiently and the Parrinello and Rahman constant-pressure algorithm [24] for a constant-enthalpy $\{N, P, T\}$ ensemble. A time step of 10^{-15} s is used, and the system is allowed to evolve over a large number of time steps to ensure equilibrium, followed by thermodynamic property averaging over a further several thousand time steps. Therefore equilibrium properties can be obtained from the statistical average of the behaviour of these particles over a simulated time period of several picoseconds. Quantum corrections were included in the simulations and were based upon the Wigner–Kirkwood expansion of free energy in term of Planck's constant, as proposed by Matsui [23]. This method naturally includes the effects of anharmonicity, in which the potential is sampled along the asymmetric well. For more details of the ensemble sizes and lengths of simulations, see Ref. [10].

3.4.3   Geophysical Applications

3.4.3.1   Thermoelastic Properties of $MgSiO_3$ Perovskite

In this subsection we report the thermoelastic properties of $MgSiO_3$ perovskite calculated from our computer-simulation approach. We focus on the compression (V/V_0), isothermal bulk modulus (K_T), thermal expansion (α), Grüneisen parameters (γ and q), and the Anderson–Grüneisen parameter (δ_T). All the results from our study are from MD simulations except for the data for γ and q.

Figure 3.4.2 shows a plot of the compression of $MgSiO_3$ perovskite as a function of pressure at four isotherms ($T = 500$ to 2000 K), where our 500 K data are being

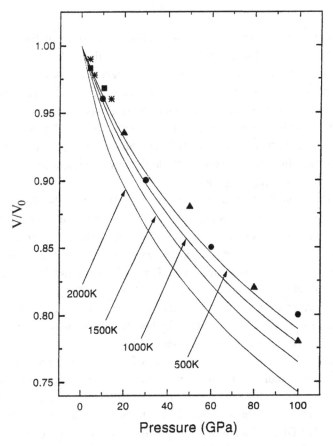

Figure 3.4.2 Plot of compression as a function of pressure at four isotherms. Other data: filled circles [25], triangles [26], stars [1], and squares [2].

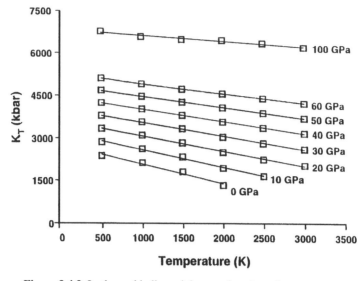

Figure 3.4.3 Isothermal bulk modulus as a function of temperature.

compared with various results from experiment and ab initio calculations. Given that most of the data compared with our results are either athermal (i.e., 0 K for the ab initio calculations) or at room temperatures, we believe that our calculations agree well with all other sources, but most notably with that of Mao et al. [1] and Wang et al. [2]. As expected, as the temperature increases, the solid becomes more compressible whereas the V/V_0 ratio becomes smaller. The classical (V_{CL}) and quantum-corrected (V_Q) molar volumes obtained from the MD simulations are very close; the quantum corrections to the volume are less than 0.5% at the lower temperatures and diminish, as expected, with increasing temperature.

Our $V(P)$ data from the MD simulations were fitted to the Birch–Murnaghan equation of state. Figure 3.4.3 shows K_T against T for various isobars. We see that as the temperature increases, K_T decreases, reflecting the fact that the solid is more compressible as it is heated, and, as expected, the converse is true for increasing pressure. Note that there is a constant value for $\partial K_t/\partial T$ at each isobar and that it decreases as P increases. The values for $\partial K_t/\partial T$ span from -0.069 (0 GPa) to -0.021 (100 GPa) GPa K^{-1}. The zero-pressure value is in good agreement with the data of Mao et al. [1]; they obtained a value of -0.063 GPa K^{-1}, higher than other measurements, whereas Wang et al. [2] found a value of -0.023 GPa K^{-1}.

The calculated thermal-expansion coefficients for a variety of pressures and temperatures are given in Table 3.4.1, where once again there is reasonable agreement with most of the observed data. In Fig. 3.4.4 we present the pressure dependence of α for six isotherms, and it is clearly seen that $(\alpha)_T$ all tend to the same value (1.4×10^{-5} K^{-1}) at high pressures (100 GPa), which is in good agreement with the value of 1.1×10^{-5} K^{-1} obtained by Anderson and Masuda [3] with their thermodynamic method. The temperature dependence of α for isobars up to 100 GPa are presented in Fig. 3.4.5, and it is

Table 3.4.1. *Comparison of thermal-expansion coefficients of* (Mg, Fe)SiO₃ *perovskite*

P(GPa)	$\alpha/\mathrm{x}10^{-5}\ K^{-1}$	T(K)	References
0.0	1.90	150–373	[27]
0.0	2.20	298–381	[28]
0.0	4.00	450–840	[29]
0.0	3.20	500	This study
0.0	1.7	300	[2]
0.0	1.7–1.9	300	[4]
0.0	4.5	300	[1]
0.0	1.7	300	[30]
0.0	1.55	300	[31]
20.5	0.80	300	[32]
20.0	2.62	500	This study
36.0	≧1.70	300	[33]
35.0	2.18	500	This study

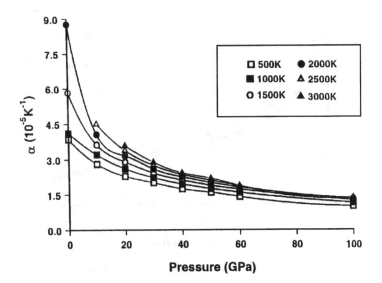

Figure 3.4.4 Thermal-expansion coefficient as a function of pressure.

clearly seen that α increases with T at all pressures, as expected. Moreover, the zero-pressure isobar shows a rather sharp increase in thermal expansion with temperature as the temperature is approaching melting. At high pressures (100 GPa) α increases by only $0.6 \times 10^{-6}\ K^{-1}$ over 2500 K.

From LD calculations it is possible to calculate the mode Grüneisen parameters γ_i that, when summed and weighted appropriately, are equivalent to the thermal

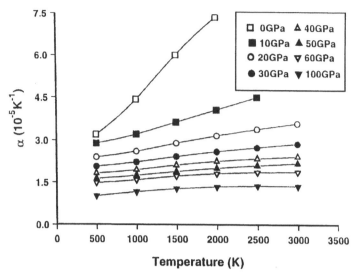

Figure 3.4.5 Thermal-expansion coefficient as a function of temperature.

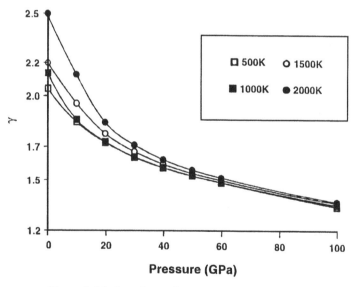

Figure 3.4.6 Grüneisen ratio as a function of pressure.

Grüneisen ratio γ_{TH}. Our LD values for γ_{TH} along four isotherms are displayed as a function of pressure in Fig. 3.4.6. γ_{TH} increases with temperature at a constant pressure and decreases with pressure. At high T and low P the quasi-harmonic approximation breaks down, and this results in an overestimation of γ_{TH} (see Ref. [34] for a more detailed discussion). Our value of $\gamma_0(T = 500 \text{ K}) = 2.05$ is in very good agreement with the experimental values of 2.20 [29] and 1.96 [35]. However, these values are higher than those obtained from other experimental or theoretical studies

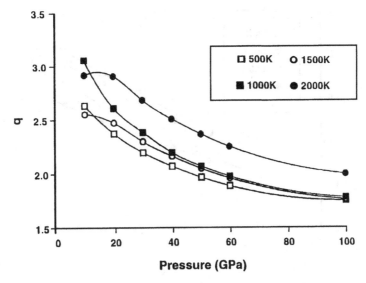

Figure 3.4.7 Second Grüneisen parameter q as a function of pressure.

(see Refs. [2, 4] and [36]), in which γ_0 is given as being in the range 1.4–1.5. At high pressures (100 GPa) each isotherm in Fig. 3.4.6 tends to the same value of 1.4.

It is generally assumed that the Grüneisen ratio has a pressure dependence given by

$$\frac{\gamma}{\gamma_0} = \left(\frac{V}{V_0}\right)^q,$$ (3.4.21)

where q is the second Grüneisen parameter and is constant. However, from our plots given in Fig. 3.4.7 we see that, for each isotherm, q decreases with pressure and tends to a value of 1.7 at 100 GPa. The $T = 2000$ K isotherm appears to give higher values of q than the other three isotherms, and once again we believe this to be due to the breakdown of the quasi-harmonic approximation at high temperatures. Our value of q_0 is 3.0, which is comparable (within quoted uncertainty limits) with the value of 2.5 adopted by Stixrude et al. [37]. The plots reveal that there is no systematic nesting of curves for the isotherms, implying that q has a rather complex temperature dependence consistent with the study on MgO by Anderson et al. [38].

The Anderson–Grüneisen parameter can be calculated with the thermodynamic definition

$$\delta_T = -\frac{1}{\alpha K_T}\left(\frac{\partial K_T}{\partial T}\right)_P.$$ (3.4.22)

From our simulations, we obtained values of $\delta_{T_0} = 7.0$ if K'_{T_0} is fixed at 4.0 and a value of 8.0 if K'_{T_0} is allowed to vary for the 500 K isotherm. These values are in agreement with those of Hemley et al. [35], but higher than the values found by Wang et al. [2] and Anderson et al. [36]. In addition, our calculations show a correlation among δ_{T_0},

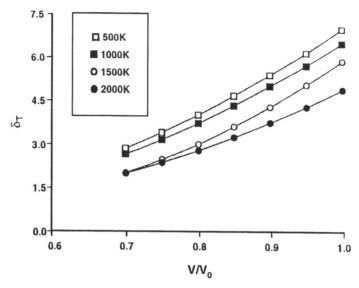

Figure 3.4.8 Compressional dependence of the Anderson–Grüneisen parameter.

K'_{T_0}, and q_0, which is totally consistent with the thermodynamic equation

$$\delta_{T_0} = \frac{\partial K_{T_0}}{\partial P} + q_0 - 1 + \left(\frac{\partial \ln C_V}{\partial \ln V}\right)_T, \qquad (3.4.23)$$

where C_V is the heat capacity at constant volume. δ_T as a function of pressure is plotted in Fig. 3.4.8, from which we obtain a value of $\sim$2.5 for κ for each isotherm when the following empirical relationship is used [39]:

$$\delta_T = \delta_{T_0}\left(\frac{V}{V_0}\right)^\kappa. \qquad (3.4.24)$$

This value is larger by $\sim$1 than the values of κ found for MgO by Anderson et al. [39], but it is in agreement with the variational-induced-breathing calculations of Inbar and Cohen [40].

Anderson [41] has shown an approximate relationship between the isothermal and the adiabatic Anderson–Grüneisen parameters:

$$\delta_T = \delta_S + \gamma. \qquad (3.4.25)$$

From seismology we believe that the lower-mantle value of δ_S is between 1.8 and 1.0. Using our values of δ_T and γ, we obtain $\delta_S = 1.05$ for $MgSiO_3$ perovskite, which is totally consistent with seismic observations.

The thermoelastic parameters calculated in our study and other theoretical and experimental approaches are summarised in Table 3.4.2, where we see, in general, excellent agreement between our predictions and the values obtained from experiments performed by Mao et al. [1] (for an analysis of their data see Ref. [35]). There is disagreement, however, between our calculated values of δ_{T_0} and γ and those inferred from the multianvil high-pressure experiments of Wang et al. [2] and the fitted values of Anderson et al. [36].

3.4.3.2 Diffusion in MgO

In this subsection we investigate the application of Vineyard theory (see the subsection entitled "Vineyard Theory") for investigating ionic diffusion in MgO and also investigating the effect on the energy surface resulting from the anharmonicity of the vibrational modes in the region of the saddle point. We show how this leads to bifurcation in the $\langle 001 \rangle$ direction, displacing the saddle point from its central position as the lowest-energy diffusion route deviates from the most direct one. This effect becomes especially noticeable at high temperatures as the lattice expands. Therefore we correct the Vineyard theory to accommodate this anharmonicity. We present results for Schottky defect and ion migration energies that enable us to calculate absolute ionic diffusion coefficients for both cation and anion diffusion in MgO.

Schottky Defect Formation

The structure of MgO is one of dense ionic packing; because of this, Frenkel defects are energetically unfavourable, and so Schottky defects are the dominant source of intrinsic defects in MgO. The Gibbs free energy, volume, and entropy of Schottky formation was calculated, with the $3 \times 3 \times 3$ MgO supercell containing 216 atoms, as a function of temperature; the results of our calculations over a simulated temperature range of 100 to 1000 K are given in Table 3.4.3 and Fig. 3.4.9. We calculate a formation entropy that tends towards a high temperature limit of $\sim 4.1k$ (Fig. 3.4.10). Having calculated a formation enthalpy and entropy for Schottky defects in MgO, the next stage is to calculate an activation enthalpy and entropy for ionic migration to enable Eq. (3.4.10) to be evaluated.

Table 3.4.2. *Thermoelastic parameters for silicate perovskite*

Parameter	Source				
	[2]	[35]	[3]	This study, $K_0' = 4$ (fixed)	This study, K_0' varies
Θ (K)	1094	1017	1017	1039	1039
K_0 (GPa)	261	263	263	250	239
K_0'	4.0	3.9	3.9	4.0	5.0
V_0 (cm^3 mol^{-1} K^{-1})	24.46	24.46	24.46	24.44	24.44
γ_0	1.3	1.96	1.5	1.97	1.97
δ_{T_0}	4.3	7.0	5.0	7.0	8.0
κ	1.5	0.8	1.5	2.5	3.0

Ion Migration

The first requirement when trying to obtain ion migration energies is to establish where in the saddle energy surface the minimum energy configuration lies. With MgO this is not as simple as it may initially seem, for although MgO has a face-centred cubic crystal structure, the saddle point bifurcates as a function of cell volume, therefore making the problem much more complex. In PARAPOCS, the cell volume can be varied with

Table 3.4.3. *Schottky formation parameters for* MgO

Temperature (K)	Formation energy (eV)	Formation volume $(cm^3\ mol^{-1})$	Formation entropy (k)
100	7.371	14.522	1.928
200	7.340	14.584	3.593
300	7.299	14.655	3.863
400	7.254	14.734	3.913
500	7.207	14.819	3.935
800	7.063	15.113	4.004
1000	6.963	15.349	4.084

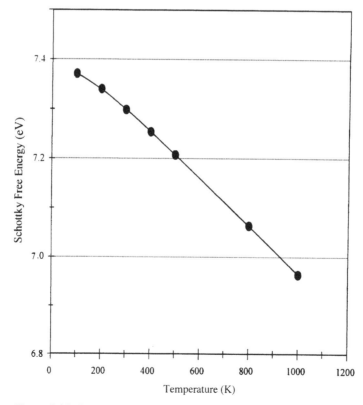

Figure 3.4.9 Free energy vs. temperature for Schottky defect formation.

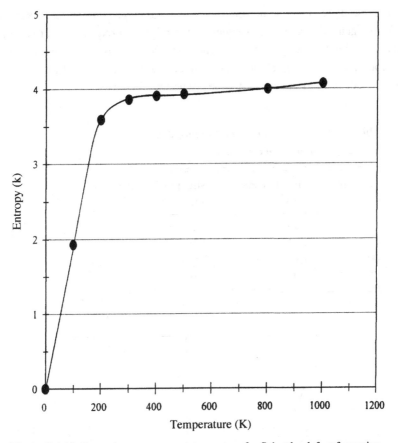

Figure 3.4.10 Formation entropy vs. temperature for Schottky defect formation.

temperature by means of the quasi-harmonic approximation [12]. An ion may be fixed at an absolute position in space to simulate the migration. At very low temperatures, at which the cell volume is at its smallest, the saddle surface is approximately harmonic in the $\langle 001 \rangle$ direction, but as the volume increases this energy surface bifurcates, evolving into a double-welled saddle surface with a maximum at its symmetric centre. In our calculations, placing an ion at this point would generate two imaginary frequencies: the first is expected, resulting from the energy maximum in the $\langle 110 \rangle$ direction, i.e., the jump direction; the second unstable vibrational mode results from the bifurcation with its principal motion being perpendicular to the jump path. The harmonic approximation fails at this point as an increasingly anharmonic, flattened saddle surface evolves with increasing volume.

We obtain the bifurcation described above by considering the potential well along the z direction of the migrating ion (which is diffusing in the $\langle 110 \rangle$ direction) and performing a fixed-volume calculation at increasing cell volumes (to simulate expansion with increasing temperature). The results of such calculations are shown in Figs. 3.4.11 and 3.4.12. These diagrams show how the potential-energy well in the z direction becomes increasingly unstable with increasing cell size and therefore temperature. It also shows that there is a local minimum at $z = 0$ for magnesium but a maximum at $z = 0$ for oxygen. In fact, as a consequence, the migrating oxygen defect would not minimise with this geometry. However, when placed at the local minimum (with the defect z coordinate of 0.075 cell units), the lattice energy was successfully minimised and only the one desired imaginary frequency was predicted.

Table 3.4.4. *Migration energy and migration volume for magnesium diffusion*

Temperature (K)	Migration energy (eV)	Migration volume (cm^3 mol^{-1})
100	1.965	2.059
200	1.948	2.261
300	1.928	2.569
400	1.907	3.169

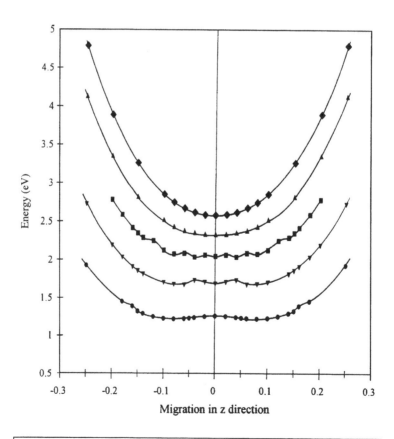

| ◆ a = 4.0112 A | ▲ a = 4.1112 A | ■ a = 4.2112 A | ▼ a = 4.3112 A | ● a = 4.4112 A |

Figure 3.4.11 Bifurcation of the magnesium saddle surface.

We found the migration free energy of Mg^{2+} and O^{2-} in MgO as a function of temperature by fixing an ion midway between two vacant lattice sites in the $x = y$ direction (see Fig. 3.4.1) and, in the case of oxygen, off centre in the z direction. The free energy was calculated up to 400 K for magnesium and 250 K for oxygen, after which the saddle surface bifurcated further and a secondary imaginary frequency appeared. Results obtained from these simulations are shown in Tables 3.4.4 and 3.4.5 and Fig. 3.4.13.

Table 3.4.5. *Migration energy and migration volume for oxygen diffusion*

Temperature (K)	Migration energy (eV)	Migration volume ($cm^3 \, mol^{-1}$)
50	1.997	2.155
100	1.997	2.163
150	1.991	2.172
200	1.988	2.179
250	1.981	2.188

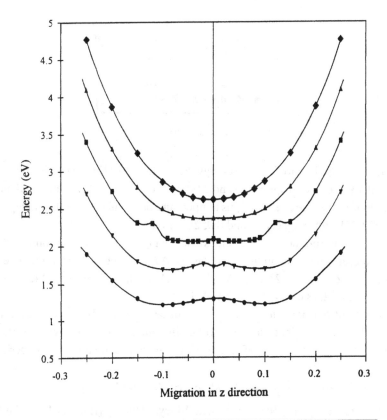

$\blacklozenge$ a = 4.0112 A $\blacktriangle$ a = 4.1112 A $\blacksquare$ a = 4.2112 A $\blacktriangledown$ a = 4.3112 A $\bullet$ a = 4.4112 A

Figure 3.4.12 Bifurcation of the oxygen saddle surface.

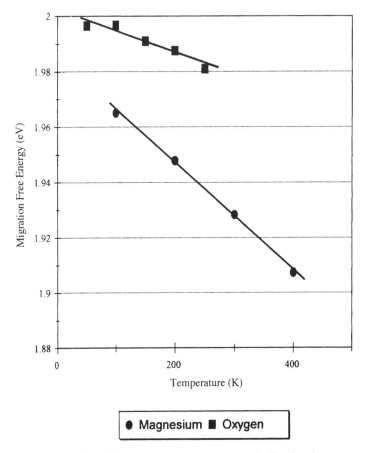

Figure 3.4.13 Free energy vs. temperature for ionic migration.

Graphical extrapolation of the free energy (Fig. 3.4.13) to 0 Kelvin yields a migration enthalpy of 1.985 eV for magnesium and 2.003 eV for oxygen. ΔS_m can be calculated from the slope of the free energy versus temperature plot, and we calculate a migration entropy of $2.251k$ for magnesium and $0.952k$ for oxygen.

The quality of the predicted ΔH_m values can be gauged when they are compared with previous experimental or theoretical estimates of ΔH_m, which include, for magnesium migration, 2.16 eV [42], 2.28 eV [43], 1.8–2.2 eV [44], 1.57–3.46 eV [45], and for oxygen diffusion, 2.38 eV [42], 1.3–2.1 eV [44].

The quality of our predicted ΔS_m values can be gauged only by comparing them with full diffusion coefficient data. In the following subsection, we use Vineyard theory to calculate absolute diffusion coefficients, which are then compared with experiment.

The Attempt Frequency by Means of Vineyard Theory

The attempt frequency may be obtained with Vineyard theory [Eq. (3.4.11)]. Vineyard theory is an harmonic theory, and therefore the eigenfrequencies of the 108 MgO unit supercell were calculated at 1 K in both the relaxed equilibrium and saddle-point configurations of the defective lattice, yielding one imaginary frequency for the unstable migrating ion as required. This unstable mode, corresponding to the motion of the defect across the saddle plane, is excluded from the denominator in Eq. (3.4.12). From Vineyard theory we found that the harmonic attempt frequencies υ^* for magnesium and oxygen diffusion were 18.99 and 9.83 THz, respectively. In fact, the attempt frequencies calculated from Vineyard theory are overestimated because of the low-frequency anharmonic mode at the saddle point associated with the displacement of the migrating ion in the $\langle 001 \rangle$ direction. From Harding et al. [46] this anharmonicity that is due to the incipient bifurcation needs to be corrected by applying a scaling factor to the results of the intrinsically harmonic Vineyard theory. The correction applied to the low-frequency mode (see Refs. [18] and [46]) reduces the calculated value of υ^*. We have to find out how much the actual saddle surface in the z direction differs from the harmonic and apply numerical integration methods to determine the rescaling factor that needs to be incorporated into Vineyard theory.

Figures 3.4.14 and 3.4.15 show the calculated migration energies at 1 K for both the cation and the anion fitted to a fourth-order polynomial compared with the harmonic

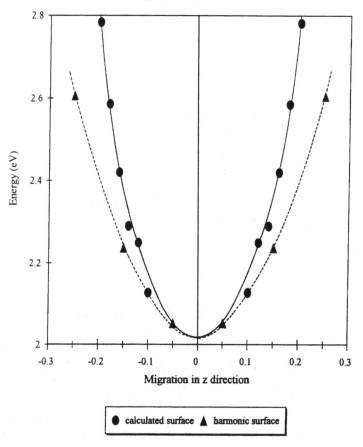

Figure 3.4.14 Anharmonicity of saddle surface for magnesium.

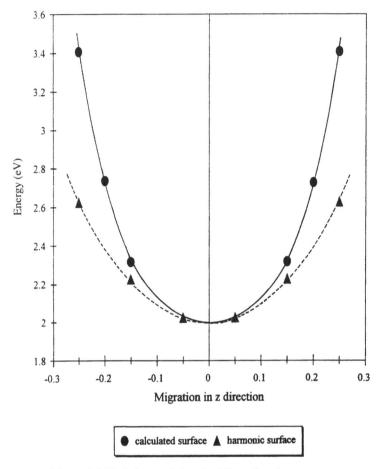

Figure 3.4.15 Anharmonicity of saddle surface for oxygen.

equivalent. From Harding et al. [46], the rescaling factor that needs to be applied to the values of υ^* predicted by Vineyard theory is given by

$$\frac{\int_0^{z_0} e^{\frac{-E(z)}{kT}} \, dz}{\int_0^{\infty} e^{\frac{-E'(z)}{kT}} \, dz}, \tag{3.4.26}$$

where $E(z)$ and $E'(z)$ are the calculated and the harmonic energies, respectively, as functions of displacement z away from the $\langle 110 \rangle$ direction to a maximum z_0, taken in this study to be $z = 0.25$. This yields an estimate of the reduction factor that must be applied to Vineyard theory in the temperature range 100–2000 K of 0.99–0.84 for magnesium and 0.97–0.87 for oxygen (see Fig. 3.4.16). When accounting for this anharmonicity by applying the maximum correction, this reduces the attempt frequencies to 16 THz for the Mg^{2+} ion and 9 THz for the O^{2-} ion. Therefore, from Eqs. (3.4.9) and (3.4.10) and taking $Z = 12$ for the diffusion of Mg and O in MgO,

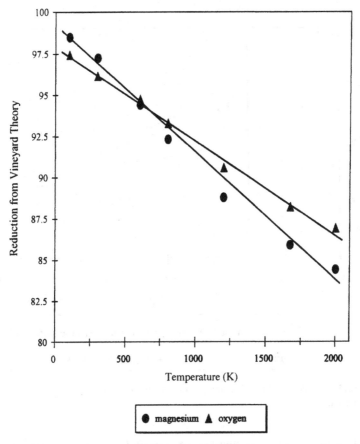

Figure 3.4.16 Preexponential correction factors for magnesium and oxygen.

we obtain the following extrinsic self-diffusion coefficients.

$$\text{For Mg:} = N_V \, 2.84 \times 10^{-6} \, \exp\left(-\frac{1.985}{kT}\right), \tag{3.4.27}$$

$$\text{For O:} = N_V \, 1.60 \times 10^{-6} \, \exp\left(-\frac{2.003}{kT}\right). \tag{3.4.28}$$

The Diffusion Coefficient

We can use the values of ΔH_f and ΔS_f to calculate the intrinsic diffusion coefficient; thus (remembering that the formation entropies and enthalpies take half the value of the Schottky defect formation energy),

$$\text{For Mg:} = 2.20 \times 10^{-5} \, \exp\left(\frac{-5.70}{kT}\right), \tag{3.4.29}$$

$$\text{For O:} = 1.24 \times 10^{-5} \, \exp\left(\frac{-5.72}{kT}\right). \tag{3.4.30}$$

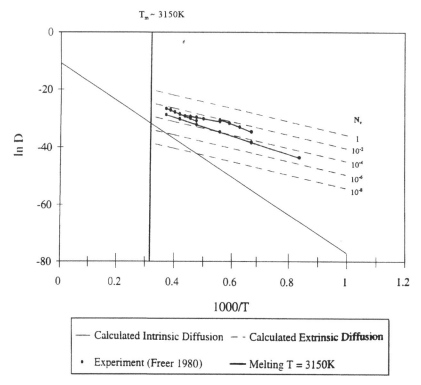

Figure 3.4.17 Calculated absolute magnesium diffusion data compared with experimental data.

Figures 3.4.17 and 3.4.18 show how our calculated diffusion coefficients vary with temperature and how they compare with some experimental data [45]. Our calculated extrinsic diffusion coefficient is plotted at different vacancy concentrations, $N_v = 1$, 10^{-2}, 10^{-4}, 10^{-6}, and 10^{-8}.

Our calculated results are in very close accord with experiment and are compatible with materials having N_V between approximately 10^{-3} and 10^{-7}. Our results suggest that the intrinsic régime is accessible at only very high temperatures, implying that all experimental data currently available give quantitave values for extrinsic diffusion only. Our results indicate that it would be exceptionally difficult to generate defects intrinsically before melting ($T_m = 3120$ K), as this would be observed only if $N_V < 10^{-6}$ and experiments were performed at >3000 K. This conclusion has also been reached by Wuensch [47], who considered diffusion in MgO in his extensive study of the results of experimental diffusion in close-packed oxides. From his research he found that both cation and anion diffusion in MgO were ill constrained with each experimental result plausible in its own right and dependent on experimental interpretation. He concluded that it was unlikely that any of the experiments he reviewed actually detected any intrinsic diffusion. It would seem apparent from this survey that computer calculations such as ours are necessary to provide a guide by which experimentalists may constrain their interpretations.

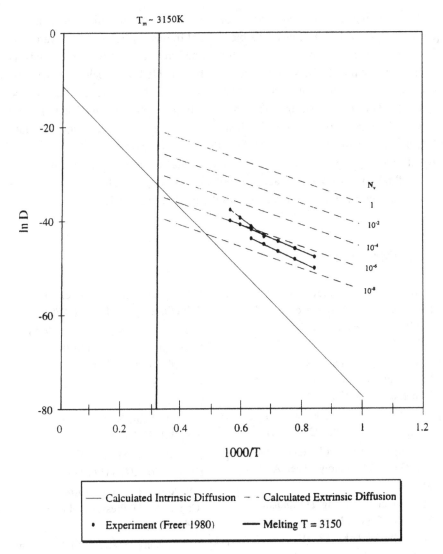

Figure 3.4.18 Calculated absolute oxygen diffusion data compared with experimental data.

3.4.4 Summary

First, we have obtained a self-consistent set of thermoelastic parameters for $MgSiO_3$ perovskite. Although a dichotomy still exists for the determination of the Grüneisen and the Anderson–Grüneisen parameters and their pressure dependence, our calculations favour the x-ray-diffraction and diamond-anvil-cell experiments of Mao et al. [1], but more experiments are required for resolving this debate. A possible explanation for the difference in the thermoelastic properties obtained from various experimental groups is due to the variation in parameterisations of the relevant equations of state used for fitting the original $V(P, T)$ data.

Second, applying an anharmonic correction to the predicted quasi-harmonic phonon frequencies allows us to calculate a feasible attempt frequency by using Vineyard theory for MgO, which is comparable with that obtained from thermodynamical analysis of the free-energy plots. The supercell method has enabled us to calculate both the extrinsic and the intrinsic diffusion coefficients for ionic diffusion in MgO that may be used as a standard for evaluating defect concentrations experimentally. These results support the findings of Wuensch [47] indicating that experimental techniques to date are only able to sample diffusion on MgO in the extrinsic regime.

Overall, we hope to have shown that the molecular dynamics and the lattice dynamics methodologies along with accurate potentials have proven to be rigorous complements to experiment and valuable tools in the field of geophysics.

References

[1] H. K. Mao, R. J. Hemley, Y. Fei, J. F. Shu, L. C. Chen, A. P. Jephcoat, Y. Wu, and W. A. Bassett, J. Geophys. Res. **96**, 8069 (1991).

[2] Y. Wang, D. J. Weidner, R. C. Liebermann, and Y. Zhao, Phys. Earth Planet. Inter. **83**, 13 (1994).

[3] O. L. Anderson and K. Masuda, Phys. Earth Planet. Inter. **85**, 227 (1994).

[4] A. Chopelas, Abstract Volume IUGG XXI General Assembly (Week A) A **354**, (1995).

[5] M. Born and K. Huang, *Dynamical Theory of Crystal Lattices* (Clarendon Press, Oxford, U.K., 1954).

[6] R. P. Ewald, Ann. Phys. **64**, 253 (1921).

[7] R. P. Ewald, Gottinger Nechr. Math-Phys. K **1**, 55 (1937).

[8] C. R. A. Catlow and M. J. Norgett, UKAEA Rep. AERE M2763 (United Kingdom Atomic Energy Authority, Harwell, U.K., 1978).

[9] B. G. Dick and A. W. Overhauser, Phys. Rev. **112**, 90 (1958).

[10] A. Patel, G. D. Price, M. Matsui, J. P. Brodholt, and R. J. Howarth, Phys. Earth Planet. Inter. **98**, 55 (1996).

[11] L. Vočadlo and G. D. Price, Phys. Chem. Miner. **23**, 42 (1996).

[12] S. C. Parker and G. D. Price, Adv. Solid State Chem. **1**, 295 (1989).

[13] G. Filippini, C. M., Gramaccioli, M. Simonetta, and G. B. Suffritti, Acta Crystallogr. A **32**, 259 (1976).

[14] A. Baldereschi, Phys. Rev. B **7**, 5212 (1973).

[15] D. J. Chadi and M. L. Cohen, Phys. Rev. B **8**, 5747 (1973).

[16] N. F. Mott and M. J. Littleton, Trans. Faraday Soc. **34**, 485 (1938).

[17] J. P. Poirier, *Creep of Crystals* (Cambridge U. Press, Cambridge, U.K., 1985).

[18] M. J. L. Sangster and A. M. Stoneham, J. Phys. C **17**, 6093 (1984).

[19] G. H. Vineyard, J. Phys. Chem. Solids **3**, 121 (1957).

[20] M. P. Allen and D. J. Tildesley, *Computer Simulation of Liquids* (Clarendon, Oxford, U.K., 1987).

[21] C. W. Gear, *Numerical Initial Value Problems in Ordinary Differential Equations* (Prentice-Hall, Englewood Cliffs, NJ, 1971).

[22] S. Nosé, J. Chem. Phys. **81**, 511 (1984).

[23] M. Matsui, J. Chem. Phys. **91**, 489 (1989).

[24] M. Parrinello and A. Rahman, Phys. Rev. Lett. **45**, 1196 (1980).

[25] R. Wentzcovitch, N. L. Ross, and G. D. Price, Phys. Earth Planet. Inter. **90**, 13 (1995).

[26] E. Knittle and R. Jeanloz, Science **235**, 668 (1987).

[27] J. B. Parise, Geophys. Res. Lett. **17**, 2089 (1990).

[28] N. L. Ross and R. M. Hazen, Phys. Chem. Miner. **16**, 415 (1989).

[29] E. Knittle, R. Jeanloz, and G. L. Smith, Nature (London) **319**, 214 (1986).

[30] N. Funamori, T. Yagi, W. Utsumi, T. Kondo, and T. Uchida, J. Geophys. Res. **101**, 8257 (1996).

[31] G. Fiquet, D. Andrault, A. Dewaele, T. Charpin, M. Kunz, and Haüsermann, Phys. Earth Planet. Inter. **105**, 21 (1998).

[32] H. Morishima, Geophys. Res. Lett. **21**, 899 (1994).

[33] N. Funamori and T. Yagi, Geophys. Res. Lett. **20**, 387 (1993).

[34] M. Matsui, G. D. Price, and A. Patel, Geophys. Res. Lett. **21**, 1659 (1994).

[35] R. J. Hemley, L. Stixrude, Y. Fei, and H. K. Mao, *High-Pressure Research: Application to Earth and Planetary Sciences*, Y. Syono and M. H. Manghnani, eds. (Terra Scientific, Tokyo, 1992), p. 183.

[36] O. L. Anderson, K. Masuda, and D. Guo, Phys. Earth Planet. Inter. **89**, 35 (1995).

[37] L. Stixrude, R. J. Hemley, Y. Fei, and H. K. Mao, Science **257**, 1099 (1992).

[38] O. L. Anderson, H. Oda, A. Chopelas, and D. Isaak, Phys. Chem. Miner. **19**, 369 (1993).

[39] O. L. Anderson, H. Oda, and D. Isaak, Geophys. Res. Lett. **19**, 1987 (1992).

[40] I. Inbar and R. E. Cohen, Geophys. Res. Lett. **22**, 1533 (1995).

[41] D. L. Anderson, Phys. Earth Planet. Inter. **45**, 307 (1987).

[42] W. C. Mackrodt and R.F. Stewart, J. Phys. C **12**, 5015 (1979).

[43] D. R. Sempolinski and W. D. Kingery, J. Am. Ceram. Soc. **63**, 664 (1980).

[44] C. R. A. Catlow and M. J. Norgett, UKAEA Rep. AERE M2936 (United Kingdom Atomic Energy Authority, Harwell, U.K., 1976).

[45] R. Freer, J. Mater. Sci. **15**, 803 (1980).

[46] J. H. Harding, M. J. L. Sangster, and A. M. Stoneham, J. Phys. C **20**, 5281 (1987).

[47] B. J. Wuensch, in *Mass Transport in Solids*, F. Bénière and C. R. A. Catlow, eds., NATO/ASI Series (Plenum, New York, 1983), p. 353.

Part IV

Transformations in Silica

Chapter 4.1

Polymorphism in Crystalline and Amorphous Silica at High Pressures

Russell J. Hemley, James Badro, and David M. Teter

Recent years have been witness to advances in our understanding of the high-pressure behavior of crystalline and amorphous silica. Experimental developments made possible by new diffraction techniques have generated new findings to megabar pressures (i.e., above 100 GPa). Theoretical advances, including increasingly accurate first-principles methods and interatomic potentials such as those first proposed by Yoshito Matsui and co-workers, have provided predictions and new understanding of experimental data. We review these theoretical developments in the context of recent experimental findings. Our analyses provides a basis for understanding the extensive metastablity of high-pressure crystalline structures, the nature of the short- and intermediate-range order of the high-pressure amorphous material, and both equilibrium and nonequilibrium transformations. Such study also provides insight into the structural basis of anomalous transport properties of the liquid predicted at high pressure.

4.1.1 Introduction

The nature of silica under pressure is a textbook example of the intersection of condensed-matter physics and mineralogy [1]. Silica is of obvious importance in mineralogy, as SiO_2 is abundant in the Earth's crust and plays a major role in the deep interior, both as a product of chemical reactions and as an important secondary phase. From the point of view of condensed-matter physics, SiO_2 presents an important system for investigating pressure-induced polymorphism, providing examples

H. Aoki et al. (eds), *Physics Meets Mineralogy* © 2000 Cambridge University Press.

of first-order reconstructive transitions [2], displacive (soft-mode-driven) transitions [3], pressure-induced amorphization [4, 5], and polymorphism and polyamorphism [4–6]. Silica is also of great technological interest in both its crystalline and glassy forms.

Since the discovery of stishovite [2], there has been significant interest in the possibility of denser phases that are stable at higher pressure. The transformation to the $CaCl_2$-type phase [7, 8] was found at $50\,GPa$ [3, 9, 10], in excellent agreement with first-principles predictions [3, 11]. Evidence for other phases, possibly stable at still higher pressures, has been reported on the basis of both static and dynamic compression of amorphous silica [12], α-cristobalite [13], and α-quartz [14, 15]. Poststishovite phases proposed in these studies or predicted theoretically include the Fe_2N [12, 15], α-PbO_2 [14, 16], $I2/a$ [17], baddeleyite [18], fluorite [19], and $Pa3$ [20–22] structures. Numerous questions also exist concerning the states of amorphous material produced at high pressure, including the distribution of ring structures in the random network, the nature of the short- and intermediate-range order, how these structures differ from crystalline (but possibly disordered) variants, and the relationship to pressure-induced amorphization (e.g., Refs. [5], and [23–25]).

In view of these problems, it is not surprising that the behavior of silica continues to be the focus of considerable experimental and theoretical study. In fact, a flurry of such studies have been reported in recent years. Many new experimental results have been made as a result of advances in synchrotron x-ray-diffraction methods. Concurrently, there have been considerable improvements in first-principles methods as well as interatomic potentials such as those proposed by Tsuneyuki et al. [26]. Such potentials have been used extensively, either as is or in a modified form (slightly different interaction parameters) [27]. Here we review these developments and examine implications for recent experimental findings. The results provide new insights into the extensive metastablity of high-pressure crystalline structures, the nature of the short- and the intermediate-range order of high-density amorphous silica, nonequilibrium transformations including amorphization, and the relationship between structure and dynamics in high-temperature liquids.

4.1.2 Equilibrium High-Pressure Phases

4.1.2.1 Energetics

The bonding in silica gives rise to extensive polymorphism. The polymorphs of silica such as quartz, cristobalite, and tridymite that are stable at low pressure can be described as spatial arrangements of corner-linked SiO_4 tetrahedra. This fourfold coordination of Si (^{IV}Si) allows a large number of connectivity sequences because of the soft, deformable, Si—O—Si linkage joining the rather rigid SiO_4 units. The properties of equilibrium phases based on tetrahedral Si (i.e., the lower-pressure polymorphs) are relatively well understood, and have been extensively reviewed elsewhere (e.g., Ref. [1]). At higher pressures, Si increases its coordination number, first in stishovite,

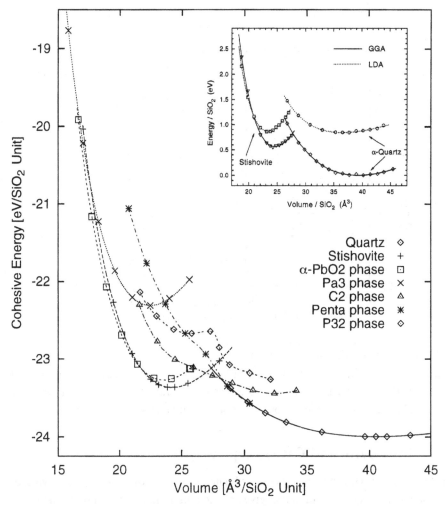

Figure 4.1.1 Calculated energy–volume curves for selected SiO₂ phases. The curves were determined from density functional theory by use of the generalized gradient approximation (GGA). The inset shows the difference between the GGA and the local-density-approximation (LDA) results for quartz and stishovite (from Ref. [28]).

which has the higher-density rutile structure and a more ionic Si—O bond. Numerous higher-pressures structures have been proposed, including those with sevenfold and eightfold coordination of Si. The energetics of representative phases considered here are given in Figs. 4.1.1 and 4.1.2.

Accurate calculation of structural properties, equations of state, and energetics of the full suite of known silica polymorphs has been a difficult problem for theory. This is true both for ionic model calculations (because of the covalency of the Si—O bond) as well as for first-principles total-energy methods [30, 31]. Figure 4.1.1 shows energy–volume relations calculated for high-pressure phases from density functional theory. Within density functional theory, the use of generalized-gradient-approximation (GGA) corrections has been shown to give a reasonably correct determination of the relative energies for quartz and stishovite compared with standard local-density-approximation

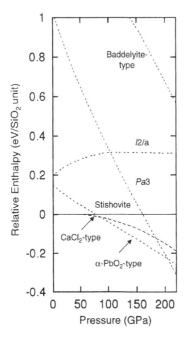

Figure 4.1.2 Enthalpies ($T = 0$ K) for a series of SiO$_2$ polymorphs calculated from the local-density approximation [29].

(LDA) theory [28] (Fig. 4.1.1). Nevertheless, structural and equation-of-state parameters calculated with the LDA are in significantly better agreement with experiment (e.g., Ref. [29]). The LDA-derived enthalpies of several hypothetical high-pressure polymorphs are shown as a function of pressure relative to stishovite in Fig. 4.1.2.

4.1.2.2 Stishovite and CaCl$_2$-Type Structures

The Si atoms (^{VI}Si) in stishovite are coordinated to six O atoms to form a network of SiO$_6$ octahedra, a configuration that gives rise to a more ionic Si—O bond. The structure can be described as a distorted *hcp* array of O ions with one half of the available octahedral interstices occupied by Si. The CaCl$_2$-type structure was first proposed as a poststishovite phase by Nagel and O'Keeffe [32] on the basis of crystal-chemical arguments; they showed that the transition from rutile is associated with an elastic instability and softening of the Raman-active B_{1g} optical phonon (Fig. 4.1.3). In this structure, the layers of O anions adopt a closer packing than is allowed by the rutile structure, resulting in greater O–O separation. This follows from the general observation that structures tend to adopt packings that maximize the anion–anion and cation–cation separations while maintaining near-ideal cation–anion bond lengths (eutaxy) [33]. Tsuchida and Yagi [7] presented evidence that the CaCl$_2$-type phase was stable at pressures of $\sim$100 GPa on the basis of in situ x-ray diffraction carried out without a pressure medium (essentially nonhydrostatic conditions). Subsequently, the transition pressure was predicted by full potential LDA methods [linearized augmented plane-wave (LAPW) calculations] at significantly lower pressures (45 GPa) [11]. The dynamics of the transition were also explored in detail at the same time by Matsui and Tsuneyuki [8].

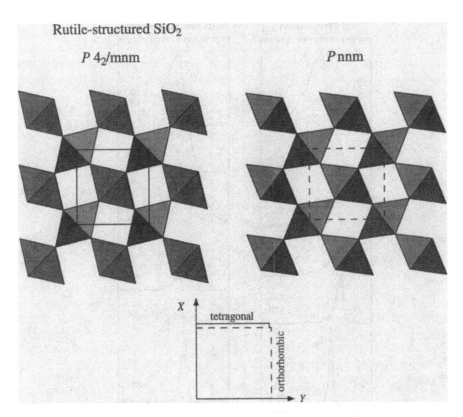

Figure 4.1.3 Relationship between the stishovite (rutile-type) and $CaCl_2$-type structures.

Raman measurements conducted on samples contained in quasi-hydrostatic pressure-transmitting media (e.g., Ne) showed that the transition occurs near 50 GPa [3]. This result was in excellent agreement with the first-principles predictions [3, 11]. The calculations of Matsui and Tsuneyuki [8] showed that the material is strongly anharmonic at the transition. This large anharmonicity is likely to be responsible for the additional features observed in at the transition (Fig. 4.1.4), as expected for a displacive transition (e.g., here associated with a $c_{11}-c_{12}$ elastic instability). Single-crystal x-ray diffraction revealed variable strains in the (domain) microcrystals formed by the breakup of the single crystal at the phase transition [34]; these variable strains appear to arise from the grain–grain contacts giving rise to a complex Raman spectrum in the vicinity of the transition.

The transition pressure of the rutile–$CaCl_2$ transition was confirmed by single-crystal x-ray diffraction, measurements that also mapped out the spontaneous strain associated with the transition [9, 34]. Angle-dispersive powder-x-ray-diffraction measurements combined with laser heating [10] are also consistent with the single-crystal study. The recent powder-diffraction study has provided evidence for the stability of the $CaCl_2$-type phase to at least 120 GPa, as discussed further below. A complete Landau model has also been developed to examine the strain/order parameter coupling in the transition [35].

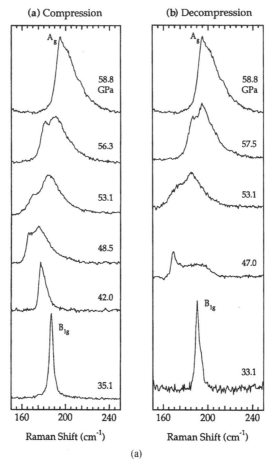

(a) Compression   **(b) Decompression**

Figure 4.1.4 (a) Low-frequency Raman spectrum
showing the transition between stishovite and the
$CaCl_2$-type structures [3, 34].

4.1.2.3 Higher-Pressure Phases

The phases described above have been established experimentally. We now discuss denser phases that are predicted theoretically to be equilibrium high-pressure phases. There have been several conflicting experimental reports for the existence of the first of these, α-PbO_2-type SiO_2 (Fig. 4.1.5). The α-PbO_2 structure type, which is observed as a postrutile phase in a number of analog systems, can also be described in terms of an *hcp* packing of O. However, in this case the Si is arranged in such a way as to generate 2×2 zigzag chains of edge-sharing octahedra. The α-PbO_2 structure type was first discussed by Pauling and Sturdivant [36], and the geophysical implications of such a structure type for SiO_2 were examined by Liu [37]. Theoretical calculations of α-PbO_2-type SiO_2 were first performed by Matsui and Kawamura [38], who used interatomic potentials.

α-PbO_2-type silica was first reported by German et al. [14] in samples recovered from the dynamic compression of quartz. Their reported refinement gave lattice

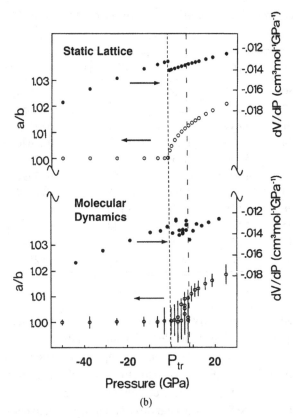

Figure 4.1.4 (b) the c/a calculated in the simulations of Matsui and Tsuneyuki [8], showing the strongly anharmonic character of the transition.

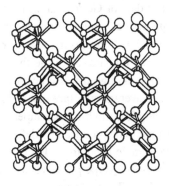

Figure 4.1.5 α-PbO$_2$ structure.

parameters of $a = 4.30$ Å, $b = 4.70$ Å, and $c = 4.50$ Å. However, there is poor agreement between their reported data and recent theoretical results obtained with LDA methods; in particular, the lattice parameters calculated at zero pressure are $a = 4.06$ Å, $b = 5.01$ Å, and $c = 4.48$ Å. More recently, Dubrovinsky et al. [39] reported the synthesis of a new phase of silica by laser heating of amorphous silica at high pressure. They refer to this structure as a *Pnc2* type based on earlier theoretical studies carried out with empirical potentials [40]. However, a simple cell transformation confirms

Table 4.1.1. *Calculated structure parameters for the α-PbO$_2$-type silica structure as a function of pressure*

Pressure	a (Å)	b (Å)	c (Å)	Si$(0, y, \frac{1}{4})$	O(x, y, z)
0	4.050	5.001	4.468	0.153	(0.2684, 0.3828, 0.4193)
20	3.967	4.914	4.395	0.153	(0.2659, 0.3838, 0.4192)
40	3.901	4.844	4.334	0.152	(0.2638, 0.3847, 0.4192)
60	3.844	4.786	4.282	0.152	(0.2620, 0.3854, 0.4194)
80	3.795	4.736	4.237	0.151	(0.2604, 0.3860, 0.4196)
100	3.753	4.691	4.196	0.151	(0.2590, 0.3865, 0.4198)
120	3.713	4.653	4.160	0.150	(0.2577, 0.3870, 0.4201)
140	3.678	4.616	4.126	0.150	(0.2565, 0.3873, 0.4203)
160	3.645	4.584	4.096	0.149	(0.2554, 0.3876, 0.4205)
180	3.616	4.554	4.067	0.149	(0.2548, 0.3879, 0.4207)
200	3.588	4.526	4.040	0.148	(0.2534, 0.3881, 0.4210)

that the *Pnc*2 structure is essentially identical to α-PbO$_2$ [29], a result also pointed out by Kanzaki et al. [41]. Also, first-principles calculations indicated that the *Pnc*2 structure adopts the higher-symmetry *Pbcn* space group at all pressures considered and that the optimized *Pnc*2 and α-PbO$_2$-type structures have the same total energies and simulated diffraction patterns [29]. These calculations also predict the transformation of CaCl$_2$-type to PbO$_2$-type SiO$_2$ to be near 80 GPa. The agreement between the experimental and the theoretically calculated diffraction patterns indicate that the α-PbO$_2$-type phase may have been synthesized in the experiment of Dubrovinsky et al. [39].

Recently, tentative evidence was presented for α-PbO$_2$-type and baddeleyite-type silica in the SNC Shergotty meteorite [16, 18]. The data of German et al. [14] were used as the basis for the identification of the new phase (i.e., similar lattice parameters obtained from limited diffraction data). There are major differences in the axial ratios, a result that persists when the experimental data are compared with the theoretical predictions at all calculated pressures (Table 4.1.1). Although the discrepancy could in principle arise from the samples having a distorted average structure that is due to a high defect concentration (e.g., as a result of quenching from high shock temperatures), the magnitude of the distortion appears to be too large to be produced by defects. It is likely that the crystal structure of the experimental sample is not the α-PbO$_2$ type.

The compatibility of the reported evidence for α-PbO$_2$-type phase versus the reported stability of CaCl$_2$-type silica to at least $\sim$100 GPa [7, 10] remains to be clarified. It should be noted that the first-principles calculations indicate that the transition to the α-PbO$_2$-type phase should occur near the reported maximum pressure of these experimental studies [7, 10]. It would be useful to examine both experimentally and theoretically the *P–T* boundary of the transition as well as the effects of nonhydrostatic stress, as the experimental studies all involved high temperatures and varying degrees of pressure inhomogeneity. Moreover, the evidence for extensive metastability of dense SiO$_2$ phases at these pressure complicates the experimental results, as discussed below.

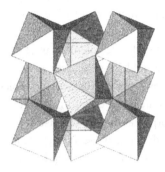

Figure 4.1.6 *Pa*3 structure.

On the other hand, the reported observation of the baddeleyite-type structure is not consistent with theoretical calculations, which indicate that this structure (with ^{VII}Si) is so unfavorable energetically (Fig. 4.1.2) that quenching to the metastable state under ambient conditions seems unlikely. Moreover, the structure is calculated to be mechanically unstable below 100 GPa [29] and transforms to the α-PbO$_2$ structure, as discussed by Tse et al. [45]. The monoclinic $I2/a$ structure is more competitive but still has a lower density and a higher enthalpy relative to CaCl$_2$- and α-PbO$_2$-type silica (Fig. 4.1.2).

A higher-pressure phase of SiO$_2$ consisting of corner-linked octahedra and *Pa*3 space group symmetry (Fig. 4.1.6) was proposed to be a possible poststishovite phase by Tressaud and Demazeau [20]. This *Pa*3 structure (earlier called modified fluorite) was apparently first observed experimentally by the compression of rutile-type PdF$_2$ [43]. It was later proposed that this structure type may in general be adopted instead of the fluorite structure as a postrutile phase for metal dioxides [22]. The structure allows increased anion–anion and cation–cation separation compared with the close-packed phases discussed above. This structure type for silica was examined theoretically by Matsui and Kawamura [38] by molecular-dynamics simulations and subsequently by first-principles methods by Park et al. [44]. Recent total-energy calculations predict that the *Pa*3-type structure becomes the stable phase above $\sim$215 GPa [29, 31]. The phase has not yet been identified experimentally.

4.1.3   Metastable Crystalline High-Pressure Phases

4.1.3.1   Quartz-II

Recent theoretical studies have provided a deeper understanding of metastable phase transitions in silica and of these transitions, those occurring in α-quartz have been the subject of continuing interest. The nature and the mechanism of pressure-induced amorphization at 20–30 GPa [4, 5] and the associated metastable crystalline–crystalline transition (to quartz-II) [34, 48] is intriguing. The latter transition was accompanied by the appearance of new diffraction lines and the loss of several strong diffraction lines belonging to α-quartz, but the relatively low resolution and limited reciprocal space coverage in the diffraction data prevented identification of its structure. Theoretical models have been used to identify possible structural models for the phase as well to locate possible underlying structural instabilities.

Theoretical calculations that employ different interatomic potentials have predicted a structural instability for α-quartz of ~20 GPa. Tse and Klug [46] predicted a mechanical (i.e., elastic) Γ-point instability in quartz at this pressure. Subsequent molecular-dynamics simulations of Somayazulu et al. [47] indicated a transformation to a monoclinic phase at these pressures (i.e., before amorphization). It was also reported on the basis of lattice-dynamics calculations that a dynamical instability occurs at the K point of the Brillouin zone ($\frac{1}{3}, \frac{1}{3}, 0$) before the onset of the mechanical instability (e.g., Ref. [48]). These transition pressures are close to that of the experimentally observed crystalline–crystalline phase transition in quartz.

Wentzcovitch et al. [49] found that if the dynamical instability at the K point is allowed to proceed within the simulation, then the transition results in a hexagonal phase with fivefold and sixfold coordinated Si ($P3_2$ structure; Table 4.1.2) [51]. However,

Table 4.1.2. *Calculated structural parameters at 22 GPa for α-quartz and the postquartz structures described in the text*

Structure	Atom	x	y	z
α-quartz (100% ^{IV}Si)	Si(1)	0.4821	0	$\frac{1}{6}$
$a = b = 4.337$ Å, $c = 5.090$ Å	O(1)	0.3742	0.3236	0.2505
$P3_221$ #154 $Z = 3$				
Penta phase (100% ^{V}Si)	Si(1)	0.5043	0.3240	0.2201
$a = b = 4.364$ Å, $c = 9.245$ Å	O(1)	0.4734	0.0759	0.0527
$P3_221$ #154 $Z = 6$	O(2)	0.8131	0.6700	0.1394
$C2$ phase ($\frac{1}{3}{}^{IV}$Si $+ \frac{2}{3}{}^{VI}$Si)	Si(1)	0	0.0540	0
$a = 8.403$ Å, $b = 3.385$ Å, $c = 5.219$ Å	Si(2)	0.3243	0.3192	0.3818
$\beta = 110.2°$	O(1)	0.1664	0.3151	0.0856
$C2$ #5 $Z = 6$	O(2)	0.1864	0.3194	0.5728
	O(2)	0.4957	0.3251	0.2779
$P3_2$ phase ($\frac{2}{3}{}^{V}$Si $+ \frac{1}{3}{}^{VI}$Si)	Si(1)	0.6953	0.1297	0.6045
$a = b = 6.951$ Å, $c = 4.837$ Å	Si(2)	0.7723	0.9304	0.0000
$P3_2$ #145 $Z = 9$	Si(3)	0.3645	0.9156	0.1464
	O(1)	0.8890	0.0976	0.2858
	O(2)	0.8736	0.4133	0.5581
	O(3)	0.9099	0.4294	0.0477
	O(4)	0.9124	0.1548	0.7900
	O(5)	0.1507	0.7517	0.3611
	O(6)	0.2175	0.7488	0.8490
Stishovite (100% ^{VI}Si)	Si(1)	0	0	0
$a = b = 4.059$ Å, $c = 2.628$ Å	O(1)	0.3035	0.3035	0.3035
$P4_2/mnm$ #135 $Z = 2$				

if the mechanical (Γ-point) instability is allowed to drive the phase transition, then the resulting phase is monoclinic with fourfold and sixfold coordinated Si ($C2$ phase; Table 4.1.2). Unfortunately, the comparison of diffraction patterns calculated for these model structures with the experimental patterns [45] does not yield a unique solution. On the other hand, vibrational spectroscopy provides an important constraint on the transformation: Raman and IR measurements suggest a minor structural change (and point to an order–disorder transformation), with no evidence for sixfold coordinated Si atoms in the quartz-II phase [52]. The theoretical results suggest that for an ideal crystal under hydrostatic stress the dynamical instability occurs first, yielding the $P3_2$ phase predicted by Wentzcovitch et al. [49]. The calculation also indicates that inclusion of the defects has an effect on the transformation [51]. The presence of twin boundaries in the simulated α-quartz lowers the instability by several gigapascals, usually resulting in the formation of the $C2$ structure. It appears that the instability nucleates at the twin boundary and grows throughout the crystal.

Recent molecular-dynamics studies for which the potential of van Beest, Kramer, and van Santen was used [27] predict that an orthorhombic phase of silica with space group $Pnc2$ becomes stable above 6 GPa [53]. This phase consists entirely of tetrahedral SiO_4 units and could be a good candidate for a postquartz phase; in fact, its simulated diffraction pattern shows some similarities to that of quartz-II. The reported calculations indicate [54] that the structure becomes mechanically unstable at 15 GPa and undergoes a transformation to a monoclinic phase with mixed fourfold and sixfold coordination. Examination of the stability of the phase with total-energy calculations as well as calculation of complete diffraction patterns and vibrational spectra as functions of pressure are needed for detailed comparison with experimental data.

4.1.3.2 Nonhydrostatic Stress Effects and Fivefold Coordination

The effects of nonhydrostatic stresses on phase transitions and structures of silica polymorphs have been the focus of several recent studies. Recent work indicates that specific nonhydrostatic stress conditions can radically alter the compressional behavior of α-quartz [55, 56]. Molecular-dynamics calculations suggest that when quartz is compressed nonhydrostatically with the stress along c greater than that along a and b ($\sigma_c > \sigma_a, \sigma_b$), new structural transformations result [55]. The resulting strain allows the Si atoms to experience a different O environment than would be encountered under hydrostatic conditions. A low-energy pathway is created that promotes a crystalline–crystalline transition from α-quartz to a phase with fivefold coordinated silicon (^VSi). The structure possesses $P3_221$ space-group symmetry, i.e., it is isosymmetrical with α-quartz. The fivefold coordination of Si by O gives rise to O atoms that have both twofold and threefold coordination by Si. The SiO_5 pentahedra form a distorted square pyramid linked together as edge-sharing and corner-sharing dimers to form a three-dimensional network (Fig. 4.1.7, Table 4.1.2).

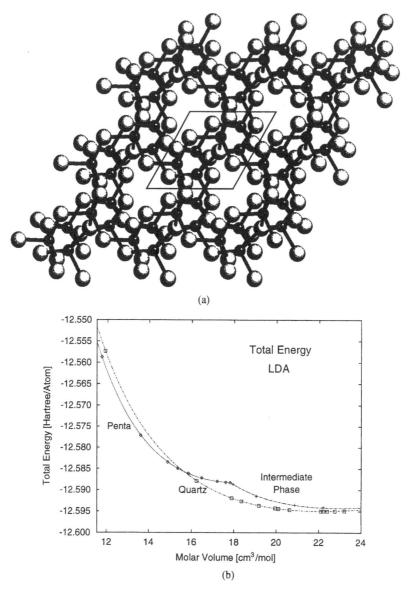

(a)

(b)

Figure 4.1.7 (a) Structure of the theoretically predicted five-coordinate Si phase, (b) energy–volume curves.

First-principles LDA calculations, as well as molecular dynamics and energy minimization with interatomic potentials, find this phase to be mechanically stable and energetically stable with respect to quartz at high pressure [55]. On decompression, the ^{V}Si phase reverts to α-quartz through an intermediate fourfold coordinated phase in an unusual isosymmetrical phase transformation. During the transformation, the coordination number of the Si changes from five to four, and the oxygens change from threefold to twofold. Taken together with the theoretical results described above for quartz-II, these results suggest the existence of extensive polymorphism and the possibility of numerous structure types constructed from combinations of fourfold, fivefold, and sixfold coordinated Si at pressures of 15–30 GPa.

4.1.3.3 Polymorphism in Octahedrally Coordinated Silica

Simulations with empirical potentials by Matsui and Kawamura [38] showed that there is a variety of structures that are close in both energy and density to the α-PbO_2 type; they called these structures defect structures, and these structures would form metastable phases. This general finding was extended by Teter et al. [29] on the basis of crystal-chemical arguments and first-principles calculations. Many crystal structures can be visualized in terms of close-packed arrays (either *hcp* or *ccp*, or superpositions of the two stacking sequences) of anions or cations. For each ion in the array, there are two available tetrahedral interstices and one available octahedral interstice. If these interstices are occupied by cations, the close-packed arrays of anions are usually distorted from ideal packing by bonded interactions between adjacent cations and anions and by cation–cation and anion–anion antibonding and nonbonding interactions. The structure types of octahedrally coordinated silica can be visualized in terms of close-packed arrays (either *hcp* or *ccp*, or combinations of the two stacking sequences) of anions. For example, stishovite can be described as a distorted *hcp* array of O ions with one half of the available octahedral interstices occupied by Si. This gives rise to straight chains of edge-sharing SiO_6 octahedra that are corner linked between ($ABAB\cdots$) layers to form a three-dimensional network.

Such an approach allows us to methodically generate hypothetical structures, which can then be examined by theoretical methods, including first-principles calculations (Fig. 4.1.8) [29]. The MX superstructures for hexagonal and cubic close packing are the NiAs and NaCl structure types. In both structures, cations occupy all of the octahedral sites between the close-packed anion layers. The selective removal of one-half of the M ions from the NiAs structure leads to the rutile and α-PbO_2 structures, whereas the anatase structure can be derived by removal of one half of the M atoms from the NaCl structure. Novel framework structures of MX_6 octahedra can be generated by construction of NiAs- or NaCl-type supercells, or combinations of their stacking variants, and removal of one half of the metal ions according to simple rules [57].

LDA calculations for numerous structures found in this way reveal that the most favorable are those constructed of edge-sharing octahedral chains with varying degrees of kinking (Fig. 4.1.8; Table 4.1.3) [29]. The densities and the enthalpies of phases having these structures correlate with the number of kinks along the chains (Fig. 4.1.9). Thus, stishovite (straight chains) is the least dense, whereas α-PbO_2 type (the most possible kinks in its 2×2 zigzag chains) is the densest, with an essentially unlimited number of intermediate structures. The results for structures with 4×4, 3×3, and 3×2 kinked octahedral chains in ($ABAB\cdots$) stacking can be compared with previously examined structures and experimental diffraction data. The favourability of these structures as high-pressure phases is determined by a delicate balance between maximizing cation–anion attractions and minimizing anion–anion and cation–cation

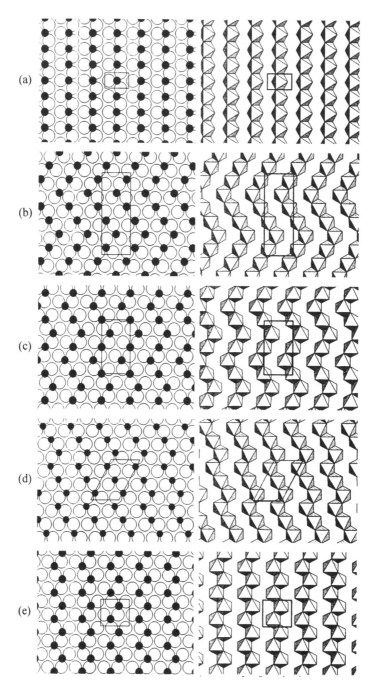

Figure 4.1.8 Representations of dense silica structures: (a) CaCl$_2$, (b) 4 × 4 SnO$_2$, (c) 3 × 3 NaTiF$_4$, (d) 3 × 2 $P2_1/c$, (e) 2 × 2 α-PbO$_2$ structure types. The left-hand figures show one layer of the $ABAB\cdots$ stacking of hcp oxygen anions (white) with one half of the octahedral interstices filled with Si ions (black). The right-hand figures show how these patterns form edge-sharing octahedral chains with various degrees of kinking.

repulsions while simultaneously maximizing the density (and at finite temperature, the entropy). α-PbO$_2$-type SiO$_2$ becomes the stable phase at 80–85 GPa at this level of calculation [29, 42]. The phase is denser than that with the CaCl$_2$ structure type and allows greater O–O separation.

As the intermediate kinked phases are both structurally similar to and enthalpically competitive with the rutile and α-PbO$_2$ structures, these may form metastable phases in high-pressure silica samples (e.g., if transformation pathways kinetically favor their formation instead of α-PbO$_2$). These theoretical results can thus be used to examine some of the contradictory high-pressure experimental data for SiO$_2$. The brookite- and anatase-type structures, as well as the previously mentioned $I2/a$-type [17] and

Table 4.1.3. *Calculated structural parameters at zero pressure, space groups, formula units/cell, and predicted bulk moduli for representative dense SiO$_2$ structures*

Structure	Atom	x	y	z
Stishovite	Si(1)	0	0	0
$P4_2/mnm$ #156 $Z = 2$	O(1)	0.3053	0.3053	0
$a = 4.154$ Å, $b = 4.154$ Å, $c = 2.667$ Å				
$K_0 = 306$ GPa, $K_0' = 4.6$				
4×4 SnO$_2$ type	Si(1)	0	0.0607	$\frac{1}{4}$
$Pbcn$ #60 $Z = 12$	Si(2)	0.9964	0.3573	0.4162
$a = 4.086$ Å, $b = 4.987$ Å, $c = 13.488$ Å	O(1)	0.2910	0.1292	0.0296
$K_0 = 311$ GPa, $K_0' = 4.0$	O(2)	0.7877	0.0882	0.1376
	O(3)	0.2295	0.3330	0.1954
3×3 NaTiF$_4$ type	Si(1)	0	0.7949	$\frac{1}{4}$
$Pbcn$ #60 $Z = 8$	Si(2)	0	$\frac{1}{2}$	0
$a = 4.067$ Å, $b = 4.993$ Å, $c = 8.975$ Å	O(1)	0.7298	0.0233	0.1670
$K_0 = 317$ GPa, $K_0' = 4.0$	O(2)	0.7843	0.7663	0.4177
3×2 $P2_1/c$ type	Si(1)	$\frac{1}{2}$	0	0
$P2_1/a$ #14 $Z = 6$	Si(2)	0.1656	0.5032	0.9657
$a = 7.619$ Å, $b = 4.063$ Å, $c = 4.996$ Å	O(1)	0.0545	0.2327	0.6557
$\beta = 118.09°$	O(2)	0.7227	0.2286	0.1823
$K_0 = 320$ GPa, $K_0' = 4.0$	O(3)	0.3897	0.2149	0.6542
2×2 α-PbO$_2$ type	Si(1)	0	0.1502	$\frac{1}{4}$
$Pbcn$ #60 $Z = 4$	O(1)	0.2576	0.3870	0.4201
$a = 4.049$ Å, $b = 5.001$ Å, $c = 4.468$ Å				
$K_0 = 329$ GPa, $K_0' = 3.9$				
HP-PdF$_2$ type	Si(1)	0	0	0
$Pa3$ #205 $Z = 4$	O(1)	0.3445	0.3445	0.3445
$a = 4.422$ Å				
$K_0 = 347$ GPa, $K_0' = 4.1$				

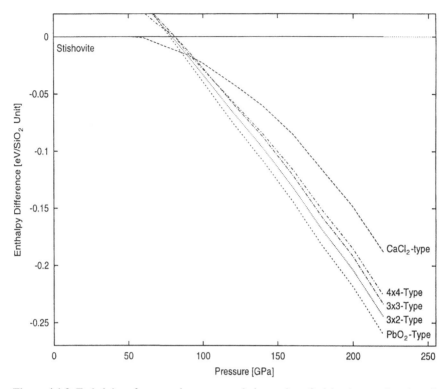

Figure 4.1.9 Enthalpies of proposed structures relative to that of stishovite as a function of pressure. First-principles pseudopotential plane-wave total-energy calculations within the LDA to electronic exchange and correlation were performed on the model structures. This method accurately reproduces many of the physical properties of silica [1, 17, 29, 58]. Vanderbilt ultrasoft pseudopotentials were generated for Si and O [59, 60]. The electronic degrees of freedom were minimized by a preconditioned conjugate-gradient method, and both the ions and cell were fully optimized at fixed volume. Because of the extremely small energy differences among the structures, a convergence of 0.001 eV/SiO₂ molecule was used for energy differences with respect to kinetic-energy cutoff and for the total energy with respect to Brillouin zone integration. A plane-wave cutoff of 395.7 eV was found to give sufficient convergence for the chosen structures and their energy differences. Calculated equations of state of α-quartz, stishovite, and the CaCl₂-type silica by the PW91–GGA [61] code show that the LDA method provides a significantly better prediction of the experimental equations of state.

baddeleyite type, are not favored at high pressure because of their low densities relative to stishovite. In contrast to these, the calculations predict that kinked phases described above are all preferred relative to the CaCl₂-type structure above ~100 GPa. The 4×4 structure type has been reported for an epitaxially modified form of SnO₂ [62]. The 3×3 structure type (Fig. 4.1.8) is adopted by α-NaTiF₄ [63]. The 3×2 structure type (apparently not previously reported) is found to be only slightly higher in enthalpy than the 2×2 α-PbO₂ type.

Yamakata and Yagi [64] (see also Chap. 4.4 of this volume) reported evidence for a new phase obtained from quasi-hydrostatic compression of cristobalite at room temperature. The structure appears to be closely related to, but distinct from, stishovite. The phase differed from stishovite by the presence of a single additional diffraction

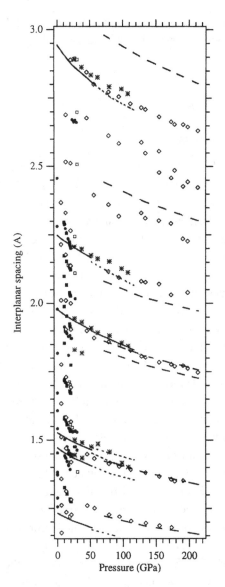

Figure 4.1.10 d spacings for silica measured in x-ray-diffraction experiments and calculated from theory to megabar pressures. The filled symbols are hydrostatic experiments: filled circles, polycrystalline quartz [45]; filled squares, single-crystal quartz [34]. The open symbols are from nonhydrostatic experiments: open squares, polycrystalline quartz run to 31 GPa [65]; open diamonds, polycrystalline quartz to 215 GPa [65]; stars, laser-heated silica [7]. Theory: solid curves, stishovite; short-dashed curves, $CaCl_2$ type; long-dashed curves, α-PbO_2.

line, which disappeared on heating (indicating metastability relative to stishovite). Kingma et al. [65] reported diffraction evidence for a dense phase from nonhydrostatic room-temperature compression of quartz powder to 213 GPa (Fig. 4.1.10). No match to the predicted equilibrium high-pressure phases was found. However, fair agreement with the measured d spacings was obtained with an orthorhombic structure derived from the observed $CaCl_2$ structure [65]. In a theoretical simulation of the experiment, compression of an 18-atom orthorhombic cell of α-quartz above 25 GPa by an energy-minimization scheme with the potential of van Beest et al. [27] results in a diffusionless transformation to the 3×2 kinked $P2_1/a$ phase. Simulated powder-diffraction patterns from LDA-derived $P2_1/a$ unit cells as functions of pressure are found to be similar to those of the experimental data (i.e., with respect to d spacings). The good agreement between theory and experiment, along with the low enthalpy of the $P2_1/a$ structure, suggests that it may have been synthesized as a metastable phase in these experiments.

4.1.4 High-Pressure Amorphous Forms

4.1.4.1 Compression of Silica Glass

Silica glass (vitreous SiO_2) consists of a fully polymerized framework of corner-linked SiO_4 tetrahedra at room pressure, with an average Si—O bond length of 1.61 Å and a wide Si—O—Si angle distribution between 120° and 180° with an average of 144° [50]. It is well established that the framework is extremely flexible in the sense that the intertetrahedral angle is strongly affected by the application of pressure. The evolution of the properties of silica glass with increasing pressure has been the subject of continued studies. High-pressure studies of the glass indicate that cooperative tilting of the SiO_4 tetrahedra accommodates all of the initial volume compression under applied pressure (e.g., Ref. [5]). The angle distribution becomes sharper and closer to 120°. In α-quartz, the O atoms approach a body-centered lattice under these conditions [25, 66]; a similar average configuration may prevail in the glass. At that point, a second densification mechanism takes over: A gradual increase of mean Si coordination occurs, as shown by x-ray diffraction and vibrational spectroscopy [23, 24, 67–71]. The mean Si—O—Si angle shifts toward lower values (asymptotically to 90°), the distribution narrows, and the average Si—O bond length increases [72]; this indicates a trend toward octahedral coordination. In principle, a third compression regime is encountered when the entire system consists of octahedra. At this point, the density increase is directly linked to linear compression of the Si—O bond (estimated to be >50 GPa).

The low-frequency boson band and the first sharp diffraction feature have been the subject of extended recent debate for silica (and related network-forming amorphous solids) (e.g., Refs. [73–78]). These in principle contain detailed information on intermediate-range order, phonon localization, and transport properties. Q_1 appears to be a signature of tetrahedral framework [70]. Strong pressure shifts in both the low-frequency Raman band identified as that of the boson peak [24, 79] and the first sharp diffraction peak of the glass [70] were measured (Fig. 4.1.11). Figure 4.1.12 shows the shift in the maximum of the boson peak v_m and the first sharp diffraction peak Q_1 obtained in situ at high pressure [24, 70, 79] and from pressure-quenched samples [77]. The linear correlation between the pressure shifts in the maxima of the two features for SiO_2 is evident, with the data for densified glass falling on a distinctly lower trend.

Notably, the width of the first sharp diffraction peak, ΔQ_1, remains essentially constant at high pressure (to at least 42 GPa [70]), despite the large shift in v_m. In situ high-pressure measurements [80–82] reveal that the transverse sound velocity v_t drops from 3.8 km/s at zero pressure to 3.3 km/s at 8 GPa; hence $v_t \Delta Q_1$ decreases rather than increases over this pressure range, in contrast to recent proposals [74]. Further, Elliot [73] has proposed that the intensity of Q_1 should decrease with pressure as

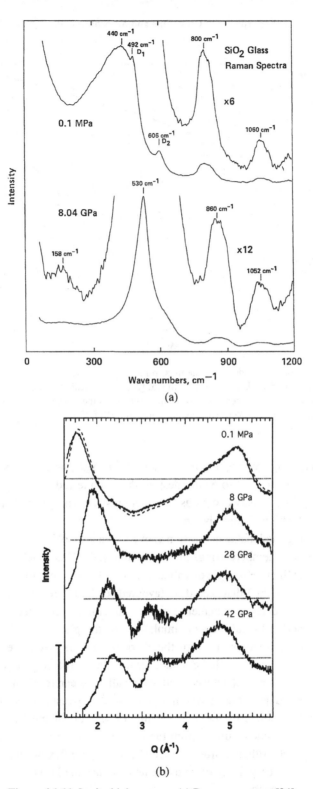

Figure 4.1.11 In situ high-pressure (a) Raman spectrum [24] and (b) x-ray-diffraction pattern [70] of SiO_2 glass at 8 GPa (room temperature).

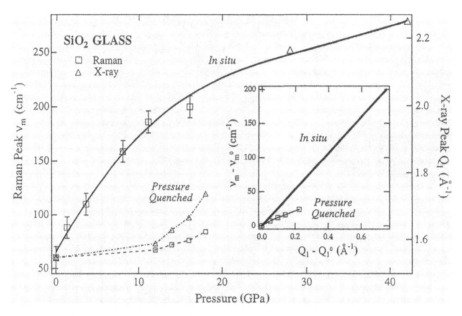

Figure 4.1.12 Pressure dependence of the maximum of the low-frequency boson peak v_m and the first sharp diffraction peak Q_1 measured in situ [24, 70, 79] and at zero pressure following quenching from the indicated pressures [77]. The inset shows the relative shifts plotted against each other where the initial zero-pressure values are (from Ref. [78]).

a result of decreasing free volume in the glass. Like ΔQ_1, however, the intensity is also unaffected by pressure to 8 GPa despite the volume reduction by $\sim$10% over this range [82]. The intensity decreases at higher pressure where densification and the gradual Si coordination changes begin.

It has been suggested that densification above 8 GPa is associated with a shift in the population of small-membered rings in the structure [24]. This conclusion was based on the shift in the Si—O—Si angle distribution inferred from vibrational spectroscopy toward values (e.g., 120°) that would be favoured for bond breaking and reforming to give a higher percentage of smaller (e.g., four- and three-membered Si rings). This proposal is supported by subsequent simulations of the glass [83, 84]. It has been proposed that increasing pressure shifts the distribution of rings to larger values [85]. This conclusion is based on a Bethe model for the glass structure and an analogy to low-density crystals, for which it is argued that small rings can result in an increase in excluded volume and therefore an overall decrease in atom density in the vicinity of the ring. This is not supported by the interpretation of vibrational spectroscopic studies of the glass. The argument suffers from the problem that the analogy to crystals is apparently incorrect: Although true for certain crystals, the flexibility of the random network glass allows easy filling of voids created around small rings, giving rise to an overall increase in density.

Less information has been available about glasses (or rather amorphous material) formed at higher pressures (e.g., >30 GPa). The behavior of amorphous silica formed at these pressures has been of interest both from a fundamental point of view and as a model for understanding the behavior of silicate melts that may form at lower-mantle conditions deep within the Earth. Under these conditions, the coordination of Si has increased, although the change (measured at room temperature) is spread out over a wide pressure interval ($\sim$10–30 GPa) [70]. As this is a metastable phase, the P–T–t trajectory is important for describing the state of the material.

In principle, three different paths may be considered: (1) room-temperature compression of glass (vitreous silica), (2) formation of an amorphous solid by rapid quenching of the melt under pressure (as for a conventional glass), and (3) formation of an amorphous solid by pressure-induced amorphization of low-pressure crystalline phases (typically a rapid process occurring in the lower-temperature regime, i.e., well below melting). If the system were at global thermodynamic equilibrium, the ergodic principle would require the final states to be equivalent because the final state has to be independent of the P–T path, but this is not necessarily the case because ergodicity does not prevail. Recent molecular-dynamics simulations [84] have explored paths (2) and (3). The results indicate that the structure and the dynamics of the system are a continuous function of the quenching rate for path (2) and a continuous function of maximum pressure for path (3) when the systems are compared at the same density. Analysis of the structural properties (coordination and ring statistics) and the dynamical properties (velocity autocorrelation function, vibrational density of state) of the glass shows that the higher the pressure the system is compressed above the amorphization threshold (and the lower the pressure for decompression amorphization), the closer the system is to that of a supercooled liquid obtained by rapid temperature quench [86]. In pressure-induced amorphization, maximum pressure then plays a role similar to that of quenching rate in standard liquid supercooling.

Primak [87] has suggested that high-pressure amorphous silica is a dense, ordered array of O atoms with disordered Si atoms. This model is consistent with the theoretical prediction of the near degeneracy of structures formed from dense, ordered arrays of O with disordered Si filling one half of the octahedral sites. Such a phase corresponds to the defect-niccolite (Fe_2N) structure, which has been reportedly synthesized from laser-heated silica [12, 15] at high pressure. The pair correlation functions for simulations of defect-niccolite structure and dense supercooled liquid are shown in Fig. 4.1.13. The sharper peaks indicate a more ordered structure for the former; thus although the high-pressure amorphous solid formed on room-temperature compression possesses considerable short- and intermediate-range order, the material still contains residual disorder associated with defects and residual strain in the O sublattice. The availability of a large number of possible sites for the Si would lower the free energy at high temperature through entropic effects, although this needs to be quantified by further calculations.

4.1.4.2   Pressure-Induced Amorphization

As described above, it is well established that quartz (and coesite) undergo [4] pressure-induced amorphization when pressurized well into the stability field of stishovite at

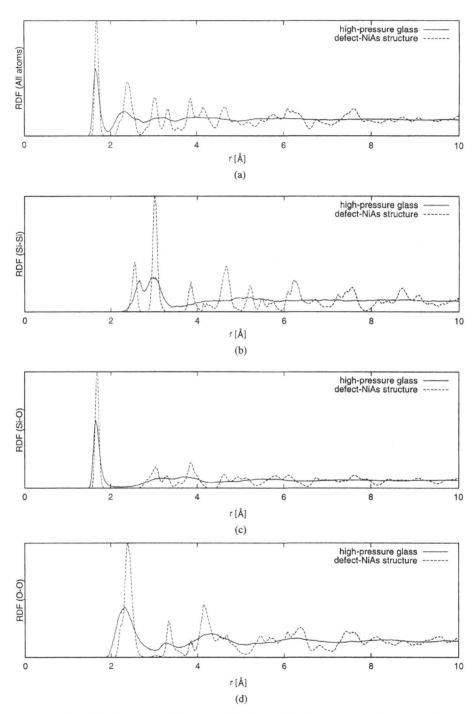

Figure 4.1.13 Radial distribution functions (RDFs) for (a) all the atoms, (b) Si—Si interactions, (c) Si—O interactions, and (d) O—O interactions.

room temperature. This class of metastable transformation was first reported for ice I [88]. Numerous experiments [5, 45, 89, 90] and simulations [46–48, 91–94] have been undertaken to understand aspects of the transformation. Early on, it was noted [4] that the amorphization threshold is modified in the presence of nonhydrostatic stress, with lower transformation pressures observed in the presence of shear stress. Subsequently, an intermediate metastable crystalline phase was reported before amorphization [45], as noted above. Different phenomena related to this transformation in quartz appear at pressures beginning at 12 GPa [25]. This depends on the magnitude of shear stress [55] and the sensitivity and length scale of order probed by different techniques (e.g., single crystal and powder x-ray diffraction, Raman and IR spectroscopy, TEM), as well as the character of the samples (e.g., single crystals or powders) [95]. Detailed discussion can be found in Refs. [30], and [52].

Several different points related to the characteristics of the amorphized samples deserve discussion. First, one of the important findings of more recent work has been the observation that as the transformation proceeds, the samples are strongly heterogeneous [45]. Raman spectra show features very similar to those of silica glass [4], and x-ray-diffraction patterns show a broadening of quartz reflections and an increase in diffuse background intensity [45]. Second, a component of the amorphization observed in the pressure-quenched samples is decompression-induced amorphization of quartz-II [30]; Raman spectra of such pressure-quenched samples show a mixture of amorphous and crystalline material. High-resolution TEM analyses of recovered samples [96] reveal lamellar features consisting of an alternation of amorphous and crystalline material. Insight into the nature of these lamellar features in situ has been provided by local structural observations by x-ray-diffraction patterns collected on a single crystal compressed to 17–25 GPa with a micro-x-ray beam [97]. In the unaltered region, Laue diffraction patterns are characteristic of a well-crystallized sample (quartz) whereas diffraction measured from the lamellae is characteristic of disordered yet crystalline material. On decompression, the same measurements reveal that the material in the lamellae becomes amorphous.

The observation of amorphous material on decompression (e.g., that in the lamellar structure) can be understood in general thermodynamically if a new high-pressure phase were formed (in this case, quartz-II); i.e., kinetics prevent recrystallization of the equilibrium crystalline polymorph [98]. For the SiO_2 system, this thermodynamic treatment is also applicable to high-temperature amorphization of high-pressure polymorphs at ambient pressure (e.g., stishovite amorphization when heated above 800 K [99, 100]), as well as the pressure-induced amorphization of the low-pressure phases discussed above [5].

Simulations of silica offer a microscopic picture of the thermodynamic and the structural changes that occur on amorphization. Both the potentials of Tsuneyuki et al. [26] and van Beest et al. [27] potentials have been used. According to simulations with the potential of van Beest et al., the amorphization process is observed to occur around 22 GPa with a disordering accompanied by an increase in mean coordination number. The systems thus consist of a mixture of fourfold, fivefold, and sixfold coordinated

Si atoms, the proportion of which is pressure dependent [27, 46–48, 55, 91–94]. In these model systems, box-sized limitations prevent the development of the hetero-geneities observed experimentally [96] and remain amorphous on pressure release. The coordination effect is largely reversible in the sense that a large proportion of Si atoms are in fourfold coordination at zero pressure [55], and is a function of the maximum pressure reached. Nonisotropic stresses lower the mean pressure of amor-phization [55], as observed experimentally [4, 5].

Quasi-harmonic lattice dynamics carried out with these potentials predict that the transformation is associated with an elastic Γ-point instability at or approximately at 22 GPa [46, 94, 101, 102], as discussed above. Moreover, similar simulations reveal a K-point (dynamical) instability [48] at lower pressures, although it has been argued that the results depend on the geometry of the simulation box [47, 48]. Simulation in a cell that is not built in a $3 \times 3 \times 1$ geometry will not undergo the K-point instability [because it occurs at the $(\frac{1}{3}, \frac{1}{3}, 0)$ point in the Brillouin zone] and will not form a superstructure associated with this instability [47]. That phase was then found to undergo amorphization, in good agreement with experimental observations [45] of the sequence quartz $\rightarrow$ quartz-II $\rightarrow$ amorphous. However, the simulated diffraction patterns do not coincide with those observed experimentally [47].

Recent molecular-dynamics simulations [84] have examined the relationship be-tween pressure-induced amorphization of quartz and the metastable extension of its melting (Fig. 4.1.14), a conjecture originally made for ice [88]. In a recent experi-mental study of ice [104], a crossover was found between the equilibrium melting (two-phase melting) and nonequilibrium amorphization (one-phase melting). Like-wise, the simulations for SiO_2 indicate a crossover between these two regimes; the pressure-induced amorphization curve follows the extrapolation of the melting curve and the structural and semidynamical properties of the amorphs show a continuous variation [84]. Raman spectroscopy of amorphized quartz shows that, with time, the high-density amorphous material relaxes toward the low-density glass, as can be seen by characteristic changes in the Raman spectrum [105]; it was also reported that temperature-amorphized stishovite at room pressure forms an amorphous material that is similar to that obtained from pressure-amorphized quartz, and then relaxes to-ward the low-density glass [100]. The combined experimental and theoretical results suggest the presence of a mechanical instability line that bounds the equilibrium and the metastable melting line, although the behavior at lower temperatures (e.g., below the glass transition at high pressure) requires additional study.

4.1.5   High-Density Liquid

The above discussion of the pressure-induced structural changes in crystalline and amorphous solids provides a basis for understanding the properties of high-temperature liquid silica under pressure, a first-order problem in the geosciences. Liquid silica is the archetypal strong tetrahedral network liquid. Historically, much of what is known

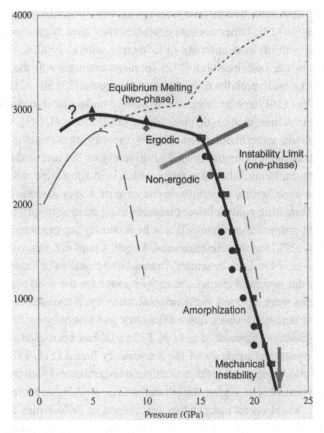

Figure 4.1.14 $P-T$ phase diagram showing the melting and amorphization of α-quartz [84]. The predicted quartz–melt (glass) coexistence curve is given by the thick curve and filled symbols. Circles and squares indicate isothermal compression, and diamonds and triangles indicate isobaric heating (two different symbols are used for each set to indicate the error bars). With decreasing temperature, the calculated transition changes over from true melting (in the ergodic regime) to amorphization (nonergodic regime). The experimental equilibrium melting curves of quartz, coesite, and stishovite are given as thin curves. The metastable extension of the experimental quartz melting curve is extrapolated (thin dashed curve) from the data of Zhang et al. [103] (see also Ref. [30]).

experimentally about the structure of the liquid at high density has been inferred by analogy from the high-pressure behavior of glass. Nevertheless, much insight into the structure of the liquid has been obtained from computer simulations, as shown by the pioneering studies of Woodcock et al. [106]. For example, more sophisticated two-body [26, 27] and three-body [107] potentials provided group-mode analysis of the vibrational spectrum in the liquid [108–110] and dynamical quantities such as diffusion and viscosity, including predictions of coordination changes even at the lowest densities [111]. Poole et al. [112] have predicted a possible second critical point in liquid silica, with two distinct liquid phases existing below T_c, as originally predicted for H_2O [113]. Other unusual characteristics have been reported, such as the existence of a third-order anomaly in isotherms around 7000 K, corresponding to a minimum in the bulk modulus [112] (perhaps analogous to the experimental observation of the bulk modulus minimum in silica glass [23, 80–82]). Belonoshko and Dubrovinsky [114] have performed a two-phase molecular-dynamics simulation of the melting of stishovite with the potential of van Beest et al. [27], which yielded good agreement with experiment. More recently, the melting curve of quartz has been examined with the same potential [84]. The difficulty of the task is due to the finite (and very short) duration of the simulation, which is incompatible with the need for thermodynamic equilibrium, especially in the case of a very viscous liquid such as silica close to its melting point at lower pressures (e.g., melting of quartz and coesite).

The transport properties of liquid silica at high density are predicted to be anomalous [106, 115–117] (see also Hemmati and Angell, Chap. 6.1, this volume). Recent predictions for the anomalous dynamical (transport) properties of liquid silica can be compared with the structural changes described above for the solid phases. Si and O atom diffusivities were reported in the original study by Woodcock et al. [106]. The Stokes–Einstein equation, which links diffusivity and viscosity by the introduction of the Stokes–Einstein diameter, $d_{SE} = k_B T / 3\pi \eta D$, has been used in more recent studies of the dynamical behavior of the low-density liquid [111, 118–120]. These studies show a good agreement (within simulation uncertainties) with the extrapolated experimental viscosities to higher temperatures ($d_{SE} = 2.5$ Å). More sophisticated methods [116], based on an independent calculation of diffusivities and viscosities (from the stress–stress autocorrelation function), show that the Stokes–Einstein relation holds better at higher densities, where d_{SE} is independent of temperature and density and has a value of 1 to 1.5 Å. Nevertheless, the temperature dependence of the diffusion constants [116, 118] agrees with the Arrhenius-like behavior of liquid SiO_2 [115] at low density (Fig. 4.1.15). The initial calculations of the system indicated that O atoms diffuse by infrequent hopping from the coordination sphere of one Si atom to that of another [106].

Recent simulations of Tsuneyuki and Matsui [117] show that the driving force for the diffusion process can be understood as arising from the differences in enthalpy between the fourfold and the sixfold coordinated crystalline phases. The model shows that diffusion is driven by a transient coordination increase. The calculations of Barrat et al. [116] predict that the diffusivity increases with pressure and then decreases, with a maximum around 20 GPa at 4000 K, an effect directly connected to the changing distribution of fourfold, fivefold, and sixfold coordinated silicon species (Fig. 4.1.15). The O diffusivity is correlated with the proportion of fivefold-coordinated Si [111, 118]. The diffusivity maximum is obtained when the fivefold coordinated species accounts for 50% of the total Si, which therefore cannot be considered as defects. The picture proposed by Tsuneyuki and Matsui [117] is broadly consistent with this view. Barrat et al. [116] have shown that not only does viscosity also reach a minimum, but there is also a transition from strong (Arrhenius-like) behavior to fragile behavior. A plausible

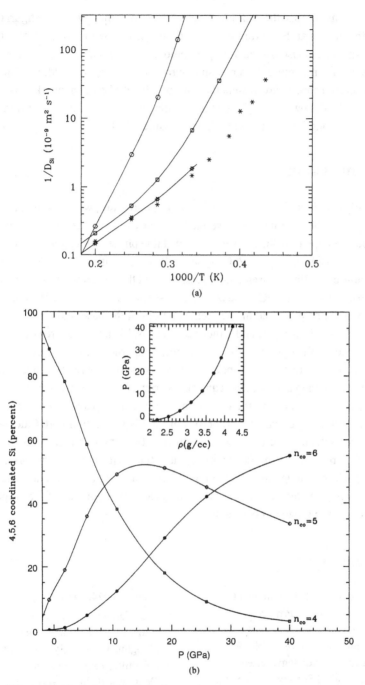

Figure 4.1.15 (a) Density dependence of transport properties of silica predicted from molecular-dynamics simulations with the potential of van Beest et al. (a) Arrhenius plots of $1/D_{Si}$ (reciprocal diffusion constant for Si); open circles, $\rho = 2.2$ g/cm^{-1}; open stars, $\rho = 3.3$ g/cm^{-1}; stars, $\rho = 3.7$ g/cm^{-1}; squares, $\rho = 4.2$ g/cm^{-1}. (b) Distribution of Si coordination numbers as a function of pressures. The curves are guides to the eye (from Ref. [116]).

explanation of this is that the minimum viscosity is obtained when configurational entropy of the melt is at its maximum [86, 116], consistent with Adam–Gibbs theory [121] for diffusion and viscosity in a simple liquid. Moreover, the activation energy for cooperative diffusion is much lower in an environment where fourfold, fivefold, and sixfold coordinated species are present than in a purely tetrahedral network; this would explain the tendency towards fragility at higher density and predict another transition from fragile to strong behavior at still higher densities [86].

4.1.6   Conclusions

Extensive polymorphism in silica is possible at high pressure, with a large number of structures than can be described as close-packed O arrays with one half of the octahedral interstices occupied by Si. The sequence of stable phases with increasing pressure is stishovite $\Rightarrow$ $CaCl_2$ type $\Rightarrow$ α-PbO_2 type $\Rightarrow$ $Pa3$-type with evidence for considerable metastable crystalline phases at high pressure. But there is conflicting evidence for the relative stability of the $CaCl_2$ and α-PbO_2-type phases in the 100-GPa range. Free SiO_2, which might be produced in chemical reactions at the base of the mantle, could have the α-PbO_2-type structure under these conditions, but a large degree of compression of amorphous silica glass at low (i.e., room) temperature can be accommodated by compression of the tetrahedral random network, with the pressure-induced coordination change occurring above 10 GPa and spread out over a wide pressure range. The high-pressure behavior of both the crystalline and the amorphous forms provides a basis for understanding dynamical properties of the high-density liquid. Calculations predict, for example, that the existence of a diffusivity maximum that arises from the existence of a broad range of coordination states (i.e., with low-energy barriers between them). The continued improvement in the accuracy of both theoretical methods and high P–T experimental techniques is providing a unifying picture of the properties of silica over a wide range of conditions.

References

[1] P. J. Heaney, C. T. Prewitt, and G. V. Gibbs, eds., *Silica: Physical Behavior, Geochemistry, and Materials Applications*, Vol. 29 of Reviews in Mineralogy Series (Mineralogical Society of America, Washington, D.C., 1994).

[2] S. M. Stishov and S. V. Popova, Geokhimiya **10**, 837 (1961).

[3] K. J. Kingma, R. E. Cohen, R. J. Hemley, and H. K. Mao, Nature (London) **374**, 243 (1995).

[4] R. J. Hemley, in *High-Pressure Research in Mineral Physics*, M. H. Manghnani and Y. Syono, eds. (Terra Scientific, Tokyo; and American Geophysical Union, Washington, D.C., 1987), p. 347.

[5] R. J. Hemley, A. P. Jephcoat, H. K. Mao, L. C. Ming, and M. H. Manghnani, Nature (London) **334**, 52 (1988).

[6] C. A. Angell, Science **267**, 1924 (1995).

[7] Y. Tsuchida and T. Yagi, Nature (London) **340**, 217 (1989).

[8] Y. Matsui and S. Tsuneyuki, in *High-Pressure Research: Application to Earth and Planetary Sciences*, Y. Syono and M. H. Manghnani, eds. (Terra Scientific, Tokyo; and American Geophysical Union, Washington, D.C., 1992), p. 433.

[9] H. K. Mao, J. Shu, J. Hu, and R. J. Hemley, Eos Trans. Am. Geophys. Union **75**, 662 (1994).

[10] D. Andrault, G. Fiquet, F. Guyot, and M. Hanfland, Science **282**, 720 (1998).

[11] R. E. Cohen, in *High-Pressure Research: Application to Earth and Planetary Sciences*, Y. Syono and M. H. Manghnani, eds. (Terrra Scientific, Tokyo; and American Geophysical Union, Washington, D.C., 1992), p. 425.

[12] L. G. Liu, W. A. Bassett, and J. Sharry, J. Geophys. Res. **83**, 2301 (1978).

[13] Y. Tsuchida and T. Yagi, Nature (London) **347**, 267 (1990).

[14] V. N. German, M. A. Podurets, and R. F. Trunin, JETP **37**, 107 (1973).

[15] T. Sekine, M. Akaishi, and N. Setaka, Geochim. Cosmochem. Acta **51**, 379 (1987).

[16] A. El Goresy, T. G. Sharp, B. Wopenka, and M. Chen, Lunar Planet. Sci. (to be published).

[17] J. S. Tse, D. D. Klug, and Y. Le Page, Phys. Rev. Lett. **69**, 3647 (1992).

[18] A. El Goresy, L. Dubrovinsky, S. Saxena, and T. G. Sharp, Meteor. Planet. Sci. **33** (Suppl. 4), A45 (1998).

[19] Y. Syono and S. Akimoto, Mater. Res. Bull. **14**, 147 (1968).

[20] A. Tressaud and G. Demazeau, High Temp. High Pressures **16**, 303 (1984).

[21] Y. Matsui and M. Matsui, in *Advances in Physical Geochemistry*, S. Ghose, E. Salje, and J. M. D. Coey, eds. (Springer, New York, 1988), Vol. 7, p. 129.

[22] J. Haines, J. M. Leger, and O. Schulte, Science **271**, 629 (1996).

[23] M. Grimsditch, Phys. Rev. Lett. **52**, 2379 (1984).

[24] R. J. Hemley, H. K. Mao, P. M. Bell, and B. O. Mysen, Phys. Rev. Lett. **57**, 747 (1986).

[25] R. M. Hazen, L. W. Finger, R. J. Hemley, and H. K. Mao, Solid State Commun. **72**, 507 (1989).

[26] S. Tsuneyuki, M. Tsukada, H. Aoki, and Y. Matsui, Phys. Rev. Lett. **61**, 869 (1988).

[27] B. W. H. van Beest, G. J. Kramer, and R. A. van Santen, Phys. Rev. Lett. **64**, 1955 (1990).

[28] D. R. Hamann, Phys. Rev. Lett. **76**, 660 (1996).

[29] D. M. Teter, R. J. Hemley, G. Kresse, and J. Hafner, Phys. Rev. Lett. **80** 2145 (1998).

[30] R. J. Hemley, C. T. Prewitt, and K. J. Kingma, in *Silica: Physical Behavior, Geochemistry, and Materials Applications*, P. J. Heaney, C. T. Prewitt, and G. V. Gibbs, eds., Vol. 29 of Reviews in Mineralogy (Mineralogical Society of America, Washington, D.C., 1994) p. 41.

[31] R. E. Cohen, in *Silica: Physical Behavior, Geochemistry, and Materials Applications*, P. J. Heaney, C. T. Prewitt, and G. V. Gibbs, eds., Vol. 29 of Reviews in Mineralogy (Mineralogical Society of America, Washington, D.C., 1994), p. 369.

[32] L. Nagel and M. O'Keeffe, Mater. Res. Bull. **6**, 1317 (1970).

[33] M. O'Keeffe, Acta Crystallogr. A **33**, 924 (1977).

[34] R. J. Hemley, J. Shu, M. A. Carpenter, J. Hu, H. K. Mao, and K. J. Kingma, Solid State Commun., in press.

[35] M. A. Carpenter, R. J. Hemley, and H. K. Mao, J. Geophys. Res., in press.

[36] L. Pauling and J. H. Sturdivant, Z. Kristallogr. **68**, 239 (1928).

[37] L. G. Liu, Phys. Earth Planet. Inter. **9**, 338 (1974).

[38] Y. Matsui and K. Kawamura, in *High-Pressure Research in Mineral Physics*, M. H. Manghnani and Y. Syono, eds. (American Geophysical Union, Washington, D.C., and Terra Scientific, Tokyo, 1987), p. 305.

[39] L. S. Dubrovinsky, S. K. Saxena, P. Lazor, R. Ahuja, O. Eriksson, J. M. Wills, and B. Johannson, Nature (London) **388**, 362 (1997).

[40] A. B. Belonoshko, L. S. Dubrovinsky, and N. A. Dubrovinsky, Am. Mineral. **81**, 785 (1996).

[41] M. Kanzaki, Y. Matsui, and M. Matsui, Am. Mineral. **82**, 1042 (1997).

[42] J. S. Tse, D. D. Klug, and D. C. Allan, Phys. Rev. B **51**, 16392 (1995).

[43] A. Tressaud, F. Langlais, G. Demazeau, and P. Hagenmuller, Mater. Res. Bull. **14**, 1147 (1979).

[44] K. T. Park, K. Terakura, and Y. Matsui, Nature (London) **336**, 670 (1988).

[45] K. J. Kingma, R. J. Hemley, H. K. Mao, and D. R. Veblen, Phys. Rev. Lett. **70**, 3927 (1993).

[46] J. S. Tse and D. D. Klug, Phys. Rev. Lett. **67**, 3559 (1991).

[47] M. S. Somayazulu, S. M. Sharma, and S. K. Sikka, Phys. Rev. Lett. **73**, 98 (1994).

[48] G. W. Watson and S. C. Parker, Phys. Rev. B **52**, 13306 (1995).

[49] R. M. Wentzcovitch, C. da Silva, J. R. Chelikowsky, and N. Binggeli, Phys. Rev. Lett. **80**, 2149 (1998).

[50] R. L. Mozzi and B. E. Warren, J. Appl. Crystallogr. **2**, 164 (1969).

[51] D. M. Teter, to be published.

[52] P. Richet and P. Gillet, Eur. J. Mineral. **9**, 907 (1997).

[53] I. M. Svishchev, P. G. Kusalik, and V. V. Murashov, Phys. Rev. B **55**, 721 (1997).

[54] V. V. Murashov and I. M. Svishchev, Phys. Rev. B **57**, 5639 (1998).

[55] J. Badro, J.-L. Barrat, and P. Gillet, Phys. Rev. Lett. **76**, 772 (1996).

[56] J. Badro, D. M. Teter, R. T. Downs, P. Gillet, R. J. Hemley, and J.-L. Barrat, Phys. Rev. B **56**, 5797 (1997).

[57] A. F. Wells, *Structural Inorganic Chemistry*, 4th ed. (Clarendon, Oxford, U.K., 1975).

[58] B. B. Karki, M. C. Warren, L. Stixrude, G. J. Ackland, and J. Crain, Phys. Rev. B **55**, 3465 (1997).

[59] D. Vanderbilt, Phys. Rev. B **41**, 7892 (1990).

[60] G. Kresse and J. Hafner, J. Phys. Condens. Matter **6**, 8245 (1994).

[61] J. P. Perdew et al., Phys. Rev. B **46**, 6671 (1992).

[62] V. E. Müller, Acta. Crystallogr. B **40**, 359 (1984).

[63] P. J. Omaly, P. Batail, D. Grandjean, D. Avignant, and J.-C. Coussiens, Acta. Crystallogr. B **32**, 2106 (1976).

[64] M. Yamakata and T. Yagi, Rev. High Pressure Sci. Technol. **7**, 107 (1998).

[65] K. J. Kingma, H. K. Mao, and R. J. Hemley, High Pressure Res. **14**, 363 (1996).

[66] H. Sowa, Z. Kristallogr. **184**, 257 (1988).

[67] R. Couty and G. Sabatier, J. Chim. Phys. **75**, 843 (1978).

[68] P. F. McMillan, B. Piriou, and R. Couty J. Chem. Phys. **81**, 4234 (1984).

[69] Q. Williams and R. Jeanloz, Science **239**, 902 (1988).

[70] C. Meade, R. J. Hemley, and H. K. Mao, Phys. Rev. Lett. **69**, 1387 (1992).

[71] M. Verhelst-Voorhees, J. Yarger, J. Diefenbacher, B. T. Poe, G. H. Wolf, and P. F. McMillan, Eos Trans. Am. Geophys. Union **75**, 63 (1994).

[72] R. A. B. Devine and J. Arndt, Phys. Rev. B **35** (1987).

[73] S. R. Elliot, Phys. Rev. Lett. **67**, 711 (1991).

[74] A. P. Sokolov, A. Kisliuk, M. Soltwisch, and D. Quitmann, Phys. Rev. Lett. **69**, 1540 (1992).

[75] L. Borjesson, A. K. Hassan, J. Swenson, L. M. Torrell, and A. Fontana, Phys. Rev. Lett. **70**, 1275 (1993).

[76] P. Benasi, M. Krisch, C. Masciovecchio, V. Mazzacurati, G. Monaco, G. Ruocco, F. Sette, and R. Verbeni, Phys. Rev. Lett. **77**, 3835 (1996).

[77] S. Sugai and A. Onodera, Phys. Rev. Lett. **77**, 4210 (1996).

[78] R. J. Hemley, C. Meade, and H. K. Mao, Phys. Rev. Lett. **79**, 1420 (1997).

[79] Q. Williams, R. J. Hemley, M. Kruger, and R. Jeanloz, J. Geophys. Res. **98**, 22157 (1993).

[80] J. Schroeder, T. G. Bilodeau, and X. S. Zhao, High Pressure Res. **4**, 531 (1990).

[81] A. Polian and M. Grimsditch, Phys. Rev. B **47**, 13979 (1993).

[82] C. S. Zha, R. J. Hemley, H. K. Mao, T. S. Duffy, and C. Meade, Phys. Rev. B **50**, 13105 (1994).

[83] J. R. Rustad, D. A. Yuen, and F. J. Spera, Chem. Geol. **96**, 421 (1992).

[84] J. Badro, P. Gillet, and J.-L. Barrat, Europhys. Lett. **42**, 643 (1998).

[85] L. Stixrude, in press.

[86] J. Badro and J.-L. Barrat (personal communication, 1999).

[87] W. Primak, *The Compacted States of Vitreous Silica* (Gordon & Breach, New York, 1975), p. 81.

[88] O. Mishima, L. D. Calvert, and E. Whalley, Nature (London) **310**, 395 (1984).

[89] A. J. Jayaraman, D. L. Wood, and R. G. Maines, Phys. Rev. B **35**, 8316 (1987).

[90] P. Cordier and J. C. Doukhan and J. Peyronneau, Phys. Chem. Minerals **20**, 176 (1993).

[91] S. Tsuneyuki, Y. Matsui, H. Aoki, and M. Tsukada, Nature (London) **339**, 209 (1989).

[92] N. Binggeli, N. Troullier, J.-L. Martins, and J. R. Chelikowsky, Phys. Rev. B **44**, 4471 (1991).

[93] J. S. Tse and D. D. Klug. Science **255**, 1559 (1992).

[94] S. L. Chaplot and S. K. Sikka, Phys. Rev. B **47**, 5710 (1993).

[95] S. A. T. Redfern, Mineral. Mag. **60**, 493 (1996).

[96] K. J. Kingma, C. Meade, R. J. Hemley, H. K. Mao, and D. R. Veblen, Science **259**, 666 (1993).

[97] P. Gillet, J. Badro, M. Hanfland, and D. Hausermann (personal communication).

[98] P. Richet, Nature (London) **331**, 56 (1988).

[99] P. Gillet, A. Le Cléac'h, and M. Madon, J. Geophys. Res. **95**, 21635 (1990).

[100] M. Grimsditch, S. Popova, V. V. Brazhkin, and R. N. Voloshin, Phys. Rev. B **50**, 12984 (1994).

[101] N. Binggeli and J. R. Chelikowsky, Phys. Rev. Lett. **69**, 2220 (1992).

[102] N. Binggeli, N. Keskar, and J. R. Chelikowsky, Phys. Rev. B **49**, 3075 (1994).

[103] J. Zhang, R. C. Lieberman, T. Gasparik, and C. T. Herzberg, J. Geophys. Res. **98**, 19785 (1993).

[104] O. Mishima, Nature (London) **384**, 546 (1996).

[105] P. Gillet and J. Badro (personal communication, 1999).

[106] L. V. Woodcock, C. A. Angell, and P. Cheeseman, J. Chem. Phys. **65**, 1565 (1976).

[107] M. J. Sanders, M. Leslie, and C. R. A. Catlow, J. Chem. Soc. Chem. Commun., 1271 (1984).

[108] S. H. Garofalini, J. Chem. Phys. **76**, 3189 (1982).

[109] R. G. Della Valle and H. C. Andersen, J. Chem. Phys. **94**, 5056 (1991).

[110] W. Jin, R. K. Kalia, and P. Vashishta, Phys. Rev. Lett. **71**, 3146 (1993).

[111] J. R. Rustad, D. A. Yuen, and F. J. Spera, Phys. Rev. A **42**, 2081 (1990).

[112] P. H. Poole, T. Grande, F. Sciortino, H. E. Stanley, and C. A. Angell, Comput. Mater. Sci. **223**, 1 (1995).

[113] P. H. Poole, F. Sciortino, U. Essmann, and H. E. Stanley, Nature (London) **360**, 324 (1992).

[114] A. B. Belonoshko and L. S. Dubrovinsky, Geochim. Cosmochim. Acta **59**, 1883 (1995).

[115] C. A. Angell, J. Non-Cryst. Solids **131–133**, 13 (1991).

[116] J.-L. Barrat, J. Badro, and P. Gillet, J. Mol. Sim. **20**, 17 (1997).

[117] S. Tsuneyuki and Y. Matsui, Phys. Rev. Lett. **74**, 3197 (1995).

[118] J. D. Kubicki and A. C. Lasaga, Am. Mineral. **73**, 941 (1988).

[119] M. Y. Frenkel and Y. A. Vasserman, Geokhimiya **8**, 1194 (1991).

[120] R. G. Della Valle and H. C. Andersen, J. Chem. Phys. **97**, 2682 (1992).

[121] G. Adam and J. H. Gibbs, J. Chem. Phys. **43**, 139 (1965).

Chapter 4.2

Shock-Induced Phase Transitions of Rutile Structures Studied by the Molecular-Dynamics Calculation

Keiji Kusaba and Yasuhiko Syono

To understand the anisotropic nature of the shock-induced phase transition of TiO_2 (rutile), the transition to postrutile phases under isotropic and several stress-field conditions was studied by molecular-dynamics calculations. The rutile structure was shown to transform to the fluorite structure by a displacive mechanism under the isotropic compression within the time scale of the shock transition. Calculations with variable stress fields showed anisotropic behavior. The rutile–fluorite transition occurred smoothly with [100] compression of rutile. In the case of the [110] compression, the rutile structure transformed to a twinned fluorite structure. However, the rutile structure transformed to the $CaCl_2$-type structure instead of the fluorite structure under the [001] compression. These results were in good agreement with actual shock experiments.

4.2.1 Introduction

4.2.1.1 Shock Compression Method for Solids

The dynamic compression method using shock wave has been utilized for high-pressure research on solids. The method can easily generate high pressures to more than 100 GPa. In principle, the nature of the dynamic compression is understood by hydrodynamical considerations [1]: When a shock wave travels in solids with a supersonic speed (U) and accelerates particles to a particle velocity (u), the shock front is generated as a discontinuity boundary of pressure and density. In real solids, the discontinuity is observed as a very steep change of pressure in solids and the transition

H. Aoki et al. (eds), *Physics Meets Mineralogy* © 2000 Cambridge University Press.

interval is only several nanoseconds. The shock-induced state can be described by Rankine–Hugoniot relations based on conservations of mass, momentum, and energy between the states ahead of and behind the shock front. In particular, thermodynamical treatments with plane shock-wave geometry lead to simple one-dimensional problems, and pressure and density can be calculated simply from observed U and u. The high-pressure state behind the shock front is considered to be hydrostatic even in solids. An interval of a high-pressure state is limited to a very short time, less than 1 microsecond in the gun method, which depends mainly on the thickness of the impactor as well as the shock-wave velocity (U) and sound velocity (c) in solids (U and c; as fast as several kilometers per second for a general experiment up to 100 GPa).

Although the shock method has been applied for high-pressure equation of state study of solid materials, the shock compression process is not examined precisely from crystallographical basis. In particular, only a few exceptional cases of shock-induced phase transitions have been investigated from the atomistic viewpoint. In order to understand the shock compression process from this viewpoint, following specific features of the high-pressure state achieved by a plane shock wave must always be kept in mind: (1) a very short interval as long as several nanoseconds of the shock transition to the high-pressure condition, (2) a short time duration of high hydrostatic pressure state, less than 1 μs, and (3) a possibility of anisotropic effects, which are expected from the uniaxial character of the plane shock wave, in particular within the shock front. The short intervals of the shock compression process [features (1) and (2)] limit the atomic diffusion in the shock-induced transition and also make microscopic observation difficult, for example with an in situ x-ray observation. The anisotropic effects [feature (3)] may be observed with only a single crystal with an anisotropic structure like the rutile structure.

4.2.1.2 Shock Compression Behavior of the Rutile Structure

The rutile structure is one of the most common structure with tetragonal symmetry (space group $P4_2/mnm$) for MX$_2$-type ionic compounds such as TiO$_2$, SnO$_2$, PbO$_2$, and MnF$_2$. There have been a large number of high-pressure investigations of various compounds with the rutile structure because of interest in the high-pressure behavior of stishovite (the rutile polymorph of SiO$_2$). The rutile structure is typically anisotropic [2]. Each M cation is surrounded by six X anions and these atoms form a slightly distorted octahedron. Edge-sharing chains of the octahedra run along the [001] direction, and each chain is jointed by corner sharing with four neighboring chains along the [110] direction, and the chains are isolated with each other along the [100] direction. In other words, the packing density along the [001] direction is higher than that along other directions, and open spaces are arranged along the [100] direction. Structural anisotropy is directly reflected in the remarkable elastic anisotropy [3].

We have already reported anisotropic behavior of the shock-induced phase transition of single crystal TiO$_2$ [4, 5] and SnO$_2$ [6] by two complementary methods by using a plane shock wave generated by the gun method; in situ measurements of the $P-V$ curve based on the Rankine–Hugoniot equations and observations of the shock-recovered specimen by use of an x-ray-powder-diffraction and electron microscope observation. The anisotropic effect of the plane shock wave on the phase transition was shown for the shock directions parallel and perpendicular to the rutile [001] direction: In in situ observation [4], the shock wave propagating perpendicularly to the [001] direction can cause a shock-induced phase transition at a much lower pressure than the shock wave parallel to the [001] direction. In the shock recovery experiments [5, 6], the yield of the recovered α-PbO_2-type phase (space group $Pbcn$) with shock loading perpendicular to the [001] direction was much larger than that recovered after shock loading parallel to the [001] direction.

We proposed a mechanism of the anisotropic shock behavior of the rutile compounds [5] by assuming that the shock-induced high-pressure phase had the fluorite or its related structure and that the high-pressure phase converted to a metastable phase with the α-PbO_2-type structure during the pressure release. This assumption was supported by an in situ x-ray observation of TiO_2 [7, 8], SnO_2, PbO_2, RuO_2 [9, 10], and MnF_2 [11] under static compression: The high-pressure phase of TiO_2 had the baddeleyite (ZrO_2) structure (space group $P2_1/c$), and modified fluorite phases (space group $Pa\bar{3}$) were found as the high-pressure phases of SnO_2, PbO_2, RuO_2 and MnF_2. These high-pressure phases were observed to convert to α-PbO_2-type phases in the pressure-release process, except for RuO_2 [10].

Our mechanism [5], based on the model proposed by Hyde et al. [12] (which could explain the rutile–fluorite transition without atomic diffusion), can be described by simple deformation of the rutile structure along the [100] direction and can explain the anisotropic shock behavior of the rutile compounds. However, it is difficult to prove the mechanism by real shock experiments because the shock compression interval is too short to examine the phase transition within the context of atomic movements, as mentioned above. The computer calculation will be most promising to throw light on the phase-transition mechanism. In fact, our mechanism has already been examined by molecular dynamics (MD) calculation for simple condition [13].

4.2.1.3 Computer Calculation of Phase Transitions under High Pressure

Application of the computer-calculation method to crystallographic problem for modeling extreme conditions has recently been recognized as one of the most powerful methods. In particular, MD calculations have demonstrated the behavior of several crystalline solids under high pressure [14–16]. In the MD method, statistical thermodynamical values of the system, for example pressure and temperature, are calculated from classical mechanics applied for a system with more than several hundred independent atoms in a cell. Such a cell is called a basic cell, which is surrounded by replica cells with three-dimensional periodic boundary conditions. Initial data for atoms are taken from the crystalline structure data, and initial vectors of atomic movements are taken at random. The atomic motion is calculated with a potential field, which is the sum of two-body potentials of atoms in the system involving the replica cells.

The computer calculation not only explains experimental results, but also predicts new phenomena under extreme conditions. For example, the calculation result had predicted a new crystal structure with a cubic symmetry (space group $Pa\bar{3}$) for a high-pressure phase of MX_2-type compounds [15, 16]. Indeed the structure was found as postrutile phases of SnO_2, PbO_2, RuO_2 [9, 10], and MnF_2 [11], as mentioned above. In particular, the high-pressure phase of RuO_2 was analyzed to have this structure by neutron powder diffraction [10].

The MD calculation can also show atomic movements in a crystal structure under high pressure. Seifert [17] proposed a displacive mechanism from the fluorite structure to the α-PbO_2-type structure, in which the cation– coordination number changed from eight to six. Matsui and Kawamura [15] reproduced the atomic movements in this transition by the MD calculation. These results show that it is possible to reproduce the phase transition of TiO_2 with the rutile structure induced by a plane shock wave.

Our aim in this study is to examine the rutile-type structure under several kinds of stress conditions by use of the MD calculation and to consider the phase-transition mechanism induced by the plane shock wave from the microscopic viewpoint. Our final goal is to compare the calculation results with our shock compression experiments on TiO_2.

4.2.2   Computational Experiments

4.2.2.1   Modified XDORTHO Computer Program

XDORTHO computer program [15] has been modified to enable the calculation for anisotropic stress-field conditions at the Institute for Study of the Earth's Interior, Okayama University, Misasa, Japan. In the calculation, pressure is defined as $3P_{total} = P_x + P_y + P_z$. The modified XDORTHO program can calculate only crystals of cubic, tetragonal, hexagonal, and orthorhombic systems because the frame angles of the basic cell are fixed to be right angles. Components of pressure are independently controlled when the length of each basic cell axis is changed. For the hydrostatic pressure condition, pressure is defined as $P_{total} = P_x = P_y = P_z$.

The present calculations were carried out under isotropic conditions and three kinds of stress conditions (the [100], [110], and [001] compressions of the rutile structure). The basic cell contained 125, 90, or 80 unit cells of the rutile structure (space group $P4_2/mnm$), for which the unit-cell parameters were $a = 4.1772$ Å and $c = 2.6651$ Å. Cations occupied the $2a$ site [(0, 0, 0) and (1/2, 1/2, 1/2)] and anions occupied the $4f$ site [$\pm(u, u, 0)$ and $\pm(u + 1/2, 1/2$-$u, 1/2)$] with $u = 0.3062$. In the calculation of compression along the rutile [110] direction, a tentative tetragonal cell ($a = 5.9076$ Å, $c = 2.6651$ Å, and $Z = 4$), obtained by rotation of the rutile unit cell by 45° around the c axis, was used, because the modified XDORTHO computer program can allow the uniaxial compression only parallel to the edge of the basic cell. Initial stress conditions and the sizes of basic cell for all calculations are listed in Table 4.2.1.

The present calculation was carried out under isothermal compression at 298 K by controlling velocities of atomic motions in the basic cell. For each calculation, the step interval of calculation was chosen to be 2 fs, and the total calculated interval was several picoseconds. The step size was short enough to reproduce the pressure increase process during the shock compression of a few nanoseconds.

4.2.2.2 Potential Function

In the present study, two-body central force type potential function was chosen [15, 16], which was the Gilbert–Ida-type function [18, 19] with the dispersion term.

$$U(r_{ij}) = q_i q_j / r_{ij} + f_o(b_i + b_j) \exp\{(a_i + a_j + r_{ij})/(b_i + b_j)\} + c_i c_j r_{ij}^{-6}.$$

The parameter set MS-1 was optimized for both the structural parameters [20] and elastic constants [21] of rutile-type SiO_2 (stishovite) by using the elasticity-constrained extended WMIN technique [22]. The optimized parameters are listed in Table 4.2.2.

The MS-1 potential function can reproduce structural and elastic properties of rutile-type SiO_2 using the XDORTHO program [15], as listed in Table 4.2.3. It also stabilizes the cubic fluorite structure under very high pressure [15]. Furthermore, the MS-1 potential function can reproduce the modified fluorite (space group $Pa\overline{3}$) structure [15, 16] and the $CaCl_2$ type structure (space group $Pnnm$) [13] under high

Table 4.2.1. *Initial condition parameters for computer calculations of the rutile–fluorite transition*

		Basic cell condition	
	P_{total}	Number of rutile cell	Sum of atoms
Isotropic condition	350 GPa	$4 \times 4 \times 5 = 80$	480
[100] compression	320 GPa	$5 \times 5 \times 5 = 125$	750
[110] compression[a]	350 GPa	$3\sqrt{2} \times 3\sqrt{2} \times 5 = 90$	540
[001] compression	350 GPa	$4 \times 4 \times 5 = 80$	480

[a]In the [110] compression simulation, a tentative tetragonal cell was used, which was obtained by rotation of the rutile cell by 45°. The notation of the basic cell corresponds to the original rutile cell as follows: x, [110]; y, [1$\overline{1}$0]; and z, [001].

Table 4.2.2. *Parameters of the MS-1 potential function*

	O	Si
q/|e|	−1.3	+2.6
a/Å	1.727	0.962
b/Å	0.150	0.050
c/Å k Jmol^{-1}	45.6	0

$f_0 = 1\,\text{kcal}\,\text{Å}^{-1}\,\text{mol}^{-1} = 6.948 \times 10^{-6}$ dyn.

pressure and the α-PbO_2 type structure (space group $Pbcn$) [15] at ambient condition in computational experiments.

As mentioned above, the combination of the XDORTHO program and the MS-1 potential function set for stishovite can reproduce many postrutile structures, which are observed in actual high-pressure experiments. The fact shows that the choice of the MS-1 potential function set is quite pertinent to reproduce the shock-induced phase transition of the rutile structure in computational experiments. However, it must be noted that the cubic fluorite structure of SiO_2 has not been found in actual experiments. Hence, the present calculation using the MS-1 potential function can be considered to simulate shock-induced phase transition of an imaginary compound of MX_2, which stabilizes the rutile structure at ambient condition and the cubic fluorite structure at very high-pressure condition. Then absolute value of pressure in the present calculation is for the imaginary compound, and the result can be used for only qualitative comparison with actual shock experiments.

4.2.3 Result

4.2.3.1 Isotropic Compression

The rutile–fluorite phase transition was observed under isotropic conditions ($P_{total} = 350$ GPa) in which the MS-1 potential function can stabilize the fluorite structure. Changes of pressure, cell parameters, and positional parameters with time are shown in Figs. 4.2.1–4.2.4. The figures show that the rutile–fluorite transition process consists of four stages. Typical crystal structures of each stage are shown in Fig. 4.2.5.

In the first stage between 0 and 1 ps, the pressure increased to 350 GPa and the rutile structure was compressed isotropically. In the second stage between 1 and 3.4 ps, the cell parameters along the [100] and [010] directions became different from each other and the single positional parameter u of the anions increased to two parameters u and v by a rotation of the MX_6 octahedron along the [001] direction. In other words, the

Table 4.2.3. *Crystallographic data and bulk modulus of MD-synthesized stishovite using MS-1 potential function*

	MD calculation	Observation[a]
a/Å	4.128(1)	4.1772
c/Å	2.758(1)	2.6651
ρ/Mgm-3	4.246(1)	4.2915
O(u)	0.305(1)	0.3062
K_0/GPa	325	298

[a]Crystal data and bulk modulus are from [20] and [21], respectively.

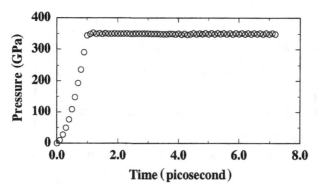

Figure 4.2.1 Variation of pressure under isotropic compression conditions with time.

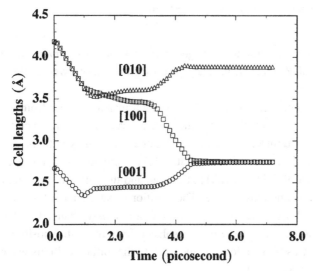

Figure 4.2.2 Variation of cell parameters under isotropic compression conditions with time.

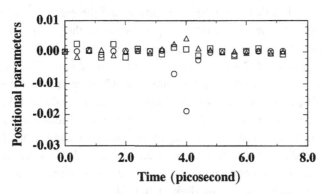

Figure 4.2.3 Time variation of positional parameters for the cation, of which the initial parameters are $x = y = z = 0$. Circles, triangles, and squares show the positional parameters of x, y, and z, respectively.

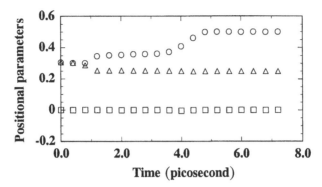

Figure 4.2.4 Time variation of positional parameters for the
anion, of which the initial parameters are $x = y = 0.3062$ and
$z = 0$. Circles, triangles, and squares show the positional
parameters of x, y, and z, respectively.

overall symmetry of the structure changed from $P4_2/mnm$ to $Pnnm$, which was one
of the subgroups of $P4_2/mnm$. In space group $Pnnm$, cations were at the $2a$ site and
anions at the $4g$ site.

The third stage between 3.4 and 4.6 ps was a transient process from the distorted
rutile structure (space group $Pnnm$) to the fluorite structure, where all cell parameters
and the positional parameters of the anions changed drastically with the displacement
of cations: The cell parameter of the rutile [100] direction was compressed to be
nearly equal to that of the rutile [001] direction, and the cell parameter ratio between
[010] and [100] became close to $\sqrt{2}$. The positional parameters of anions, u for the
[100] direction and v for the [010] direction, continuously changed to 0.5 and 0.25,
respectively. Cations, which were once displaced from the initial positions when the
MX_6 octahedron was broken down, came back to the initial positions when the MX_8
cubic polyhedron was built up.

In the fourth stage between 3.4 and 7.2 ps, the cubic fluorite structure was formed
and it was stable throughout the stage: The cell parameter of the [100] direction
became exactly equal to that of the [001] direction, and the parameter of the [010]
direction was just $\sqrt{2}$ times larger than the parameters of the [100] and the [001]
directions. In the cell, cations were in the $2a$ site and anions in the $4g$ site with
positional parameters $u = 1/2$ and $v = 1/4$ of space group $Pnnm$. When axes were
converted, the structure was found to have a cubic cell, whose cell parameter was equal
to that of the [010] direction. Cations and anions occupied the $4a$ and the $8c$ sites
respectively, of space group $Fm3m$. The crystal-orientational relationship between
the rutile and fluorite structures was envisaged as though the compressed [100] and
the expanded [010] directions of the rutile structure were parallel to the fluorite [110]
and the [001] directions, respectively. The outline of the present result was consistent
with the transition model proposed by Hyde et al. [12].

The present calculation under isotropic compression shows some important results:
(1) The rutile–fluorite transition can occur without atomic diffusion, i.e., the transition

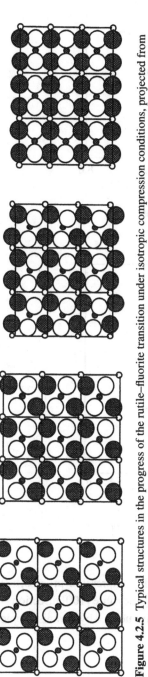

Figure 4.2.5 Typical structures in the progress of the rutile–fluorite transition under isotropic compression conditions, projected from the rutile [001] direction. From left to right, the rutile structure at 0.0 ps, an intermediate phase at 2.8 ps, a transient state to the fluorite structure at 4.0 ps, and the fluorite structure at 7.2 ps, are shown. Small and large circles indicate cations and anions, respectively. Open and hatched marks indicate $z = 0$ and $z = 1/2$, respectively.

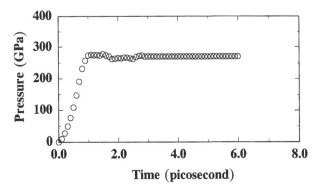

Figure 4.2.6 Variation of pressure under the [100]
compression condition with time.

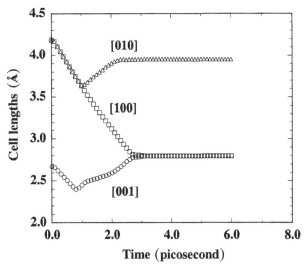

Figure 4.2.7 Variation of cell parameters under the [100]
compression condition with time.

may be induced in a very short time such as in a shock compression process; (2)
even under isotropic conditions, drastic compression along the rutile [100] direction is
caused in the transition. This suggests that the transition is influenced by the anisotropic
stress condition.

4.2.3.2 [100] Compression

The rutile–fluorite phase transition was also observed in the [100] compression. Changes of pressure and cell parameters with time are shown in Figs. 4.2.6 and 4.2.7,
respectively. The transition process could be considered as having four stages, and the
crystal structures of each stage are shown in Fig. 4.2.8.

In the first stage between 0 and 1 ps, the pressure increased to 270 GPa and the
rutile structure was compressed isotropically. Pressure was held at ~270 GPa in the

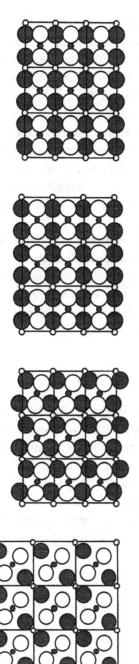

Figure 4.2.8 Typical structures in the progress of the rutile–fluorite transition under the [100] compression condition, projected from the rutile [001] direction. From left to right, the rutile structure at 0.0 ps, an intermediate phase at 1.2 ps, a fluorite-related structure at 2.6 ps, and the fluorite structure at 6.0 ps, are shown. The symbols are the same as those of Fig. 4.2.5.

second stage between 1 and 1.6 ps, although the initial pressure value was $P_{total} = 320$ GPa. The cell length of the [100] direction decreased by application of the stress along the rutile [100] direction, whereas the cell length of the [010] and the [001] directions increased. The overall symmetry of the structure changed from $P4_2/mnm$ to *Pnnm*, similar to the second stage in the isotropic calculation. The cell length of the [100] direction smoothly decreased in the third stage until 2.6 ps, and a slightly distorted fluorite structure was finally formed. In the fourth stage, the stress condition was switched from [100] compression to isotropic compression with a total pressure of 270 GPa. The distorted fluorite structure was converted to the cubic fluorite structure, and it was shown stable under pressure. The crystal-orientational relationship between the rutile and the fluorite structures was the same as that under the isotropic condition.

When this calculation result is compared with that under the isotropic condition, it is found that the rutile–fluorite transition is induced by the compression from the rutile [100] direction and relative expansion along the [010] and the [001] directions. The [100] high-stress condition indicates the transition more smoothly and directly than the isotropic condition. The present calculation could reproduce our transition model [5] by applying the simple deformation from the rutile [100] direction.

4.2.3.3 [110] Compression

Changes of pressure and cell parameters with time are shown in Figs. 4.2.9 and 4.2.10. This result was made up of two stages. The first stage, until 2.4 ps, was controlled under rutile [110] compression. In this stage the rutile structure could not transform to the simple fluorite structure but to a twinned fluorite structure, as shown in Fig. 4.2.11. The structure was found to be stable under isotropic condition with $P_{total} = 350$ GPa.

The twinned structure can be considered as a unit-cell-order twinned fluorite structure, in which the average coordination number of cations is 7.6. Similar unit-cell-order twinned structure has been reported as a high-pressure phase of ZrO_2 [23], although

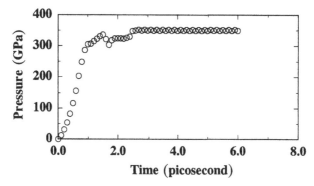

Figure 4.2.9 Variation of pressure under the [110] compression condition with time.

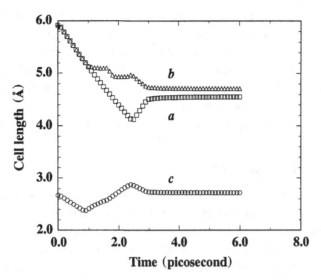

Figure 4.2.10 Variation of cell parameters under the [110] compression condition with time.

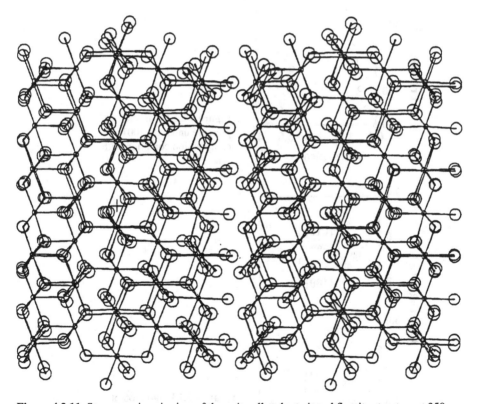

Figure 4.2.11 Stereoscopic pair view of the unit-cell-order twinned fluorite structure at 350 GPa and 298 K. Small and large circles indicate cations and anions, respectively.

the present structure has never been observed in real high-pressure experiments: The crystal structure of the high-pressure phase of ZrO_2 stable above 3 GPa has been determined by the neutron-powder-diffraction method. The crystal structure (space group $Pbca$, $Z = 8$) [23] can be explained by the unit-cell-order twinned structure of the baddeleyite (ZrO_2) type structure (space group $P2_1/c$, $Z = 4$). Both structures are explained as modified fluorite structures, in which each cation is surrounded by seven anions. The result for ZrO_2 suggests the possibility that the present twinned structure will be found in actual high-pressure experiments of other MX_2-type compounds.

4.2.3.4 [001] Compression

This process consisted of two stages. In the first stage, until 2.4 ps, the rutile structure was anisotropically compressed from the [001] direction with $P_{total} = 350$ GPa, and transformed to the $CaCl_2$-type structure (space group $Pnnm$) instead of the fluorite-type or modified fluorite-type structures, as shown in Figs. 4.2.12 and 4.2.13. The $CaCl_2$-type structure could be explained as an orthorhombically distorted rutile

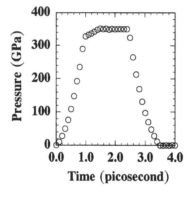

Figure 4.2.12 Variation of pressure under the [001] compression condition with time.

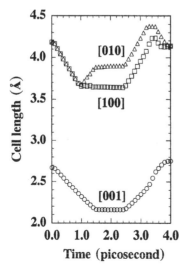

Figure 4.2.13 Variation of cell parameters under the [001] compression condition with time.

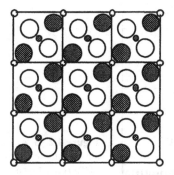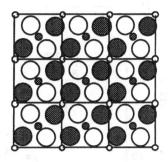

Figure 4.2.14 Typical structures in the progress of the rutile–CaCl$_2$ type transition under the [001] compression condition, projected from the rutile [001] direction. The rutile structure at 0.0 ps is shown on the left-hand side, and the CaCl$_2$-type structure at 2.4 ps is shown on the right-hand side. The symbols are the same as those of Fig. 4.2.5.

structure, in which each cation was coordinated by six anions, similar to those of the rutile structure (Fig. 4.2.14). The CaCl$_2$-type structure completely reverted to the rutile structure in the second stage of the isotropic pressure release process from 2.4 ps.

Although the CaCl$_2$-type structure has the same overall symmetry (space group *Pnnm*) as that of the intermediate structure between the rutile and the fluorite structures in the isotropic and the [100] compressions, the rotation direction of the MX$_6$ octahedra from the rutile structure is reversed with respect to each other. In the intermediate structure, as shown in Figs. 4.2.5 and 4.2.8, the displacement can be understood as an increase in the coordination number of the cation from six to eight. On the other hand, the displacement in the CaCl$_2$-type structure, as shown in Fig. 4.2.14, can be explained as a construction of the hexagonally close-packed arrangement for the anions. There is another difference between these transitions. In the transition to the intermediate structure, the cell parameter along the [001] direction increases, whereas it decreases in the case of the rutile–CaCl$_2$ transition. These differences suggest that the rutile–fluorite transition is infavorable under [001] compression.

4.2.4 Comparing Calculation Results with High-Pressure Experiments

The present calculations showed two important properties of the rutile–fluorite transition. First is that the transition can occur without atomic diffusion. This fact indicates that the transition can be induced in the pressure increase in shock compression as long as several nanoseconds. The diffusionless nature of the shock-induced transition can be supported by real shock experiments on ZrSiO$_4$ [24, 25] and FeTaO$_4$ [6].

ZrSiO$_4$ has the zircon-type structure under ambient conditions and transforms to the scheelite-type structure under shock compression [24]. The scheelite phase is 9.9%

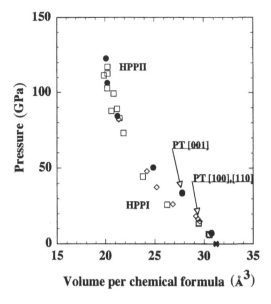

Figure 4.2.15 Anisotropic shock compression curves of TiO_2 [4]. Open square, open rhombic, and solid circle indicate the rutile [100], [110] and [001] shock directions, respectively. PT, phase transition; HPPI, high pressure phase I; HPPII, high pressure phase II.

denser than the zircon phase at ambient condition [24]. The zircon and scheelite structure can be considered as a cation-ordered rutile and fluorite structure, respectively [26]. In the shock-induced transition of $ZrSiO_4$, the relative configuration of cations is conserved, following the rutile–fluorite transition [25]. In the case of $FeTaO_4$, which has the cation-disordered rutile structure, no evidence of the cation ordering is observed in the shock-recovered specimen [6].

Another important point is the anisotropic effect on the transition: The rutile–fluorite transition occurs by the compression perpendicular to the rutile [001] direction, whereas the [001] compression induces the transition to the $CaCl_2$-type structure instead of to the fluorite structure. The anisotropic results are in very good agreement with the shock compression measurements of TiO_2 [4].

Figure 4.2.15 shows the shock compression curves measured along three different shock directions [4]. The compression curves show the anisotropic effect of the plane shock wave above the phase-transition pressure, although isotropic compression is achieved below the phase transition. In the case of the shock directions perpendicular to the [001] direction, two high-pressure phases [high-pressure phase (HPPI) and high-pressure phase II (HPPII)] are clearly observed, and the transition pressures to HPPI are determined to be 13.7 and 17.0 GPa for the shock directions of [100] and [110], respectively. The volume of HPPI under ambient conditions is estimated to be about 15% smaller than that of the rutile phase. The HPPI is considered to have the fluorite structure. The baddeleyite structure, which is a monoclinically distorted fluorite structure, is confirmed to be stable above ~15 GPa in static high-pressure experiments on TiO_2 [7, 8]. The baddeleyite structure can explain the volume change between the rutile phase and the HPPI under ambient conditions. However, the [001] compression curve suggests that the rutile structure never transforms to the baddeleyite structure.

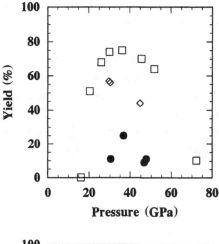

Figure 4.2.16 Anisotropic yields of the shock-induced phase of TiO$_2$ [5]. The symbols are the same as those of Fig. 4.2.15.

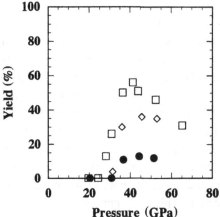

Figure 4.2.17 Anisotropic yields of the shock-induced phase of SnO$_2$ [6]. The symbols are the same as those of Fig. 4.2.15.

The anisotropic calculation results are also consistent with the yield of shock-recovered phases of TiO$_2$ [5] and SnO$_2$ [6]. In the shock recovery experiments, the α-PbO$_2$-type phase was found to coexist with the rutile phase. The volume of the shock-induced α-PbO$_2$-type phase is calculated to be only 3% smaller than that of the rutile phase. The α-PbO$_2$-type TiO$_2$ is considered as a conversion phase from the HPPI in the pressure-release process, because the baddeleyite structure of TiO$_2$ is observed to revert to the α-PbO$_2$-type structure in the pressure-release process during static high-pressure experiments [7, 8]. The anisotropic effects of TiO$_2$ and SnO$_2$ are also confirmed in the yield of the shock-induced α-PbO$_2$-type phase, as shown in Figs. 4.2.16 and 4.2.17, respectively. Maximum yield of the shock-induced phase is observed in the rutile [100] shock compression in both TiO$_2$ and SnO$_2$. The yield in the shock compression perpendicular to the rutile [001] direction is always several times larger than that for compression parallel to the [001].

By combining the present results and the MD calculation of the fluorite–α-PbO$_2$-type transition by Matsui and Kawamura [15], we derive the crystal-orientational relationships among the rutile-, fluorite-, and the α-PbO$_2$-type structures as follows.

The compressed rutile [100] direction, which corresponds to the shock direction, is parallel to the [110] direction of the fluorite structure at high pressure and the [110] direction of the α-PbO$_2$-type structure at ambient condition. The derived relationship is confirmed by observation with transmission electron microscope observation of a shock-recovered specimen of TiO$_2$ [5]. A lamellae pattern of the rutile- and α-PbO$_2$-type phases several tens of micrometers in width was frequently found in the [100] shocked specimen. The electron diffraction pattern of the lamellae indicated a definite crystal-orientational relationship between the two phases; the [100] direction of the rutile structure is parallel to the [110] direction of the α-PbO$_2$-type structure, and it is also parallel to the shock compression direction.

4.2.5 Summary

The present MD calculation has shown that the rutile–fluorite transition can occur without atomic diffusion under an isotropic pressure condition and that the transition could be more smoothly induced by uniaxial compression along the rutile [100] compression. The present calculation also showed a remarkable anisotropic effect on uniaxial compression directions: The rutile structure transformed to fluorite or a modified structure by shock compression perpendicularly to the rutile [001] direction. On the other hand, the rutile structure transformed to the CaCl$_2$-type structure instead of the fluorite structure by shock direction parallel to the rutile [001] direction. This anisotropic behavior was consistent with our real shock experiments on TiO$_2$. The calculation result has a revealed orientational relationship among relevant crystal structures and the shock direction, which was confirmed by electron diffraction observation of shocked TiO$_2$ specimen. Furthermore, the MD calculation clearly showed the detail of the reconstruction from MX$_6$ octahedron to MX$_8$ cubic polyhedron in the transition. Such atomic movements have never been taken in the previous considerations.

The present results indicate that the MD calculation method provides a great deal of information about shock-induced phase transitions in terms of crystallography, and leads us to full understanding of shock-induced phase transition by combining thermodynamic considerations and static compression experiments.

In the future, when the shock-induced transition is directly observable with a synchrotron x-ray source, the MD method may become one of the most powerful methods for helping to design the in situ x-ray observation experiments properly.

Acknowledgment

The authors wish to express their sincere thanks to Professor Yoshito Matsui, Institute for Study of the Earth's Interior, Okayama University, who kindly modified the XDORTHO computer program for the anisotropic shock-induced transition, and rendered them many helpful suggestions and advice throughout the present study.

References

[1] As a general introduction, see R. Kinslow, ed., *High Velocity Impact Phenomena* (Academic, New York, 1970).

[2] As a general introduction, see F. D. Bloss, *Crystallography and Crystal Chemistry* (Mineralogical Society of America, Washington, D.C., 1994).

[3] I. J. Fritz, *J. Phys. Chem. Solids* **35**, 817 (1974).

[4] Y. Syono, K. Kusaba, M. Kikuchi, K. Fukuoka, and T. Goto, in *High-Pressure Research in Mineral Physics*, M. H. Manghnani and Y. Syono, eds. (Terra Scientific and American Geophysical Union, Tokyo, Washington, D.C., 1987), p. 385.

[5] K. Kusaba, M. Kikuchi, K. Fukuoka, and Y. Syono, *Phys. Chem. Miner.* **15**, 238 (1988).

[6] K. Kusaba, K. Fukuoka, and Y. Syono, *J. Phys. Chem. Solids* **52**, 845 (1991).

[7] H. Sato, S. Endo, M. Sugiyama, T. Kikegawa, O. Shimomura, and K. Kusaba, *Science* **251**, 786 (1991).

[8] S. Endo, H. Sato, J. Tang, Y. Nakamoto, T. Kikegawa, O. Shimomura, and K. Kusaba, in *High-Pressure Research: Application to Earth and Planetary Science*, Y. Syono and M. H. Manghnani, eds. (Terra Scientific and American Geophysical Union, Tokyo, Washington, D.C., 1992), p. 457.

[9] J. Haines, J. M. Léger, and O. Schulte, *Science* **271**, 629 (1996).

[10] J. Haines, J. M. Leger, M. W. Schmidt, J. P. Petitet, A. S. Pereira, J. A. H. Dajornada, and S. Hull, *J. Phys. Chem. Solids* **59**, 239 (1998).

[11] N. Hamaya (private communication).

[12] B. G. Hyde, L. A. Bursill, M. O'Keeffe, and S. Andersson, *Nature* **237**, 35 (1972).

[13] K. Kusaba, Y. Syono, and Y. Matsui, in *Shock Compression of Condensed Matter-1989* (North-Holland, Amsterdam, The Netherlands, 1990), p. 135.

[14] Y. Matsui, K. Kawamura, and Y. Syono, in *High-Pressure Research in Geophysics*, S. Akimoto and M. H. Manghnani, eds. (Center for Academic Publications Japan, Tokyo, 1982), p. 511.

[15] Y. Matsui and K. Kawamura, in *High-Pressure Research in Mineral Physics*, M. H. Manghnani and Y. Syono, eds. (Terra Scientific and American Geophysical Union, Tokyo, Washington, D.C., 1987), p. 305.

[16] Y. Matsui and M. Matsui, in *Vol. 7 of Structural and Magnetic Phase Transitions in Minerals, Advances in Physical Geochemistry*, S. Ghose, J. M. D. Coey, and E. Salje, eds. (Springer, New York, 1988), p. 129.

[17] K. F. Seifert, *Fortschr. Miner.* **45**, 214 (1968).

[18] T. L. Gilbert, *J. Chem. Phys.* **49**, 2640 (1968).

[19] Y. Ida, *Phys. Earth Planet. Inter.* **13**, 97 (1976)

[20] W. Sinclair and A. E. Ringwood, *Nature* **272**, 714 (1978).

[21] D. J. Weidner, J. D. Bass, A. E. Ringwood, and W. Sinclair, *J. Geophys. Res.* **87**, 4740 (1982).

[22] W. R. Busing and M. Matsui, *Acta Cryst.* **A40**, 532 (1984).

[23] O. Ohtaka, T. Yamanaka, S. Kume, N. Hara, H. Asano, and F. Izumi, in *High-Pressure Research: Application to Earth and Planetary Science*, Y. Syono and M. H. Manghnani, eds. (Terra Scientific and American Geophysical Union, Tokyo, Washington, D.C., 1992), p. 463.

[24] K. Kusaba, Y. Syono, M. Kikuchi, and K. Fukuoka, *Earth Planet Sci. Lett.* **72**, 433 (1985).

[25] K. Kusaba, T. Yagi, and Y. Syono, *J. Phys. Chem. Solids* **47**, 675 (1986).

[26] M. O'Keeffe and B. G. Hyde, *J. Solid State Chem.* **44**, 24 (1982).

Chapter 4.3

Lattice Instabilities Examined by X-ray Diffractometery and Molecular Dynamics

Takamitsu Yamanaka and Taku Tsuchiya

X-ray diffractometry and molecular dynamics calculation examine the existence of lattice instabilities under the condition of homogeneous or inhomogeneous stress. The atomic positional displacement in each unit cell and fluctuation of the crystallographic translation operation creates lattice deformations that affect both line broadening and diffraction intensities. Pressure-induced amorphization is attributed to these lattice instabilities. Reversible and irreversible pressure-induced amorphization comes about from elastic deformation, and irreversible amorphization is caused by plastic deformation. These amorphous states are precursors to some equilibrium phase transformations.

4.3.1 Introduction

Much attention has been paid to the following phase changes of Earth interiors under high pressure and temperature; (1) phase transformation (transition, melt, amorphization); (2) decomposition, (exsolution, dissociation, phase separation); (3) chemical reaction (solid–solid, solid–molten salt); (4) electronic changes, (electronic excitation, charge transfer, high–low spin state). Many of these changes are associated with lattice instabilities accompanied by changing pressure and/or temperature conditions.

Pressure-induced phase transformation including amorphization is a significant subject not only for phase transition but also for high-pressure studies of industrial materials. Pressure-induced amorphizations of H_2O (Ih) and SiO_2 were previously regarded as thermodynamically metastable phases, which correspond to the supercooled liquid phase at room temperature [1, 2]; since these discoveries, several crystalline

H. Aoki et al. (eds), *Physics Meets Mineralogy* © 2000 Cambridge University Press.

substances have been found to transform into amorphous states under compression at sufficiently low temperatures (Table 4.3.1). Some retain the amorphous state on the release of pressure to ambient, (i.e., showing an irreversible behaviour). The others reversibly recover initial structures under ambient conditions. This reversible recrystallization from the amorphous state has a morphological memory, as in $Ca(OH)_2$ [10]. Because $Mg(OH)_2$, in spite of having the same structure as $Ca(OH)_2$, does not appear to undergo amorphization but directly transforms to the high-pressure stable form, the mechanism of pressure-induced amorphization is not necessarily dictated by crystallographic symmetry.

Dynamical lattice instability, because of its elastic and plastic deformations under pressure by shear and stress, induces reversible and irreversible amorphizations, respectively, as shown in Figure 4.3.1 [14]. The reversible mode is produced by lattice fluctuations. The irreversible model is attributed to the nucleation of high-pressure form in the parent lattice but thermal energy is not kinetically high enough to provide the large crystallite size for x-ray-diffraction coherence. The small grains can be defined as x-ray amorphous substance. These reversible and irreversible transformations arise from a hindrance to sufficient atomic mobility. These pressure-induced amorphizations are precursors of phase transformations to high-pressure stable polymorphs. Samples that show the reversible and the irreversible transformations listed in Table 4.3.1 have neither compositional nor structural restrictions.

This chapter presents a discussion of lattice instability at high pressure that can be observed by x-ray diffractometry. Successive structure studies of amorphization under

Table 4.3.1.

Sample	Pressure (GPa)	Reference
Irreversible amorphization		
$H_2O(Ih)$ (77 K)	10	Mishima et al. (1984)[1]
SiO_2(a-quartz)	25	Hemley et al. (1988)[2]
SiO_2(coesite)	30	Hemley et al. (1988)[2]
GeO_2	6.5	Yamanaka et al. (1992)[3]
Fe_2SiO_4	39	Williams et al. (1990)[4]
Mg_2GeO_4	25	Nagai et al. (1994)[5]
$CaAl_2Si_2O_8$(300 K)	22	Williams et al. (1989)[6]
$Pb_5Ge_3O_{11}$	14.6	Lin et al. (1993)[7]
Reversible amorphization		
SnI_4	20	Fujii et al. (1985)[8]
α-$AlPO_4$	18	Kruger et al. (1990)[9]
$Ca(OH)_2$	11	Meade et al. (1992)[10]
serpentine	9.7	Meade et al. (1992)[10]
$C_2(CN)_4$		Chaplot et al. (1986)[11]
$LiKSO_4$	13	Sankaran et al. (1988)[12]
$Ca(NO_3)_2/NaNO_3$	9.4	Winters et al. (1992)[13]

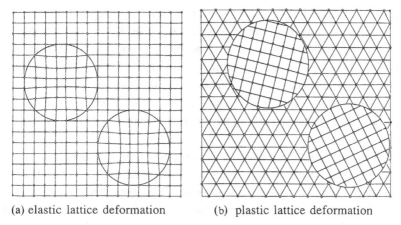

(a) elastic lattice deformation   (b) plastic lattice deformation

Figure 4.3.1 Nucleation having a tetragonal symmetry comes out in the hexagonal lattice in the plastic deformation. The orientations of the crystallites are randomly distributed.

various pressures and temperatures are investigated by molecular dynamics (MD) calculation.

4.3.2   Lattice Instability Under Pressure

Phase transitions induced by fluctuations of magnetic spin, electron or phonon, belong to the class of second-order transitions. Structural transitions (regarding atomic displacement or ordering) have been extensively studied both theoretically and experimentally. However, pressure-induced transitions appear to be first-order transitions or reconstructive transformations. There are relatively few published analyses of their experimental observations and theoretical interpretation.

In general, the transformations involving diffusion require a large thermal energy. The pressure-induced phase transition at ambient temperature is extremely sluggish compared with the thermally induced transitions. At kinetically low temperatures compared with the Debye temperature, reversible amorphization on compression is characterized by a short mean free path with a small atomic displacement and partial lattice deformation without long-range ordering. On decompression, the lattice slips back, the deformation is relaxed, and finally the amorphous state changes back to the original structure. Hence it is regarded as an elastic deformation and makes a memory glass, as seen with $Ca(OH)_2$ and α-$AlPO_4$. In these cases kinetically low temperatures often hinder the transition process to the stable phases under compression alone and bring about metastable amorphization.

The plastic lattice deformation, rather than the elastic deformation under pressure, gives rise to a lattice instability by increasing the defect, dislocation density, and atomic-ordering fluctuations. The disturbance of the periodicity of the atomic array or lattice causes the amorphization by compression. The amorphous state can be irreversibly recovered on decompression. This case can be regarded as x-ray amorphous. Nucleation of the high-pressure form can occur, but crystal growth cannot progress after nucleation with pressure alone with no furnishing of thermal energy. Randomly oriented fine crystallites insufficient for x-ray coherence show no diffraction peak but a halo pattern. The reorientation of crystallites or domains needs enormously large energy, compared with that of the nucleation. In fact, it has never been reported that high-pressure crystallization from the amorphous state is performed by only pressurization.

Amorphous materials represent a subset of noncrystalline materials in the broad sense. The amorphous state in this chapter indicates the samples that provide no sharp powder pattern (only, at most a halo pattern). X-ray amorphous materials include fine crystallites smaller than $\sim 1000\,\text{Å}$, which are incoherent for x-ray wavelength. Pressure-induced amorphous substances differ from glasses rapidly quenched from high temperatures. In general, a crystalline state of condensed matters has a higher density than the amorphous state. However, it has been often found that some crystalline substances change to amorphous states with higher density under compression. Homogeneous compression produces a uniform strain in a crystalline lattice in the pressure range of the elastic deformation. However, compression of polycrystalline substances can produce nonhydrostatic pressure and random strains in neighboring grains, resulting in anisotropic distortion. Each crystallite suffers from nonuniform stress in intercrystalline and intracrystalline lattices.

4.3.3   Homogeneous Three-Dimensional Strain

From statistical physics, external pressure P_{ext} is equivalent to the inner pressure of a crystal or a molecule in which all particles are under the thermal-equilibrium condition. Based on the virial theorem, P_{ext} of the time average is

$$
\begin{aligned}
P_{ext} &= \frac{Nk_BT}{V} - \frac{1}{3V}\left\langle \sum_{i=1}^{N}\sum_{j>i}^{N}\left(-\frac{\partial \psi_{ij}}{\partial \mathbf{r}_{ij}}\right)\mathbf{r}_{ij}\right\rangle \\
&= \frac{Nk_BT}{V} - \frac{1}{3V}\left\langle \sum_{i=1}^{N}\sum_{j>i}^{N}\mathbf{F}_{ij}\,\mathbf{r}_{ij}\right\rangle
\end{aligned}
\tag{4.3.1}
$$

where V is volume, N is the number of particles, ψ_{ij} and $\mathbf{r}_{ij}$ are the interatomic potential and distance between i and j particles, respectively, and $\mathbf{F}_{ij}$ is the force between particles j and i. The stress induced from the external pressure is represented by symmetrical tensors:

$$
P = \begin{pmatrix} P_{xx} & P_{xy} & P_{xz} \\ & P_{yy} & P_{yz} \\ & & P_{zz} \end{pmatrix},
\tag{4.3.2}
$$

where the component P_{xx} is expressed by

$$P_{\alpha\beta} = \frac{1}{V}\left(Nk_BT - \left\langle \sum_{i=1}^{N}\sum_{j>i}^{N} \frac{|\mathbf{F}_{ij}|}{|\mathbf{r}_{ij}|}r_{ij\alpha}\,r_{ij\beta}\right\rangle\right). \qquad (4.3.3)$$

The variation of the displacement $\Delta\mathbf{r}_{ij}$ with an initial bond distance $\mathbf{r}_{ij0}$ is a linear function for homogeneous compression that is due to the external force within the elastic range (i.e., following Hook's law). Linear compression with applied pressure is assumed with the energy determined by parabolic and harmonic interatomic potentials. In the more general cases of nonlinear compression, the actual displacement is $\Delta\mathbf{r}_{ij} + \Delta\mathbf{u}_{ij}$.

4.3.4   Effect on the Diffraction Intensity

As shown in Fig. 4.3.2, $\mathbf{r} = \mathbf{r}_0 + \mathbf{t}_i + \mathbf{R}_j$. Lattice instabilities are caused by several factors: the disorder of the displacement of the j atom in the unit cell, which denotes the fluctuation in $\Delta\mathbf{R}_j$, irregular arrangement of the i cell ($\Delta\mathbf{r}_i$), and lack of periodicity of the lattice indicated by the translation $\Delta\mathbf{t}_i$.

$$\mathbf{r} = \mathbf{r}_0 + \Delta\mathbf{r}_0 + \mathbf{t}_i + \Delta\mathbf{t}_i + \mathbf{R}_j + \Delta\mathbf{R}_j \qquad (4.3.4)$$

The electron-density distribution at the j atom position is given by

$$\rho(\mathbf{r}) = \rho(\mathbf{r}_0 + \Delta\mathbf{r}_0 + \mathbf{t}_i + \Delta\mathbf{t}_i + \mathbf{R}_j + \Delta\mathbf{R}_j). \qquad (4.3.5)$$

The atomic displacement is induced from a stress or shear force, but it is not large enough to form a higher-pressure structure because of the interference to sufficient atomic mobility at kinetically low temperatures. X-ray diffraction is an effective technique for observing these lattice instabilities.

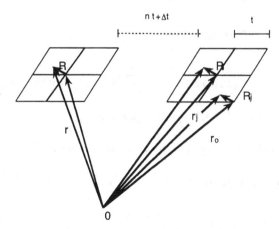

Figure 4.3.2 Atomic and unit cell positional vectors. Atomic displacement of the j atom in the i unit cell: $(\mathbf{R}_j + \Delta\mathbf{R}_j)$ and fluctuation of crystallographic translation: $(\mathbf{t}_i + \Delta\mathbf{t}_i)$. Non-periodic translation between particles is illustrated.

The entire x-ray scattering power of an ideal crystallite with periodic $\mathbf{t}_i$ can be expressed by

$$C(K) = \int_{\text{cryst}} \rho(\mathbf{r}) \exp(2\pi i \mathbf{Kr}) \, d\tau$$

$$= \sum \exp[2\pi i \mathbf{K}(\mathbf{r}_0 + \mathbf{t}_i)] \int_{\text{cryst}} \rho(\mathbf{R}) \exp(2\pi i \mathbf{KR}) \, d\tau, \tag{4.3.6}$$

where $\mathbf{K}$ is a scattering vector.

This structure factor of the unit cell in the ideal crystal can be written as

$$F(\mathbf{K}) = \int_{\text{unit}} (\mathbf{R}_j \exp(2\pi i \mathbf{KR}) \, d\tau. \tag{4.3.7}$$

This structure factor is rewritten in atomic coordinates (x_j, y_j, z_j) of the j atom, together with the atomic scattering factor f_j and thermal atomic vibrational parameters T_j:

$$F(\mathbf{K}) = \sum f_j T_j \exp(2\pi i \mathbf{KR}_j). \tag{4.3.8}$$

$\mathbf{R}_j$ is represented by atomic positional coordinates x_j, y_j, and z_j that are parallel to the unit-cell vectors $\mathbf{a}_0$, $\mathbf{b}_0$, and $\mathbf{c}_0$, respectively. Then $\mathbf{R}_j = x_j \mathbf{a}_0 + y_j \mathbf{b}_0 + z_j \mathbf{c}_0$.

If we apply atomic coordinates (x_j, y_j, z_j), the structure factor at the reciprocal lattice point (hkl) is

$$F(h)_{\text{unit}} = \sum f_j T_j \exp(2\pi i h x_j l)$$

$$= \sum f_j T_j \exp 2\pi i (h x_j + k y_j + l z_j) \tag{4.3.9}$$

The diffraction intensity $I(h)$ calculated from the structure factor is

$$I(h)_{\text{unit}} = c|F(h)_{\text{unit}}|^2. \tag{4.3.10}$$

The set of all atoms in the crystal is represented by $x'_j = x_j + dx_j$ in consideration of the atomic displacement $(x_i + dx_{ij}, y_j + dy_{ij}, z_j + dz_{ij})$ of the j atom in the i unit cell in the crystal composed of N unit cells. Equation (4.3.10) changes to the formula given below.

The structure factor of a crystal composed of N unit cells will be

$$F(h)_{\text{deform}} = \sum_i \sum_j^N f_j T_j \exp\{2\pi i [\underline{h}(x_j| + \delta x_{ij}|)]\}$$

$$= \sum_j f_j T_j \exp[2\pi i (\underline{h} x_j|)] \left\{ 1 + \sum \sum^N f_{ji} T_j \exp[2\pi i (\underline{h} \delta x_{ij}|)] \right\}$$

$$= F(h)_{\text{unit}} \left\{ 1 + \sum \sum f_j T_j \exp 2\pi i [\underline{h}(\delta x_{ij}|)] \right\} \tag{4.3.11}$$

The diffraction intensity determined by the above structure factor is changed to

$$I(h)_{\text{deform}} = c|F(h)_{\text{deform}}|^2 = c|F(h)_{\text{unit}}|^2 \left(2 + 2 \sum_i \sum_j^N \cos 2\pi \delta x_{ij} \right)$$

$$= I(h)_{\text{unit}} \left(2 + 2 \sum_i \sum_j^N \cos 2\pi \delta x_{ij} \right). \tag{4.3.12}$$

Structure factor in the whole crystal $F(h)_{\text{cryst}}$ includes the fluctuation of the translation operation of $N\mathbf{t}+\Delta\mathbf{t}_N$ besides the summation of the deformed cell. $N\mathbf{t}$ indicates a periodical translation at the N cell and $\Delta\mathbf{t}_N$. is the deviation from $N\mathbf{t}$. With $\exp(2\pi i N\mathbf{t}) = 1$ taken into consideration, $F(h)_{\text{cryst}}$ is expressed by

$$F(h)_{\text{cryst}} = \sum_i \sum_j^N f_j T_j \exp\{2\pi i[\underline{h}(x_j| + \delta x_{ij}| + \Delta t_i)]\}$$

$$= \sum_j f_j T_j \exp[2\pi i(\underline{h}x_j|)]\left(1 + \sum_i \sum_j^N f_{ji} T_j \exp\{2\pi i[\underline{h}(\delta x_{ij}| + \Delta t_i)]\}\right)$$

$$= F(h)_{\text{unit}}\left\{1 + \sum_i \sum_j^N f_j T_j \exp[2\pi i(\underline{h}\delta x_{ij}|)]\exp[2\pi i(\underline{h}\Delta t_i)]\right\}. \quad (4.3.13)$$

Then Eq. (4.3.11) and Eq. (4.3.12) yield the relation of $F(h)_{\text{cryst}} \leqq F(h)_{\text{deform}} \leqq F(h)_{\text{unit}}$. If I_O indicates the incident source intensity, then the intensity for the bulk crystal is

$$I(hkl)_{\text{cryst}} = I_O|F(h)_{\text{cryst}}|^2$$

$$\fallingdotseq I(hkl)_{\text{unit}}\left\{2 + 2\sum\sum f_j \cdot T_j \cos[2\pi h\,\delta x_{ij} + \Delta t_i)]\right\}. \quad (4.3.14)$$

Equation (4.3.14) indicates that the diffraction intensity is more reduced with an increase in the fluctuation of lattice showing $\mathbf{t}_i$ and atomic displacement (dx_j, dy_j, dz_j).

4.3.5   Effect of the Diffraction Profile on the FWHM

Interplaner distances randomly change with fluctuation of the lattice, which is equivalent to the random of crystallographic translation $\mathbf{t}_i$. This lattice deformation is one of the reasons for diffraction line broadening.

Under inhomogeneous stress, the variations $\Delta\mathbf{a}$, $\Delta\mathbf{b}$, and $\Delta\mathbf{c}$ in the unit cell edges of the crystal comprise the summation of the components of displacements parallel to the cell edges, which are introduced from ε_{11}, ε_{22}, and ε_{33}. The displacements are homogenized in a whole crystal, which is composed of the periodic repetition of the unit cell. However, an inhomogeneous stress generates anisotropic strains. In this case the basal spacings in the direction of $\langle 100\rangle$, $\langle 010\rangle$, and $\langle 001\rangle$ are not equivalent to each other, even in the symmetrically cubic crystal, resulting in $d_{100} \neq d_{010} \neq d_{001}$. This local deformation can be observed by single-crystal x-ray diffractometry through very accurate diffraction optics with a highly collimated x-ray source such as synchrotron radiation.

Lattice deformation relates not only to crystallographic symmetry but also to valence-electron orbitals. Space-group symmetry of a crystal may be locally collapsed by the anisotropy of the deformation. If crystallographic site symmetries with screw axes or glide planes of symmetry operations are present, the anisotropy of these crystallographic translations is easily generated from the inhomogeneous inner pressures in a crystal. Further, when the constituent atoms of the crystal are composed of nonspherical valence electrons such as p or d electrons rather than s electrons, these atoms located at the noncubic site symmetry easily induce the anisotropic deformation by external force.

A fluctuation of the crystallographic translation (Δt) or ($n\Delta t$) (Fig. 4.3.2) as well as the atomic random displacement (Δu) can be sufficient to brake periodicity of the translation (n should be an integer in the ideal crystal). Imperfections of the crystal such as mosaicity or dislocation also produce a lack of the lattice periodicity. The lattice deformations present in a bulk crystal introduce a broadened diffraction profile. Powder diffractometry can also detect the variation of the interplaner distances d_{hkl} of crystallographically equivalent indices hkl. The superposed diffraction peaks induce broadened diffraction profiles. Because an increasing inhomogeneous stress can make the lattice deformation more extended [as indicated Eq. (4.3.3)], the diffraction profile broadens.

4.3.6   Observations of Lattice Instability

Investigations of the lattice instability under pressure can be conducted by x-ray diffraction, x-ray absorption such as near-edge-structure or extended x-ray-absorption fine-structure techniques, and Raman spectroscopy by means of diamond anvil cell (DAC) or a multianvil high-pressure apparatus, which are effective techniques for in situ observations. We devised an oil-pressure-controlled DAC with an internal electric resistance heater for kinetic study as a function of pressure [15].

Lattice instability of crystal under static compression causes line broadening in the diffraction profile. Profile analysis can be made from energy- or angle-dispersive diffraction spectra taken under hydrostatic pressures with a DAC. Line broadening originates from increasing lattice strain and reduced average crystallite size with increasing pressure, besides the deviatoric stress in the DAC. Extensive deformation and pulverization under compression presents a halo pattern in addition to the broadening. To quantify and analyze the broadening, several profile analytical methods have been proposed to consider the effects on the diffraction profile.

The integrated intensity $y(\theta_H)$ of diffraction H is calculated for the powder diffraction by

$$y(\theta_H) = y_0(\theta_H)p(LP)|F(H)|^2 T(\theta_H)A(\theta_H), \tag{4.3.15}$$

where $F(H)$ is the structure factor, θ_H angular position of diffraction H, p multiplicity parameter, LP Lorentz and polarization factor, and $A(\theta_H)$ absorption correction. The observed Bragg intensity $y(\theta_i)$ at θ_i is expressed by

$$y(\theta_i) = \sum y(\Delta\theta_{iH}) + B(\theta_i), \tag{4.3.16}$$

where $y(\Delta\theta_{iH}) = y(\Delta\theta_{iH})P(\Delta\theta_{iH})$, with $\Delta\theta_{iH} = \theta_i - \theta_H$, $B(\theta_i)$ is the background intensity.

For the line profile, based on a classic method for determining the effect of distortion on x-ray patterns [16] and of the crystallite size distribution [17–19]. The Voigt function

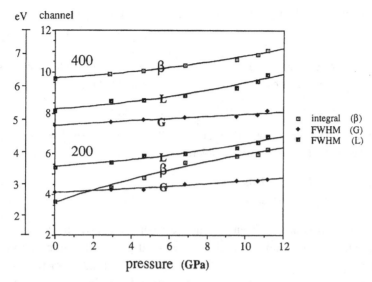

Figure 4.3.3 Diffraction profile change of NaCl as a function of pressure. Diffraction was measured using multianvil high pressure apparatus (MAX80) in Photon Factory. Γ_G, Γ_L and integral width of diffraction peaks of 200 and 400 are shown with index G, L and β.

of the convolution of Gaussian and Lorentzian functions has been applied for the quantitative discussion of the strain energy and crystallite size distribution from the profile fitting [20, 21].

The intensity distribution in profile is written as

$$P(\Delta\theta_{iH}) = c[h\exp(-X^2) * (1-h)(1+Y^2)^{-1}], \tag{4.3.17}$$

where $X = \Delta\theta_{iH}/\Gamma_G$ and $Y = \Delta\theta_{iH}/\Gamma_L$, Γ_G and Γ_L are full width at a half maximum (FWHM) of a profile of the Gaussian and the Lorentzian functions, respectively.

The FWHM Γ_L of the Lorentzian component continuously increases with pressure compared with Γ_G of the Gaussian. Conventional integrated half-width $B(hkl)$ is expressed by b related to the size effect and β of instrumental broadening. This relation is expressed by $B = b + \beta$. The mean crystallite size $\langle D \rangle_a$ and mean-squared strain $\langle \varepsilon^2 \rangle$ are derived from the observed Γ_G and Γ_L. As seen from an experimental example of NaCl, diffraction changes with increasing pressure (Fig. 4.3.3), and a larger Γ_L indicates reduced crystallite size. Lattice deformation and intrinsic strain in the lattice are not much changed. Several other factors such as asymmetric component and background fitting giving an effect to profile are also considered in the calculation. The corrected diffraction intensity $I(x)$ is composed of $h(x)$ the intrinsic crystalline parameters, such as crystallite size, size distribution, and lattice deformation, and $g(x)$, and the instrumental and optical conditions.

$$I(x) = \int h(y) * g(x - y)\, dy, \tag{4.3.18}$$

After deconvolution, the crystallite size effect $f(s)$ is expressed by

$$f(s) = h(s) - g(s). \tag{4.3.19}$$

The FWHM of the total profile, Lorentzian and Gaussian components have individual relations to $h(x)$ and $g(x)$, or $f(x)$ and $g(x)$ by the following equations:

$$\Gamma_{\text{total}} = 2w/\beta, \quad \Gamma_L = g_L * f_L \quad \text{and} \quad \Gamma_G = g_G * f_G.$$

Line broadening is written with the relation of $B = I(x)/\mathrm{d}x I(0)$, $b = f(x)\mathrm{d}x/f(0)$, and $\beta = g(x)\mathrm{d}(x)/g(0)$ taken into consideration.

$$B^2 = \beta^2 + b^2 \quad \text{(Gaussian)}, \tag{4.3.20}$$

$$B = \beta + b \quad \text{(Lorentzian)}. \tag{4.3.21}$$

The detailed procedure has been presented elsewhere. It is worthwhile to apply the above procedure for the profile analysis to a certain degree of the lattice deformation. However, the procedure is not applicable to the extremely broad diffraction peaks just before amorphization. Probably Voigtian, the profile fitting function is limited in its application to this broad peak.

4.3.7 Simulation of Pressure-Induced Amorphization by Molecular Dynamics

Lattice instability is one of the significant applications of MD from the viewpoint of the study of the phase transition under pressure. Pressure-induced amorphization has attracted much attention in the simulation of experimental observations.

The first-principle model or pseudopotential calculation with the self-consistent local-density approximation is used as the functional form of the interatomic potentials for the simulation of crystal and high-pressure structure modifications, amorphization and calculation of bulk moduli, volume expansion coefficients, and enthalpies [22, 23]. The quasi-classical MD method is commonly used for pressure-induced structure modification because of its advantages of simplicity and applicability for heavy ions. Pressure-induced amorphizations of α-AlPO$_4$-berlinite and SiO$_2$ [24] have been simulated with ab initio pseudopotentials for elastic instability and soft modes in the pressure-induced amorphization of quartz [25].

In our MD calculation of the phase transformation of GeO$_2$ including its pressure-induced amorphization [26], Coulomb-, van der Waals-, and Gilbert-type repulsion energy terms were taken into account in the ionic soft-shell model of the MD equation. The energy parameters of the effective charge, the repulsive radius, the softness parameter, and the van der Waals coefficients were empirically optimized to reproduce the structures, bulk moduli, and thermal expansion coefficients of the polymorphs under desired conditions.

Elastic properties of the α-quartz have been studied as functions of pressure with both classical interatomic potential and first-principles pseudopotential approaches

[22]. Binggeli and Chelikowsky [27] found that the structure becomes mechanically unstable at ~30 GPa and suggested that amorphization is triggered by the onset of lattice shear instability. A change in the Si coordination is intimately related to the shear instability. Tse [28] simulated the pressure-induced crystalline-to-amorphous transformation in ice-Ih by MD calculation and concluded that the actual mechanism is due to a mechanical instability in the ice framework. This is contrary to the supercooling model, i.e., transition on the extrapolated melting curve [1]. MD calculation [29] revealed that a pressure-induced amorphized solid is disordered and has an anisotropic property in the structure. However, it is not structurally related to an isotropic melting [3] that showed the mechanism of the structure memory effect by decompression after the amorphization of α-AlPO$_4$ at higher pressure than 31 GPa. The recoverable feature comes from the PO$_4$ tetrahedron that apparently exists even at over 80 GPa like a rigid body in the distorted lattice.

4.3.8   MD-Dynamics Simulation Techniques

The functional form of the interatomic potentials $\psi(\mathbf{r}_{ij})$ is expressed by the following conventional equation as the partially ionic model:

$$\psi(\mathbf{r}_{ij}) = \frac{q_i q_j}{r_{ij}} - \frac{C_i C_j}{r_{ij}^6} + f(B_i + B_j) \exp\frac{A_i + A_j - r_{ij}}{B_i + B_j}, \qquad (4.3.22)$$

where the first, second, and third terms on the right-hand side represent Coulomb-, van der Waals- [30], and Gilbert-type [31] repulsion energies, respectively. $\mathbf{r}_{ij}$ is the interatomic distance between the ith and the jth ions, and f is a standard force constant, 4184 kJ/Å mol. The effective charge q, the repulsive radius A, the softness parameter B, and van der Waals coefficient C are the energy parameters. These parameters were empirically optimized to reproduce the structure, bulk moduli, and thermal expansivities.

The energy parameters can also be optimized from the observed crystal structure, if the static lattice energy is minimized. An empirical two-body interatomic potential has been determined to simulate the structure change of crystalline GeO$_2$. Based on this potential, two polymorphs of GeO$_2$, an α-quartz type and a rutile type, have been reproduced by MD simulation techniques. Crystal structures, bulk moduli, volume thermal-expansion coefficients, and enthalpies of these polymorphs of GeO$_2$ were simulated. The bulk moduli and the thermal expansivities are in good agreement with the reliable experimental data on both polymorphs previously reported. With this potential, MD simulation was further used to study the structural changes of GeO$_2$ under high pressure. We investigated pressure-induced amorphization. Under hydrostatic compression, α-quartz-type GeO$_2$ transformed to a denser amorphous state at 7.4 GPa with a change in the packing of oxygen ions and an increase in the Ge coordination. At still higher pressure over 80 GPa, rutile-type GeO$_2$ transformed to CaCl$_2$-type structure.

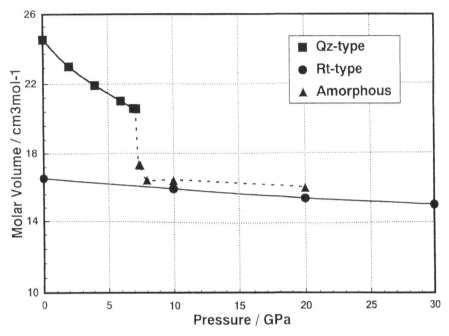

Figure 4.3.4 Molar volume of GeO$_2$ polymorphs as a function of pressure. MD basic cell: 48 unit cell ($4a \times 4a \times 3c$ including 432 atoms) for α-quartz type; basic cell: 80 cell ($4a \times 4a \times 5c$ including 480 atoms) for rutile type; iteration: 10,000 steps (12 ps) after equilibration for α-quartz type and 10,000 steps (6 ps) for rutile-type.

For the simulation of pressure-induced amorphization with a coordination change in GeO$_2$, we applied both the scaling method to control pressure and temperature, and the quantum correction of free energy to simulate the equilibrium structural and physical properties of the GeO$_2$ crystals under desired temperature T and pressure P conditions. We used the periodic boundary conditions and the Ewald sum method to converge the calculation of the Coulomb and the dispersion interactions efficiently. Our MD calculation has been published [26].

The variation of the molar volume with pressure (Fig. 4.3.4) shows that the α-quartz-type GeO$_2$ structure continuously but noticeably reduces its cell volume, and a discontinuous volume drop at 7.4 GPa testifies to a transition to the disordered crystalline state. The volume of the amorphous state is a little larger than that of the rutile-type structure, and the former state never changes to the latter. Although there is no long-range order in this state, the Ge coordination changes from fourfold to sixfold. Figure 4.3.5 shows that the coordination number of the amorphous state is close to that of the rutile-type structure but a coordination lower than sixfold results from the disordered Ge–O configurations, which include fourfold and probably fivefold links. Because the amorphous state has a slightly higher enthalpy (-1724 kJ/mol) than the rutile type (-1727 kJ/mol) but lower than the α-quartz type (-1711 kJ/mol), this noncrystalline state is possibly formed as a precursor in the series of α-quartz-type-to-rutile-type transitions under the present simulated condition (300 K, 7.4 GPa).

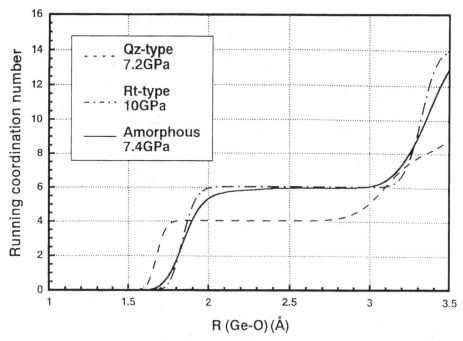

Figure 4.3.5 Coordination number change of the α-quartz-, rutile-type and amorphous state of GeO_2 as a function of average Ge–O interatomic distance. The first two crystalline states have a rigid fourfold and sixfold coordination. MD calculations were conducted in the same basic cell as in Figure 4.3.4.

The critical pressure of 7.4 GPa is a little higher than the experimental result of 6.5 GPa [3]. The population of the simulated octahedrally coordinated cluster is constant and independent of pressure, but depends on the volume of the MD basic cell. The inability of simulation techniques to model kinetic phenomena, like nucleation in a phase transition, may cause this discrepancy between the simulation and the experiment.

As mentioned above, some olivine-type structures, such as Fe_2SiO_4 [4] and Mg_2GeO_4 [5], show pressure-induced amorphization as an intermediate metastable state in the course of the olivine-to-spinel transition. The high-resolution electron microscope lattice image by Guyot and Reynard [32] suggests that the amorphous structure consists of an ideal oxygen sublattice and a cation glassy sublattice. MD simulation of the pressure-induced amorphization of Mg_2SiO_4 [33] exhibits the same feature as that of the high-resolution electron microscope observation mentioned above. In the amorphous state found between 35 GPa and 40 GPa, hexagonal close-packed oxygen layers are a little fluctuated, as seen from Fig. 4.3.6, and the fluctuated spacings are preserved to the same as those before amorphization. However, the position of both Si tetrahedral and Mg octahedral cations located in the interlayers of oxygen are noticeably randomized. Hence these cations are no more regular fourfold or sixfold coordination. This MD calculation suggests a mechanism of pressure-induced modification, which is a precursor atomic movement to the transition process. Some sublattices are preserved, and the other sublattices are deformed in a random fashion.

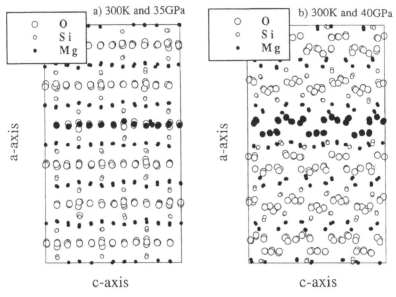

Figure 4.3.6 Atomic arrays of forsterite along its *b* axis taken from the MD calculation at 35 GPa and 40 GPa at 300 K. In the amorphous state, the oxygen layers are barely preserved but the cation positions are randomly scattered. Basic cell: $4a \times 2b \times 3c$ including 672 atoms, time increment: $2.0 \times 10{-}15$ s, iteration: 2000 step.

If the interatomic potential can be precisely given in the MD calculation, the amorphous state may transform to a stable high-pressure form after much larger steps and a larger MD basic cell for calculation, which are beyond the capabilities of our present computer. However, the amorphization may be an intrinsic precursor to phase transformation and the metastable state in the process of transition. In this case a large activation energy hinders sufficient atomic movement and can preserve an unstable state. The elucidation of this process is beyond the capability of our computer simulation.

4.3.9   Mechanism of Pressure-Induced Amorphization

There are two types of amorphization mechanisms, as described in Section 4.3.8. The amorphous state that changes from the crystalline state under compression cannot be defined as a thermodynamically stable phase but rather as a state in the kinetic process. Thus the amorphous state under pressure does not have a true stability field. However, it is a metastable state and a precursor to a phase transition, decomposition, or exsolution under pressure.

The transition energy can be defined as $\Delta G^* = V^* \Delta P - S^* \Delta T$, where V^* and S^* are activation volume and entropy, respectively, of the transition process. Pressure-induced amorphization at ambient temperature must consider only the volume change in the activation state. Generally, high-pressure transitions at low temperatures are extremely sluggish compared with thermal transition because of the exceedingly small volume change. Consequently, an individual state in the transition process can be observable over laboratory time scales.

Irreversible amorphization generates the nucleation of high-pressure form in the lattice. However, the crystal growth is not as well undertaken, as thermal energy is too small to provide a large crystallite. The small crystallite size is incoherent to the x-ray wavelength. Critical activation total energy ΔG_{total} of nucleation is possibly explained by a conventional equation for a spherical nucleus model; $\Delta G_{total} = 4/3\pi r^3(\Delta G_{bulk} + \varepsilon) + 4\pi r^2\sigma$, where ε and σ are strain energy per unit volume in lattice and surface energy per unit area, respectively, and r denotes the average radius of nuclei. Then ΔG_{total} is defined by the nucleus size. However, the combination of surface energy, strain energy, and bulk energy controls the critical size of nucleus. Practically powdered grains in the DAC or multianvil apparatus may suffer from more stress induced from the frozen pressure media with increasing pressures. This stress generates a larger strain in the lattice.

If ε becomes large under pressure, the critical energy of nucleation will become large. This is the case for pressure-induced amorphization. However, thermal energy that is due to heating enables the crystallite to increase. At kinetically low temperatures, crystalline growth is barely accelerated and the crystallite sizes are in the incoherent region, in contrast to the x-ray radiation. This feature shows the x-ray amorphous state. At higher temperatures, larger nuclei are produced and then larger crystallites will be grown. These crystallite sizes are large enough to cause x-ray diffraction.

The activation energy of the transition can be related to the coordination change, as seen in α-quartz-to-rutile transition. Tsuneyuki et al. [22] simulated the structures of four SiO_2 polymorphs by using first-principle interatomic pair potential and reported the bulk-modulus, density, and relative energies of the polymorphs. Energy differences among these polymorphs are very small (e.g., only a few kcal/mol). But the amorphous phase was not examined. From the experiment on GeO_2 shown in Fig. 4.3.5, the rutile structure cannot be crystallized from the amorphous at temperatures as high as 300°C under 12 GPa; then the crystallization from the amorphous state to the rutile-type structure is almost impossible from the pressure work alone. The long-range order of the rutile structure may require a considerably large thermal activation energy for obtaining a large atomic movement.

Quantitative analysis of the pressure-induced transformation in principle requires calorimetric study of the amorphous phase as a function of pressure. As mentioned above, the transformation on compression is extremely sluggish compared with thermal transition. If any metastable or intermediate state is present during high-pressure transition, it can be easily confirmed by in situ observation through x-ray diffraction. The Debye temperatures of many inorganic substances are exceedingly high (i.e., well above room temperature). The temperature is low enough to inhibit the transformation to a thermodynamically stable phase under high pressure. Even if nucleation of the high-pressure phase occurs, the crystallite growth at room temperature may not be sufficient to provide measurable x-ray diffraction. When the crystallite does not grow to the coherent size for x-ray wavelength, diffraction intensity cannot be detected. Accordingly it seems to be an amorphous phase and it is defined as x-ray amorphous. However, local structures in those amorphous states can often be confirmed by an extended-x-ray-absorption fine-structure technique or Raman spectroscopy. In this case the x-ray amorphous material is composed of microcrystallites of the high-density phase.

Thus lattice-instability-induced amorphization under pressure can be linked to lattice vibrations, an increase in dislocation density, and defect and entropy effect associated with atomic ordering [34].

References

[1] O. Mishima, L. D. Calvart, and E. Whalley, Nature (London) **310**, 393–395 (1984).

[2] R. J. Hemley, A. P. Jephcoat, H. K. Mao, L. C. Ming, and M. H. Manghnani, Nature (London) **334**, 52–54 (1988).

[3] T. Yamanaka, T. Shibata, S. Kawasaki, and S. Kume, *High-Pressure Research: Application to Earth and Planetary Sciences*, Y. Syono, M. H. Manghnani, eds. (Terra Scientific, Tokyo, Japan 1992), pp. 493–501.

[4] Q. Williams, E. Knittle, R. Reiichlin, S. Martin, and R. Jeanloz, J. Geophys. Res. **95**, 21549–21563 (1990).

[5] T. Nagai, K. Yano, M. Dejima, and T. Yamanaka, Mineral. J. **17**, 151–157 (1994).

[6] Q. Williams and R. Jeanloz, Nature **338**, 413–415 (1989).

[7] Y. Lin, G. Lan, and H. Wang, Solid State Commun. **86**, 99–101 (1993).

[8] Y. Fujii, M. Kowaka, and A. Onodera, J. Phys. C. **18**, 789–797 (1985).

[9] M. B. Kruger and R. Jeanloz, Science **249**, 647–649 (1990).

[10] C. Meade, R. Jeanloz, and R. J. Hemley, *High-Pressure Research: Application to Earth and Planetary Sciences*, Y. Syono and M. H. Manghnani, eds. (Terra Scientific, Tokyo, Japan, 1992), pp. 485–492.

[11] S. L. Chaplot and R. Mukhopadhyay, Phys. Rev. B **33**, 5099–5101 (1986).

[12] H. Sankaran, S. K. Sikka, S. M. Sharma, and R. Chidambaram, Phys. Rev. B **38**, 170–173 (1988).

[13] R. R. Winters, G. C. Serghiou, and W. S. Hammack, Phys. Rev. B **46**, 2792–2797 (1992).

[14] T. Yamanaka, Z. Kristallogr. **212**, 401–410 (1997).

[15] T. Yamanaka, S. Kawasaki, and T. Shibata, *Advances in X-Ray Analysis*, C. S. Barrett et al., eds. (Plenum, New York, 1992), **Vol. 35**, pp. 415–423.

[16] B. E. Warren and B. L. Averback, J. Appl. Phys. **21**, 595–599 (1950).

[17] A. J. C. Wilson, Nature (London) **193**, 568–569 (1962).

[18] A. Bienenstock, J. Appl. Phys. **32**, 187–189 (1961).

[19] A. Bienenstock, J. Appl. Phys. **34**, 1391 (1963).

[20] Th. H. de Keijser, J. I. Langford, E. J. Mittermeijer, and A. B. P. Vogls, J. Appl. Crystallogr. **15**, 308–314 (1982).

[21] D. Louer, J. P. Auffredie, J. I. Langford, D. Ciosmak, and J. C. Niepce, J. Appl. Crystallogr. **16**, 183–191 (1983).

[22] S. Tsuneyuki, M. Tsukada, H. Aoki, and Y. Matsui, Phys. Rev. Lett. **61**, 869–872 (1988).

[23] J. R. Chelikowsky, N. Troullier, J. L. Martins, and H. E. King, Phys. Rev. B **44**, 489–497 (1991).

[24] J. S. Tse and D. D. Klug, Science **255**, 1559–1561 (1992).

[25] N. Binggeli, J. R. Chelikowsky, and R. M. Wentzcovitch, Phys. Rev. B **49**, 9336–9340 (1994).

[26] T. Tsuchiya, T. Yamanaka, and M. Matsui, Phys. Chem. Miner. **25**, 94–100 (1998).

[27] N. Binggeli and J. R. Chelikowsky, Phys. Rev. Lett. **69**, 2220–2223 (1992).

[28] J. S. Tse, J. Chem. Phys. **96**, 5482–5487 (1991).

[29] J. S. Tse and D. D. Klug, Phys. Rev. Lett. **70**, 174–177 (1993).

[30] J. E. Mayer, J. Chem. Phys. **1**, 270–279 (1933).

[31] T. L. Gilbert, J. Chem. Phys. **49**, 2640–2642 (1968).

[32] F. Guyot and B. Reynard, Chem. Geol. **96**, 411–420 (1992).

[33] M. Dejima, M. S. thesis (University of Osaka, Osaka, Japan, 1994).

[34] H. J. Fecht, Nature (London) **356**, 133–135 (1992).

Chapter 4.4

Effect of Hydrostaticity on the Phase Transformations of Cristobalite

Takehiko Yagi and Masaaki Yamakata

Cristobalite was compressed at room temperature up to 30 GPa under quasi-hydrostatic and nonhydrostatic conditions. The structure of the high-pressure phase differs completely, depending on the hydrostaticity of applied pressure. Under quasi-hydrostatic conditions, a structure very similar to that of stishovite was formed above ~20 GPa, which can be quenched to ambient conditions. When this phase was heated to several hundred degrees centigrade under pressure, normal stishovite was formed. Under nonhydrostatic conditions, on the other hand, cristobalite transformed into the unidentified X-I phase at ~15 GPa, in accordance with previous study. This X-I phase transforms into stishovite on heating to above 1000 °C. The apparent compressibility of cristobalite also differs considerably, depending on the nature of applied pressure. The nature of the high-pressure phases formed by the room-temperature compression of cristobalite is discussed.

4.4.1 Introduction

Silica is one of the most intensively studied materials under a wide range of pressure and temperature conditions because of its importance in earth science as well as in material science. In spite of its simple chemical formula, many different polymorphs are formed, depending on the pressure and the temperature conditions. Moreover, recent room-temperature compression studies of various polymorphs clarified the formation of variety of metastable phases [1–5], and the situation is complicated. It is likely that the formation of some of the metastable phases is related to the nonhydrostatic nature

H. Aoki et al. (eds), *Physics Meets Mineralogy* © 2000 Cambridge University Press.

of applied pressure. Here we report a new example of the formation of completely different high-pressure structures that depend on the hydrostaticity of applied pressure.

Experiments were performed with cristobalite as the starting material; cristobalite is a high-temperature polymorph of silica stable above 1470 °C at atmospheric pressure and can be quenched to ambient condition. This structure is known to have very loose packing of atoms, and the molecular-dynamics calculation [6] predicted the transition directly, or by means of intermediate phase, to stishovite by room-temperature compression. A previous experiment [2], however, clarified the formation of a completely different, unknown structure above ~10 GPa. Tsuchida and Yagi [2] further reported the formation of two other unknown structured phases above ~40 GPa and on the release of pressure. Their study was made by direct compression of powdered cristobalite into a diamond-anvil apparatus; thus strong nonhydrostatic pressure was applied to the specimen. In the present study, various different pressure-transmitting media, including solidified rare gases that are expected to give nearly hydrostatic (quasi-hydrostatic) pressure, were used, and the effect of hydrostaticity on the phase transformation of cristobalite was considered. A brief report of this work has been published already [7] and full detail is given in this chapter. This may help us to understand the complexity of the phase transformations of silica reported by various studies and to understand the effect of nonhydrostatic stress on phase transformations in general.

4.4.2   **Experimental**

Starting cristobalite was prepared by a firing reagent grade-silica powder (Wako Chemical) at 1550 °C for 5 h and then quenched in liquid nitrogen. The material thus prepared was examined by powder x-ray diffraction and more than 30 diffraction lines were well explained by cristobalite, although there were a few very weak diffraction lines that could not be explained by cristobalite. Samples recovered from high-pressure experiments were examined by electron-probe microanalysis, and no elements other than silicon and oxygen were found. The starting material used for the YAG laser-heating experiment was mixed with small amount of platinum powder, which works as an absorber of the laser energy.

High-pressure experiments were performed with a modified Mao–Bell-type diamond anvil [8]. To keep the sample chamber as thick as possible, a rhenium gasket was used throughout the study. A 200-μm-diameter hole was drilled in the center of a preindented gasket; then the hole was very lightly filled with a cristobalite powder, so that enough space was left for the pressure-transmitting medium. A few ruby chips were placed in the sample chamber as pressure markers. During high-pressure experiments, we could clearly observe the powder sample immersed in the pressure-transmitting medium; thus it was clear that quasi-hydrostatic pressure was applied to the specimen.

Pressure-transmitting media used are argon, neon, hydrogen, and 4:1 mixture of methanol and ethanol. The gases used here solidify at approximately 1–5 GPa at room temperature but the solidified gases are soft enough to give a quasi-hydrostatic environment. The degree of hydrostaticity decreases of the order of hydrogen, neon, and argon [9]. The alcohol mixture remains hydrostatic up to 10 GPa but above that it hardens and is expected to have relatively large nonhydrostaticity compared with that of other gases. The gas medium was loaded at room temperature at ~ 0.2 GPa by a gas-loading apparatus [10].

Some of the experiments were made by increasing the amount of specimen and reducing the amount of pressure-transmitting medium considerably, so that some of the grains of the specimen were directly in contact with each other. This achieved the condition between quasi-hydrostatic and uniaxial compression. For comparison, nonhydrostatic compression was also made by directly compressing the powdered specimen without any pressure-transmitting medium. This way, a highly uniaxial stress is applied to the specimen. Pressure was measured by ruby fluorescence technique with the pressure scale of Mao and Bell [11].

Most of the experiments were made at room temperature. To see the effect of temperature, we heated one sample in an argon medium by using a CO_2 laser. The temperature of the sample was estimated to be only several hundred degrees centigrade from the temperature of the diamond anvil, which was measured by a thermocouple, and from the fact that no incandescent light was observed during heating. Another sample was heated to above $1000\,^\circ$C with a YAG laser. In this case, cristobalite was mixed with small amount of platinum powder and directly compressed without a pressure medium.

All the x-ray experiments were made at the Photon Factory, National Laboratory for High Energy Physics of Japan (KEK). Monochromatized x-rays with a wavelength of 0.6888 Å were collimated to a 80–200-μm-diameter beam and then irradiated the sample. Diffracted x-rays were recorded by an imaging plate (IP) detector, which was placed 150 mm from the sample. Exposure time was 1 to 4 h, depending on the pressure range and the sample size. The details of the analysis have already been described [12].

4.4.3   Results

Examples of the IP images at atmospheric pressure and 28 GPa are shown in Fig. 4.4.1, and diffraction profiles obtained by the integration of these patterns are shown in Fig. 4.4.2. With increasing pressure, a smooth compression of the tetragonal unit cell and a gradual broadening of the diffraction lines were observed up to 17.4 GPa. Previous x-ray studies have clarified that cristobalite transforms from a tetragonal structure to a lower symmetry phase at ~ 1.5 GPa under hydrostatic conditions [13, 14]. A similar phase transformation was also seen with Raman spectroscopy [15, 16]. In the present study, however, no splitting of the x-ray-diffraction lines as reported in the above studies was observed when argon was used as the pressure-transmitting medium. When other pressure media were used, the sample was compressed to much higher pressures from the beginning and we could not confirm the existence of a phase transformation. It is likely that such a distortion is very sensitive to the existence of nonhydrostatic stress and that the transition was smeared out when the sample was compressed in the solidified argon. When the pressure was increased from 17.4 to 21.9 GPa,

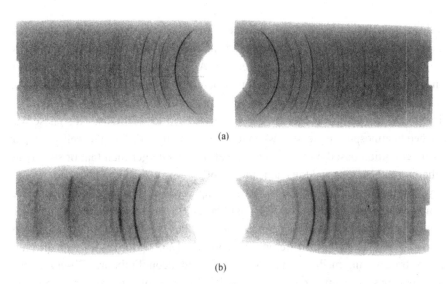

Figure 4.4.1 Examples of diffraction profiles recorded on the IP detector. (a) Starting material (cristobalite) at 0.1 MPa, (b) stishovitelike phase plus argon pressure medium at 28 GPa.

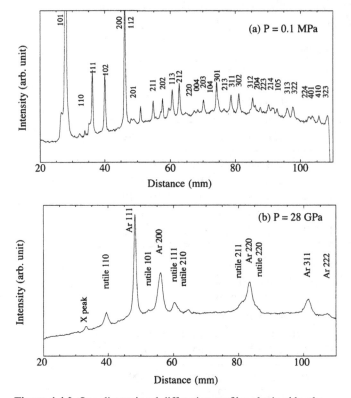

Figure 4.4.2 One-dimensional diffraction profiles obtained by the integration of the intensities of the diffractions shown in Fig. 4.4.1. The horizontal axis is the distance of the imaging plate from the direct beam position. (a) Starting material (cristobalite) at 0.1 MPa, (b) the stishovitelike phase plus argon pressure medium at 28 GPa. X peak is the diffraction that cannot be indexed by the rutile unit cell.

new diffraction lines appeared. Although the diffraction was relatively broad, as shown in Figs. 4.4.1(b) and 4.4.2(b), it can be indexed on a rutile-type unit cell similar to that of stishovite. One diffraction line near 3.5 Å [X peak in Fig. 4.4.2(b)], however, cannot be indexed with the rutile structure. There was no further transition when this new stishovitelike phase, which was so termed because of the strong similarity of its diffraction to the stishovite spectrum, was compressed to ~30 GPa. This phase was quenched to atmospheric pressure when it was decompressed. The unit-cell volume at ambient condition based on the rutile unit cell is 1.2% larger than that of stishovite. Furthermore, the a axis and the c axis are slightly elongated and compressed, respectively, resulting in a smaller c/a ratio (0.63) compared with 0.64 for normal stishovite. Examples of the unit-cell calculation at 28 GPa and 0.1 MPa are shown in Table 4.4.1.

This stishovitelike phase was formed whenever the pressure-transmitting medium was used, and no meaningful difference was observed among the results with different pressure-transmitting media such as Ar, Ne, H_2, and alcohol mixture. Observed unit-cell dimensions are summarized in Fig. 4.4.3. Because of the broad nature of the diffraction lines, calculated unit-cell dimensions are not as accurate as those calculated

Table 4.4.1. *Observed and calculated d values of stishovitelike phases at 28 GPa and 0.1 MPa*

$P = 28$ GPa
$\quad a$ (Å) $= 4.0751 \pm 0.0057$
$\quad c$ (Å) $= 2.6161 \pm 0.0064$
$\quad V$ (Å^3) $= 43.44 \pm 0.16$

h	k	l	d_{obs} (Å)	d_{cal} (Å)	$d(o)/d(c) - 1$
1	1	0	2.8768	2.8815	−0.0016
1	0	1	2.2049	2.2015	0.0016
1	1	1	1.9333	1.9369	−0.0019
2	1	0	1.8264	1.8224	0.0022
2	1	1	1.4949	1.4953	−0.0003

$P = 0.1$ MPa
$\quad a$ (Å) $= 4.2185 \pm 0.0036$
$\quad c$ (Å) $= 2.6597 \pm 0.0038$
$\quad V$ (Å^3) $= 47.33 \pm 0.10$

h	k	l	d_{obs} (Å)	d_{cal} (Å)	$d(o)/d(c) - 1$
1	1	0	2.9807	2.9829	−0.0007
1	0	1	2.2512	2.2498	0.0006
1	1	1	1.9852	1.9852	0.0000
2	1	1	1.5367	1.5388	−0.0013
2	2	0	1.4936	1.4915	0.0014

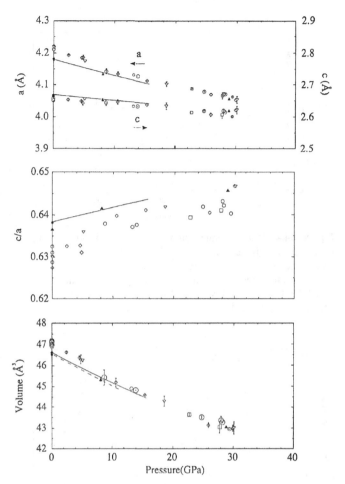

Figure 4.4.3 Lattice parameters of the stishovitelike phase formed with various pressure-transmitting mediums. Open symbols are for the phases formed at room temperature ($\bigcirc$: Ar, $\Diamond$: Ne, $\square$: H_2, ∇: Mth+Eth). Solid symbols are those of stishovite formed by heating the stishovitelike phase by CO_2 laser (triangle) and YAG laser (circle). Solid lines represent the compression curve measured on the single crystal [23] and powder [24] of stishovite.

for single crystals, but there was no systematic difference among them. Relative compression data for a and c axes, together with those of the unidentified peak near 3.5 Å, are summarized in Fig. 4.4.4. It is clear from this figure that the a axis is much more compressible than the c axis, and the compression of the unidentified peak is almost identical to that of the a axis. The linewidth of the pattern for the stishovitelike phase formed at room temperature was very broad, probably because of poor crystallinity; to improve crystallinity, the sample was heated to several hundred degrees centigrade in an argon pressure medium by irradiation with a CO_2 laser. X-ray-diffraction profiles obtained after heating at 29 GPa and after recovery to ambient conditions are shown in Figs. 4.4.5(a) and 4.4.5(b) and the unit cell calculated from the diffraction profile

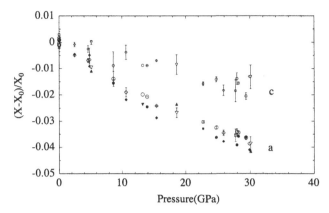

Figure 4.4.4 Relative compressions of stishovitelike phase along the *a* and *c* axis (open symbols) compared with that of the unidentified peak (solid symbols). The different symbols correspond to different pressure-transmitting medias as in Fig. 4.4.3. Compressibility of the *d* spacing of the unidentified peak is indistinguishable from that of the *a* axis.

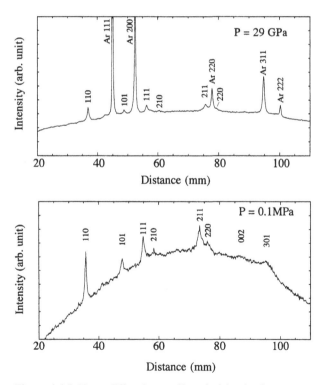

Figure 4.4.5 X-ray-diffraction profiles of stishovite that was formed by heating stishovitelike phase to approximately several hundred degrees centigrade by a CO_2 laser at (a) 29 GPa and (b) after recovered to ambient condition.

of Fig. 4.4.5(b) is given in Table 4.4.2. Compared with those of Fig. 4.4.2(b), diffraction lines from argon became much sharper after heating whereas the sharpening of the diffraction lines from the sample were less clear. The remarkable change, however, is that the unidentified peak near 3.5 Å has disappeared. Moreover, the dimensions of the unit cell became indistinguishable from those of stishovite, as shown in Table 4.4.2 and Fig. 4.4.3.

When the cristobalite was compressed directly without a pressure medium, diffraction lines became very broad, even at low pressures. An example pattern, obtained at 6 GPa, is shown in Fig. 4.4.6(a) and it was difficult to tell whether or not the phase transformation had occurred at ~1.5 GPa. Three intense and broad diffraction lines of cristobalite were clearly observed up to 15 GPa. When the pressure was increased above 15 GPa, new diffraction lines appeared. These new lines were completely different from those of the stishovitelike phase, as shown in Fig. 4.4.6(b), but were very similar to those of the X-I phase reported in the previous study under similar non-hydrostatic conditions [2]. Some additional weak lines were observed in the present study, probably because the x-ray system used in this study has much higher resolution and sensitivity. When this X-I phase was heated above 1000 °C at ~29 GPa by a YAG laser, well-crystallized stishovite was formed, as shown in Fig. 4.4.7. The unit-cell parameters of stishovite thus formed are in good agreement with those obtained in the CO$_2$ laser heating run, (Fig. 4.4.5) and with the literature data, as shown in Fig. 4.4.3.

The tetragonal unit-cell volumes of cristobalite observed under uniaxial compression, calculated from only three intense lines (101, 111, 102), are plotted in Fig. 4.4.8 and are compared with those observed under quasi-hydrostatic conditions. It is evident that the apparent compressibility differs considerably between these two conditions and cristobalite appears to be much more incompressible under the uniaxial condition.

Table 4.4.2. *Observed and calculated d values of stishovite at 0.1 MPa that was formed by the heating stishovitelike phase at 29 GPa*

$P = 0.1$ MPa
$a(\text{Å}) = 4.1838 \pm 0.0018$
$c(\text{Å}) = 2.6660 \pm 0.0020$
$V(\text{Å}^3) = 46.66 \pm 0.054$

h	k	l	d_{obs} (Å)	d_{cal} (Å)	$d(o)/d(c) - 1$
1	1	0	2.9572	2.9584	−0.0004
1	0	1	2.2496	2.2484	0.0006
1	1	1	1.9794	1.9805	−0.0006
2	1	0	1.8723	1.8710	0.0007
2	1	1	1.5311	1.5315	−0.0003

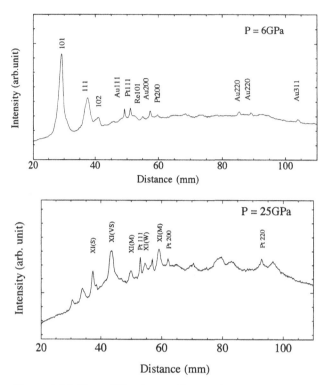

Figure 4.4.6 X-ray-diffraction profiles of cristobalite observed under nonhydrostatic conditions (a) at 6 GPa and (b) at 25 GPa transformed to the X-I phase.

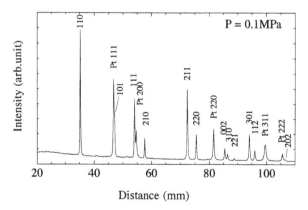

Figure 4.4.7 X-ray-diffraction profile of stishovite obtained by heating the sample shown in Fig. 4.4.6(b) to above 1000 °C by a YAG laser at ∼25 GPa and then recovered to ambient condition. Platinum was mixed as an absorber of laser energy.

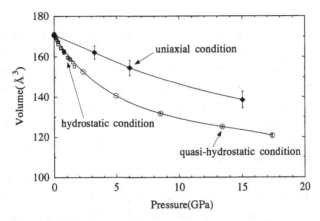

Figure 4.4.8 Apparent volume compression curves of cristobalite observed under hydrostatic (open symbols) and nonhydrostatic (solid symbols) conditions. Open squares are after [25].

When the compression was made with only a very small amount of argon pressure medium, coexistence of the X-I phase and cristobalite was observed from 15 to 20 GPa, and a mixture of the stishovitelike phase and the X-I phase was observed above 20 GPa. This means that the portion of the sample surrounded by the argon pressure medium behaved like the sample compressed under quasi-hydrostatic condition whereas the area of grain–grain contact behaved like the sample under uniaxial compression.

4.4.4 Discussion

This is one of a few experiments showing that the structure of a high-pressure phase changes completely, depending on the nature of the applied pressure. Another similar example so far reported is the transition in $CuGeO_3$, which occurs at ~7 GPa [17]. In this case, it was reported that the structure was more sensitive to the existence of nonhydrostaticity. One of the high-pressure phases was formed only when liquid-state ethanol–methanol mixture or helium was used as a pressure medium, and the other phase was formed when solidified neon or argon was used. Numerous room-temperature compression experiments on silica have been so far made under both quasi-hydrostatic and nonhydrostatic conditions. In the case of silica, however, the effect of nonhydrostatic stress is not so evident in most cases. Hemley et al. [1] compressed quartz under quasi-hydrostatic conditions by using a neon pressure medium. They observed that, above 25 GPa, diffraction peaks began to broaden and to decrease markedly in intensity and the diffraction from crystalline quartz became very weak by 30 GPa. They observed very similar amorphization when coesite was used as a starting material, although the transition pressure was slightly higher. The recovered samples remained amorphous in both cases.

Kingma et al. [4, 5, 18] made a detailed study of the amorphization of quartz under both quasi-hydrostatic and nonhydrostatic conditions. They studied the microstructure of the amorphous phase by using both electron microscope observations of recovered samples and high-pressure in situ x-ray observations. They clarified that the amorphization starts at ~15 GPa. Above that pressure, a heterogeneous sample was formed, which consists of a mixture of the amorphous phase and one or two crystalline phases. They found some differences for the sample compressed under quasi-hydrostatic and nonhydrostatic conditions above ~21 GPa. In the former case, the remaining quartz in the amorphous phase transformed into quartz II. In the latter case, in addition to the occurrence of a quartz I–II transition, they also found new sharp diffraction lines. Two crystalline phases, together with the amorphous phase, coexist from 21 to 43 GPa. Above that pressure, the diffraction from quartz II disappears and another crystalline phase persists to above 200 GPa. The behavior of the new diffraction lines is similar to that of the $CaCl_2$ structure, but the real nature of this diffraction remains unclear. Tsuchida and Yagi [19] reported the transition of stishovite into a $CaCl_2$ structure between 80 and 100 GPa under nonhydrostatic conditions. They further reported [2] the formation of poorly crystallized stishovite when quartz was compressed without a pressure medium at room temperature to above 60 GPa. Mao et al. [20] made a single-crystal x-ray diffraction study of stishovite by using a hydrogen pressure medium and observed a transition into the $CaCl_2$ structure at ~56 GPa, in agreement with the result of Raman spectroscopy under quasi-hydrostatic conditions [21].

As is clear from these examples, when the denser phases of silica such as quartz, coesite, and stishovite are used as starting materials, the results in quasi-hydrostatic and nonhydrostatic compression are similar to each other, although some differences are found in the transition pressures. On the other hand, when cristobalite is used, completely different phases are formed, depending on the hydrostaticity of applied pressure. Moreover, the stishovitelike phase reported in the present study is formed only from cristobalite. When these high-pressure phases are heated, they transform into stishovite. This means that the high-pressure phases formed by the room-temperature compression of silica are metastable, and their structures vary depending on the structure of the starting material. At room temperature, the diffusion of atoms is limited by the lack of thermal energy. The density of cristobalite at ambient condition is only 59% and 54% of that of quartz and coesite, respectively. This means that cristobalite has a very open structure, and stishovite can be formed by collapsing the structure without diffusion [6]. In quartz and coesite, on the other hand, there is no such open space and it is difficult to change the structure drastically without diffusion.

The real structure of the stishovitelike phase remains unknown. In the beginning, we thought that the pressure-transmitting medium dissolved into the structure and increased its unit-cell volume. This idea was denied by the fact that no meaningful difference was observed among the results obtained when different pressure media were used and the fact that no elements other than silicon and oxygen were found in the recovered samples. The fact that the compressibility of the unidentified diffraction line at $d = 3.5$ Å is almost identical with that of the a axis, which is the most compressible axis of this structure, strongly suggests that this line is formed by a superstructure in this direction. We therefore tried to explain the observed diffraction lines by elongating the unit cell of stishovite. When the *a* axis is tripled from the original unit cell, it is possible to index the extra line, but the difference of the observed and the calculated *d* values are still large compared with the uncertainty of the experiment [7]. It is known that the stacking faults sometimes alter the unit-cell dimension slightly. The stishovitelike phase changes into normal stishovite when it is heated up to only several hundred degrees, which means it can change into stishovite without considerable diffusion. This fact is easily understood if the structure is similar to that of stishovite and contains many imperfections. The broad nature of both x-ray- and electron-diffraction patterns of this phase made the further analysis difficult.

It is not easy to understand the mechanism of the formation of different structures as a function of the hydrostaticity of applied pressure. Figure 4.4.8 shows the compressibility of cristobalite observed under quasi-hydrostatic and nonhydrostatic compressions. This indicates that the apparent compression curves differ considerably, depending on the nature of the applied pressure. This apparent difference can be understood by the elastic strain theory of the uniaxial stress field [22]. With the geometry of a conventional diamond-anvil experiment in which the incident x ray is parallel to the main compression axis, the specimen always appears to be less compressible under nonhydrostatic conditions. The degree of difference depends on the elastic properties of the specimen, and compressible materials such as cristobalite show large differences. It is evident that a large uniaxial stress component must exist when the sample is compressed in the diamond-anvil cell without use of a pressure medium. The orientation of the grains of the sample, however, is random, and it is difficult to determine how each crystal is deformed in a uniaxial stress field. To discuss the relative stability of various phases under uniaxial compression, it is necessary to consider the free energy of each phase under a given stress field.

It is quite interesting to compare the present experimental result with that predicted by the molecular-dynamics calculation. It is clear that the present quasi-hydrostatic compression data are in close agreement with the molecular-dynamics calculation of Tsuneyuki et al. [6], whereas the result under nonhydrostatic compression is in disagreement. This is reasonable because, in the calculation, the applied pressure is assumed to be hydrostatic. The development of computer simulation has made it possible to predict the behavior of materials under extreme conditions, and it will be interesting to apply such calculations to nonhydrostatic conditions. This may help us understand the formation of the many metastable phases formed by room-temperature compressions of silica.

4.4.5 Conclusion

A structure very similar to that of stishovite was formed by room-temperature compression of cristobalite to above 20 GPa. This result is in good agreement with the predication of the molecular-dynamics calculation and is in disagreement with the result obtained by nonhydrostatic compression. This fact indicates that hydrostaticity affects not only the transition pressure but also the structure of the high-pressure phases. Many such metastable phases are formed by the room-temperature compression of various polymorphs of silica, and it may be helpful to understand these phases if theoretical calculations under nonhydrostatic conditions are made.

Acknowledgments

We are grateful to T. Kondo and S. Tsuneyuki for helpful comments and discussions throughout the study and to T. Kikegawa and O. Shimomura for various support for the experiments. The x-ray experiments were performed under the approval of the Photon Factory Advisory Committee (94G149).

References

[1] R. J. Hemley, A. P. Jephcoat, H. K. Mao, L. C. Ming, and M. H. Manghnani, Nature (London) **334**, 52 (1988).

[2] Y. Tsuchida and T. Yagi, Nature (London) **347**, 267 (1990).

[3] C. Meade, R. J. Hemley, and H. K. Mao, Phys. Rev. Lett. **69**, 1387 (1992).

[4] K. J. Kingma, R. J. Hemley, H. K. Mao, and D. R. Veblen, Phys. Rev. Lett. **70**, 3927 (1993).

[5] K. J. Kingma, H. K. Mao, and R. J. Hemley, High Pressure Res. **14**, 363 (1996).

[6] S. Tsuneyuki, Y. Matsui, H. Aoki, and M. Tsukada, Nature (London) **339**, 209 (1989).

[7] M. Yamakata and T. Yagi, Proc. Jpn. Acad. **73B**, 85 (1997).

[8] T. Yagi and S. Akimoto, in *High-Pressure Research in Geophysics*, S. Akimoto and M. H. Manghnani, eds. (CAPJ, Tokyo and Reidel, Dordrecht, The Netherlands, 1982), pp. 81–90.

[9] P. M. Bell and H. K. Mao, Carnegie Inst. Yearb. **80**, 404 (1981).

[10] T. Yagi, H. Yusa, and M. Yamakata, Rev. Sci. Instrum. **67**, 2981 (1996).

[11] H. K. Mao and P. M. Bell, Carnegie Inst. Yearb. **77**, 904 (1978).

[12] T. Yagi, Y. Uchiyama, M. Akaogi, and E. Ito, Phys. Earth Planet. Inter. **74**, 1 (1992).

[13] D. J. Palmer and L. W. Finger, Am. Mineral. **79**, 1 (1994).

[14] D. J. Parise, A. Yaganeh-Haeri, and D. J. Weidner, J. Appl. Phys. **75**, 1361 (1994).

[15] H. Sugiura and T. Yamadaya, in *Progress and Abstracts of the 30th High Pressure Symposium, Conf. in Japan* (Japan Soc. of High Pressure Sci. and Tech., Kyoto, 1989), p. 162.

[16] Y. Yahagi, T. Yagi, H. Yamawaki, and K. Aoki, Solid State Commun. **89**, 945 (1994).

[17] A. Jayaraman, S. K. Sharma, S. Y. Wang, and S.-W. Cheong, Curr. Sci. **71**, 306 (1996).

[18] K. J. Kingma, C. Meade, R. J. Hemley, H. K. Mao, and D. R. Veblen, Science **259**, 666 (1993).

[19] Y. Tsuchida and T. Yagi, Nature (London) **340**, 217 (1989).

[20] H. K. Mao, J. Shu, J. Hu, and R. J. Hemley, EOS **75**, 662 (1994).

[21] K. J. Kingma, R. E. Cohen, R. J. Hemley, and H. K. Mao, Nature (London) **374**, 243 (1995).

[22] T. Uchida, N. Funamori, and T. Yagi, J. Appl. Phys. **80**, 739 (1996).

[23] N. L. Ross, J. F. Shu, and R. M. Hazen, Am. Mineral. **75**, 739 (1990).

[24] Y. Sato, Earth Planet. Sci. Lett. **34**, 307 (1977).

[25] R. T. Downs, and D. C. Palmer, Am. Mineral. **79**, 9 (1994).

Part V

Novel Structures and Materials

Chapter 5.1

Opportunities in the Diversity of Crystal Structures – A View from Condensed-Matter Physics

Hideo Aoki

The diversity in crystal structures of materials of interest in condensed-matter physics and mineral sciences is reviewed from the viewpoint that they are an outcome of multiple ways in which building blocks (polyhedra, clusters, chains, layers, etc.) can be arranged in ambient or high pressures. Opportunities for interesting electronic properties expected from unconventional as well as conventional crystal structures are outlined, which may even include electron correlation engineering when those properties arise from the electron correlation.

5.1.1 Introduction

Although quantum mechanics laid a solid foundation for condensed-matter physics earlier this century, we are now witnessing a new era in that branch of physics in which we begin to understand crystal structures in nonempirical ways. As stressed in the Preface of this volume, this has been particularly fruitful in understanding rock-forming minerals. It is indeed fascinating if we can predict crystal structures of minerals or other materials in general from a knowledge of the chemical composition alone. This also opens up a way to search for novel crystal structures, either in denser or open-structured forms.

The purpose of this chapter is twofold. One is to take silica as an example for elaborating on various crystal structures, obtained from a nonempirical computer-simulation study. Silica comprises the two most abundant elements on the Earth, i.e., silicon and oxygen. Silica, along with silicates, is an important ingredient in rock-forming minerals. Crystal-to-crystal phase transformations, as well as pressure-induced amorphisation, are described.

H. Aoki et al. (eds), *Physics Meets Mineralogy* © 2000 Cambridge University Press.

Second, a wider overview is given of crystal structures in general, which are of interest from the condensed-matter physics point of view. One important goal to attain is how and why interesting electronic properties such as magnetism or superconductivity arise from peculiar crystal structures. For instance, high-T_C superconductivity, one of the highlights of condensed-matter physics in the 1980s and 1990s, arises from a characteristic crystal structure (i.e., layered perovskite). Indeed, a recent realisation in condensed-matter physics is that novel structures go hand in hand with novel electronic properties. Modern crystallography [1–3] and modern quantum chemistry are the basis for this.

Besides the chemical trend, an in situ way of modifying the structure is to apply high pressure. In fact, it was realised as early as the 1920s that pressure can cause a rearrangement of atomic structures [4]. Silicon, a typical semiconductor, becomes metallic when we apply a pressure >12 GPa; the material becomes even superconducting at low temperatures. At the opposite extreme lie open, zeolitelike structures, which may be thought of as being obtained by the assembling of large molecules into superstructures.

Thus the modern progress in mineralogy and crystallography can be, I believe, even reverted to facilitate condensed-matter physics – i.e., wisdom obtained in crystal-structure studies may be fed back to control electronic structures in condensed-matter physics. As it is impossible to give a comprehensive overview of such a vast subject in a short chapter, we concentrate here on some specific examples, with electronic properties briefly touched upon.

5.1.2 Polymorphism – A Case Study in Silica

Crystal-to-crystal phase transitions are ubiquitous. The transition from white (β) tin to grey (α) tin was known even to Aristotle and Plutarch. An established way of looking at the problem nowadays is the first-principles total-energy calculation of various crystal structures [5], which typically reproduces the transition from the diamond to the β-tin structure in Si (Fig. 5.1.1). Another way is the molecular-dynamics (MD) study, which has a virtue of being able to predict novel crystal structures.

We start with a case study in silica with the MD method. Silica, a mineral in itself, is interesting not only because it is a constituent of the majority of rock-forming minerals on the Earth, but also from the viewpoint of condensed-matter physics. Namely, silica has a surprisingly rich variety of crystal structures, which comes from the fact that the structures comprise SiO_4 tetrahedra, in which differences in structures correspond to different ways in which the tetrahedra are connected (Fig. 5.1.2, right panels). In high pressures, we can have SiO_6 octahedra as well, which diversify the structures even further.

A MD calculation, a collaboration among Tsuneyuki, Matsui, Tsukada, and the present author [7, 8, 14], has not only reproduced known polymorphs from non-empirical potentials [9], but has also predicted novel high-pressure forms [10].

The MD calculation starts from a nonempirical determination of interatomic potentials from a total-energy calculation for a cluster (Fig. 5.1.3). The potential may be

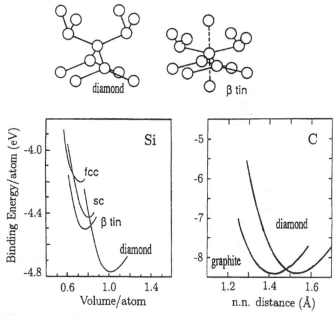

Figure 5.1.1 Example of the total-energy calculation for various polymorphs of Si or C[6]. The top panel depicts the diamond and the β-tin structures; the next-nearest neighbours are indicated by dashed lines.

accurately fitted with a pairwise interaction that comprises the Coulomb interaction, Born–Mayer-type repulsion between ionic cores, and the dispersive (van der Waals) interaction [15].

Then the interatomic (Si—Si, Si—O) potentials are fed into the MD calculation in the constant-pressure algorithm that is due to Parrinello and Rahman along with the constant-temperature algorithm that is due to Nosé. The number of atoms is naturally finite (typically 700–1000) in a computer simulation, but the shape of a unit cell is allowed to deform in the simulation under high pressures. As we look into the temporal evolution of the system, we can test the dynamical stability of structures in an MD study, unlike the static calculations such as the total-energy study.

The structures obtained from the MD study may be summarised in the descending order of density (MD result for low-temperature phases where applicable [16]) as in the following table.

Polymorph	Density (g/cm^3)	Fig. 5.1.2
C2 silica	4.24	(g)
Pa3 silica	4.20	(e)
α-PbO$_2$ silica	4.01	(h)
Stishovite	3.97	(d)
Cmcm silica	3.58	(f)
Coesite	2.72	(c)
Quartz	2.47	(a)
Cristobalite	2.41	(b)
Melanophlogite	1.67	(i)
C$_{60}$ zeolite silica	1.54	(j)

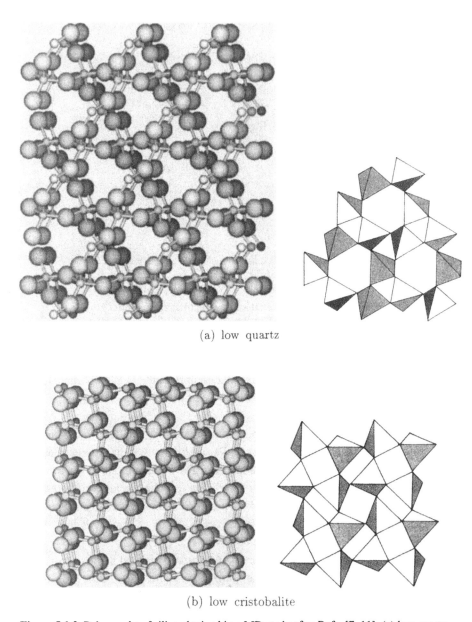

(a) low quartz

(b) low cristobalite

Figure 5.1.2 Polymorphs of silica obtained in a MD study after Refs. [7–11]: (a) low quartz, (b) low cristobalite, (c) coesite, (d) stishovite, (e) Pa$\bar{3}$ silica, (f) Cmcm silica, (g) C2 silica, (h) α-PbO$_2$ silica, (i) melanophlogite, (j) C$_{60}$ zeolite silica. Left panels are ball-and-stick representations (small spheres, Si; large spheres, O) of the time-averaged MD results; right panels schematically represent the structures with polyhedra (SiO$_4$ tetrahedra and SiO$_6$ octahedra) or skeltons of the bonding. Polyhedron representation for cristobalite is after Ref. [12]. The skeletal representation for C$_{60}$ zeolite is after Ref. [13], in which a slice of the bcc packing of the balls is depicted.

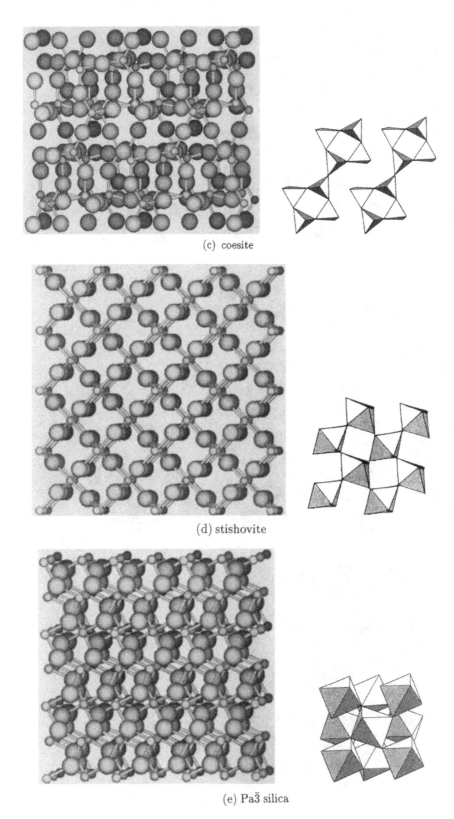

(c) coesite

(d) stishovite

(e) Pa$\bar{3}$ silica

Figure 5.1.2 (continued)

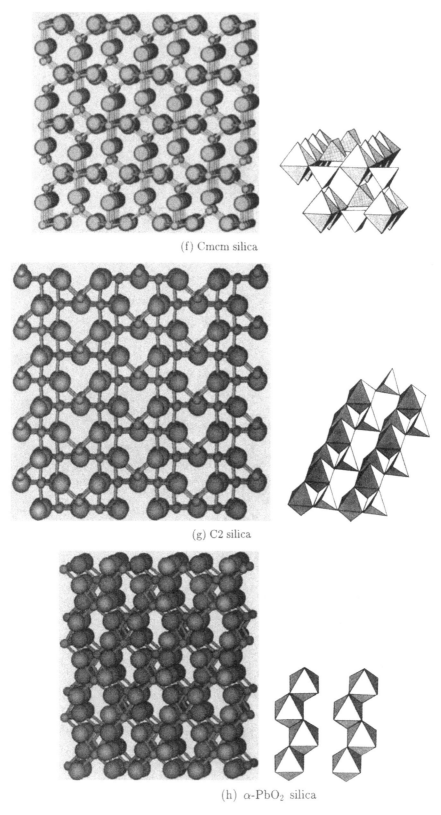

(f) Cmcm silica

(g) C2 silica

(h) α-PbO$_2$ silica

Figure 5.1.2 (continued)

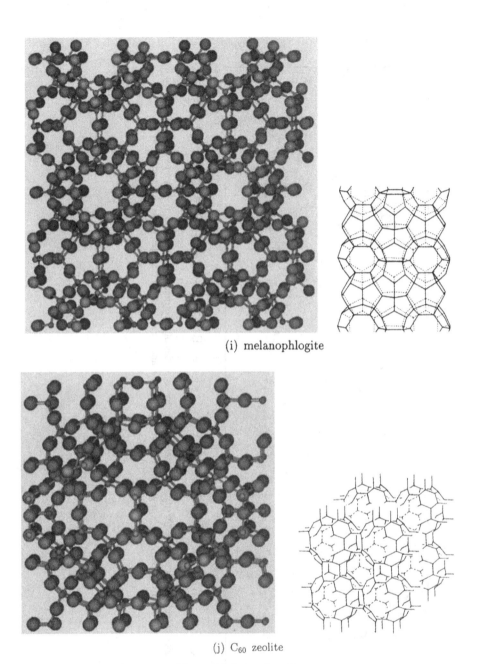

(i) melanophlogite

(j) C_{60} zeolite

Figure 5.1.2 (continued)

5.1.2.1 Conventional Structures

The MD study first reproduced the known polymorphs [17] of silica, i.e., low (α) and high (β) quartz [Fig. 5.1.2(a)], low and high cristobalite [Fig. 5.1.2(b)], coesite [Fig. 5.1.2(c)], and stishovite [Fig. 5.1.2(d)]. To help identify the idealised crystal structures, the MD result for the atomic positions averaged over a long time is shown

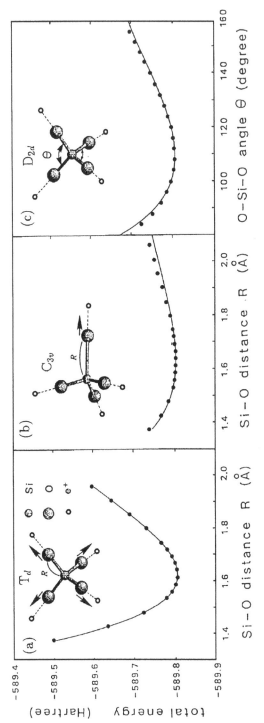

Figure 5.1.3 Total energies calculated for a cluster of silica as functions of various modes of deformation. Curves are fit with a pairwise interaction that comprises the Coulomb interaction, Born–Mayer-type repulsion between ionic cores, and the dispersive (van der Waals) interaction [9].

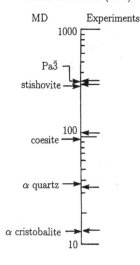

Figure 5.1.4 Bulk modulus for various polymorphs. MD results are indicated on the left. Experimental results are on the right, except for the Pa$\bar{3}$ structure where we quote a result of the full-potential linearised plane-wave calculation. Melanophlogite lies well below on this plot.

in Fig. 5.1.2. A remarkable fact is that these polymorphs are so diverse that the range of their elasticity (bulk modulus) exceeds a factor of 15 (Fig. 5.1.4).

The MD study has also revealed dynamical behaviours of polymorphs at finite temperatures. For instance, the well-known (but not too well understood) transition between low quartz and high quartz has been identified from MD to be neither the order–disorder type nor the displacive type – the transition has in fact a dynamical character, in which the high-temperature phase fluctuates between the two equivalent α_1 and α_2 phases (top panels in Fig. 5.1.5) with a temperature-dependent correlation length/time [18].

5.1.2.2 High-Pressure Forms

On the dense-polymorph side of the spectrum lies stishovite, a well-known high-pressure form of silica with sixfold Si—O coordination in a rutile-type crystal structure [Fig. 5.1.2(d)]. This polymorph is naturally occurring, as in Barringer Crater in Arizona.

In more general terms, AO$_2$-type compounds are known to take three types of structures (Fig. 5.1.6), given in the following table.

A coordination	O coordination	AO$_2$ structure	Examples
4	2	Silica structures	Quartz, cristobalite
6	3	Rutile	Stishovite, TiO$_2$
8	4	Fluorite	ZrO$_2$

Thus silica covers two rows of this table. The question, then, is this: Is rutile-type

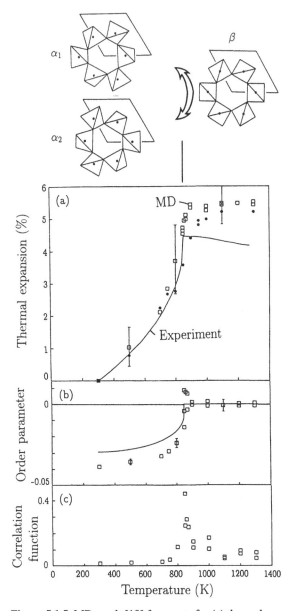

Figure 5.1.5 MD result [18] for quartz for (a) thermal
expansion (which becomes negative above T_C), (b) order
parameter (shift of the internal coordinate $\langle \bar{u} \rangle - 0.5$), and
(c) correlation function shown against temperature. The
top panel depicts the schematic structures of low quartz
and high quartz (= dynamical mixture of α_1 and α_2).

silica (stishovite) the only high-pressure phase that has coordination equal to or greater
than six?

The answer is no: Another sixfold coordinated structure, Pa$\bar{3}$, has been obtained
in an MD study by Matsui and Kawamura [19]. This structure, $\sim$6% denser than
stishovite with tilted packing of SiO$_6$ octahedra [Fig. 5.1.2(e)], was predicted to occur

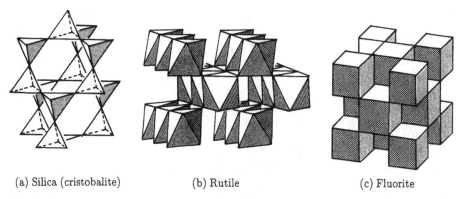

(a) Silica (cristobalite)   (b) Rutile   (c) Fluorite

Figure 5.1.6 (a) Silica structure (cristobalite here) with fourfold coordination, (b) rutile structure with sixfold coordination, and (c) fluorite structure with eightfold coordination, all represented by polyhedra.

as a phase more stable than stishovite. The total-energy calculation with the first-principles method by Park et al. [20] subsequently confirmed the stability, in which the fluorite-type silica is shown to have a higher energy. Hemley et al. [21] then found a sign for an instability of the Pa$\bar{3}$ structure in a softening of a phonon mode from a Raman experiment. Tsuchida and Yagi [22] then showed experimentally that a $CaCl_2$-type structure (having the stishovite structure with slightly different a, b sizes of the unit cell) is the stable structure. Thus the transition, if any, to the Pa$\bar{3}$ silica would have to occur at an even higher pressure.

CaCl$_2$ silica itself is of interest, as $CaCl_2$-structured dioxides have been known to occur only in β-PtO$_2$ so far. The existence of $CaCl_2$ silica is confirmed from a first-principles study by Cohen [23]. This is also the case with an MD study [24], in which the softening predicted by Hemley et al. of a B_{1g} optic phonon (along with an acoustic phonon coupled to it) is also confirmed. An interesting observation in MD is that the system around the critical pressure fluctuates between the two equivalent states (one with $a > b$, another with $a < b$), just as the high-temperature quartz fluctuates between the two equivalent α_1 and α_2 states, as described above [18]. Further, Tse et al. [25] have shown, from another MD calculation, that even richer α-PbO$_2$-type structures (see Subsection 5.1.2.4 below) are conceivable in which slabs of SiO$_6$ octahedra are stacked in a periodic manner. They conclude that the structure is comparable with the CaCl$_2$ structure and therefore may likely coexist.

5.1.2.3   Open Structures

On the low-density end of the spectrum, the MD calculation shows that the polymorph melanophlogite with a very open, zeolitelike structure is stable [Fig. 5.1.2(i)] [8, 11]. Melanophlogite, first found in the nineteenth century in Sicily, has received some attention subsequently [26]. Its structure, having a cubic space group of Pm$\bar{3}$n, is iso-morphic to a hydrate structure, as we shall come back to in Subsection 5.1.5.1, and is considered to originate from some inclusion in the material [27]. Melanophlogite's

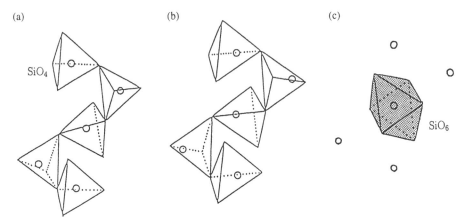

Figure 5.1.7 Elementary conversion process between fourfold and sixfold coordinations in nondiffusive crystal-to-crystal compression, exemplified here for the cristobalite to Cmcm silica transition [14]. (a) Corner-linked tetrahedra in low cristobalite, (b) deformed configuration under pressure, (c) an octahedron indicated for the same atomic positions as in (b).

density (1.67 g/cm^3 in the MD result) is more than a factor of 2.5 lower than that of Pa$\bar{3}$ silica. We can also show that another open structure [Fig. 5.1.2(j)], which is topologically the same as C$_{60}$ zeolite (discussed in Subsection 5.1.5.3 below), is stable.

5.1.2.4 Novel Silica Polymorphs

In high-pressure simulations, novel polymorphs hitherto unknown are obtained [Figs. 5.1.2(f)–5.1.2(h)]. A curious fact is that in some of high-pressure forms we have a mixture of tetrahedral (SiO$_4$) and octahedral (SiO$_6$) coordinations [Cmcm silica and C2 silica, Figs. 5.1.2(f) and 5.1.2(g)]. We can consider that such a situation becomes possible because the chemical bond in silica has both covalent and ionic characters, so that a change in the coordination number can occur at high pressures.

This has a spin-off implication for melts in the Earth's interior, as is mentioned in the subsection below on C2 silica. The ability of pair potentials to describe various phases can indeed be examined for glassy and liquid states as well. Hemmati and Angell in Chap. 6.1 of this volume [28] compare various pair potentials for liquid SiO$_2$.

These crystal-to-crystal transformations are martensitic, i.e., atomic motions are nondiffusive and reversible. The elementary process in the conversion between fourfold and sixfold coordinations is shown in Fig. 5.1.7 [14], which is rather similar to the Stolper–Ahrens mechanism considered for amorphous silica. The starting (low-pressure) polymorphs are in fact recovered by decompression in the MD simulations.

Now we take a closer look at the novel polymorphs one by one.

Cmcm Phase – An Example of Mixed Coordinations

The MD study has shown that low cristobalite exhibits a crystal-to-crystal transition at high pressures [10]. The novel high-pressure phase, obtained for the first time in MD, has a space group of Cmcm, which shares an orthorhombic subgroup C222$_1$ with low cristobalite, which has $P4_12_12$. The structure has equal numbers of fourfold and sixfold Si—O coordinations. The structure represented by tetrahedra and octahedra in Fig. 5.1.2(f), presented here for the first time, exhibits a neat, yet ingenious, array of tetrahedra and octahedra. To be precise, the arrays are somewhat corrugated, as seen in the figure.

C2 Silica – A Frustrated Crystal

Low quartz is also shown from the MD calculation to exhibit a crystal-to-crystal transition at high pressures. The high-pressure phase has a monoclinic space group of C2. This phase is of specific interest for the following reason. First, this phase again contains both fourfold and sixfold coordinations, but the fourfold Si and the sixfold Si have a ratio of 1:2, unlike the ratio of 1:1 in the Cmcm phase. We can in fact see in Fig. 5.1.2(g) that the octahedra and the tetrahedra are arrayed in yet another ingenious way.

The unusual arrangement of octahedra and tetrahedra a demands a considerable distortion of bond lengths and angles – in other words the structure is highly frustrated. This may be thought of as a frozen state on a pathway between a fourfold low-pressure phase and a sixfold high-pressure phase. The mixed geometry of tetrahedra and octahedra is also reminiscent of the intermediate state in an atomic rearrangement evoked by Tsuneyuki and Matsui [29] in their explanation of the key activation processes in silica melt. They have identified that key activation processes in a melt are rearrangements among fourfold coordinations sixfold coordinations. Thus the detailed identification of the barrier in the reaction path (e.g., whether fivefold coordinations are involved) is an interesting problem.

A telltale feature of frustration in C2 silica in fact appears in randomly distributed missing bonds with a finite density, in which we regard a Si—O bond as dislodged when its length exceeds 2.16 Å, a chemically acceptable bound. The configuration of the missing bonds fluctuates in time, and the defect density decreases with pressure as the crystal is more compressed and finally disappears above $p \sim 50$ GPa. Figure 5.1.2(g) displays this structure at higher pressures.

From Fig. 5.1.2(g) it is evident that the structure has a definite axis (along which tetrahedra are arrayed), which occurs, depending on the MD run, in one of the three equivalent crystallographic axes of the starting low quartz. Thus it is conceivable that in larger (or macroscopic) samples, an assembly of microcrystals of the C2 structure having different orientations with some domain structures can appear, which is in effect a kind of amorphous structure.

Despite the frustrated nature, the structural transition is again diffusionless. The sixfold coordinations in fact change back to fourfold almost completely on decompression.

α-PbO_2 Silica

Curiously, another pressure-induced phase appears, depending on the MD run, from low quartz, which has the α-PbO_2 structure that contains only sixfold Si—O coordinations [Fig. 5.1.2(h) and Fig. 5.1.11 below]. The way in which the structure is produced in a nondiffusive manner is illustrated in pictures of the MD run in Fig. 5.1.8. α-PbO_2 silica is slightly denser than stishovite, another structure containing sixfold coordinations only, and has been proposed as a high-pressure form by Matsui and Kawamura [19].

PbO$_2$ itself is a compound homologous to SiO_2 with a larger cation radius. α-PbO_2 is shown experimentally to transforms into a Pa$\bar{3}$ phase for $P > 7$ GPa [30].

5.1.2.5 Packing of Oxygen – A Key Factor

What factors dominate these pressure-induced transitions? We can start from observing that representing a SiO_4 unit as a tetrahedron can be misleading. Namely, in the space-filling model the ionic radius of O^{2-} (1.3Å) is much larger than that of Si^{4+} (0.4Å), so oxygen atoms occupy most of the space. The high-pressure forms must then be primarily determined by the arrangement of the bulkiest atoms.

Oxygen Packing in Ordinary Structures

Originally, ordinary structures at ambient pressures have been interpreted from the packing of oxygen ions when the oxides are analysed in terms of the ionic energy (Pauling's second principle). In this idea we classify the structures, among which are

1. fcc packing of oxygen ions as in spinel structures (Fig. 5.1.9)
2. hcp packing as in corundum structures (Fig. 5.1.10)
3. bcc packing as in rutile structures [Fig. 5.1.6(b)]

Spinel structures include silicates such as Mg_2SiO_4, a major constituent in the Earth's mantle for pressure around $p \simeq 20$ GPa, as well as a typical magnetic transition-metal oxide, magnetite Fe_3O_4 [or more precisely, in the inverse spinel form with the valence configuration $(Fe^{2+}Fe^{3+})Fe^{3+}O_4$]. The silicate comprises SiO_4 tetrahedra and MgO_6 octahedra, which are edge sharing. The Mg sites themselves (rather than the unit polyhedra) form a network of corner-shared tetrahedra, a topology the same as that of oxygen atoms in cristobalite [Fig. 5.1.6(a)]. This observation has motivated Anderson to consider the ordering of magnetic ions in ferrites ($MO \cdot M_2O_3$, M = Mn, Fe, etc.) [31].

The corundum structures, which include Al_2O_3 and hematite (Fe_2O_3), are based on the hcp packing, another close packing of spheres. The basal sheet indeed comprises a network of octahedra (Figs. 5.1.10 and 5.1.11 below).

The rutile structures, such as TiO_2 or stishovite, are based on the bcc packing, in which we have a three-dimensional (3D) staggered array of octahedra [Fig. 5.1.6(b)].

Oxygen Packing in High-Pressure Forms

The way of looking at the structure in terms of oxygen packing has a wider application, including the case of high-pressure forms. For instance, α-PbO_2 silica obtained in high-pressure MD has closely packed sheets of oxygen atoms, for which the arrangement of

(a)   (b)

(c)   (d)

(e)   (f)

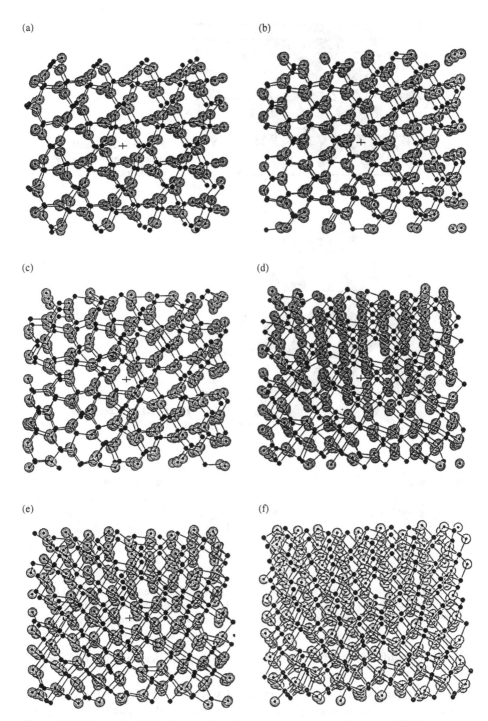

Figure 5.1.8 Example of MD pictures when low quartz is compressed at $p = 5$ GPa into α-PbO$_2$ silica (after Ref. [14]). Filled circles represent Si atoms; grey circles represent O atoms.

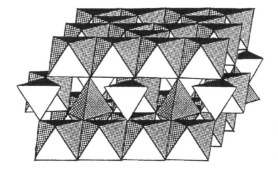

Figure 5.1.9 Spinel structure represented by octahedra and tetrahedra.

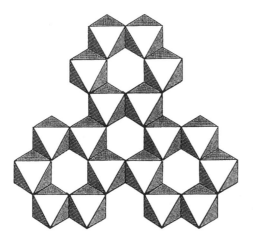

Figure 5.1.10 Corundum structure.

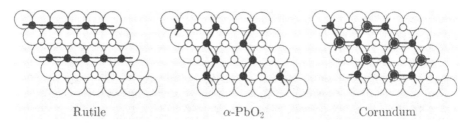

Rutile α-PbO$_2$ Corundum

Figure 5.1.11 Various arrangements of octahedral sites (Si in silica) are shown on a basal plane of the closely packed oxygen atoms (open circles). Filled circles represent one layer of the octahedral sites (lines are a guide to the eye), and small open circles represent the adjacent layer.

SiO$_6$ octahedra differs from that in stishovite, a rutile-type silica. Figure 5.1.11 shows this.

How is oxygen packed in the frustrated C2 silica [Fig. 5.1.2(g)]? Binggeli and Chelikowsky [32] have proposed an elegant way to interpret these mixed fourfold and sixfold coordinations at high pressures. They start from an observation that corner-shared SiO$_4$ tetrahedra cannot explain the large displacements of oxygen atoms even in ordinary quartz, in which oxygen atoms in fact approach the bcc stacking, as pointed out by Sowa. If we start from the bcc packing of oxygen atoms in Fig. 5.1.12

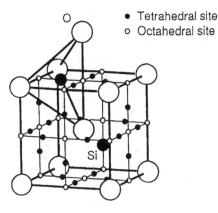

Figure 5.1.12 For the bcc packing of oxygen atoms (large circles), octahedral and tetrahedral sites for silicon are displayed (after Binggeli and Chelikowsky [32]).

to introduce Si atoms subsequently, there are many, closely spaced octahedral and tetrahedral sites for Si, which accounts for the diversity in structures in general, and the C2 silica structure in particular.

5.1.3 Polymorphs in General

5.1.3.1 Units and Layers

We can make a general question for a given material: How many polymorphs, at least in a metastable fashion, are there? For polytypes that differ in stacking sequence, a large number of polytypes (for instance, ~200 identified layered polytypes of ZnS [33]) are known to occur. Even for ordinary 3D materials a possibility of polytypes that differ in stacking sequence is now studied for low-pressure methods of synthesis [34]. In general the manner in which polymorphs occur should naturally depend on the atomic species and on the nature of their chemical bond in particular. We have a spectrum ranging from the ionic limit down to the covalent one. For the latter the transformation between diamond and graphite is the most well-known example [35].

The total-energy calculation [6] (Fig. 5.1.1) is one basis for considering the stability of polymorphs. In energetics, phonon free energy has to be considered as well [36]. For materials that are constructed from well-defined unit molecules, an attempt to analyse the vibration in terms of the rigid unit modes has been introduced by Heine and co-workers [12, 37].

Another interesting thermal effect is that some open structures have negative thermal expansion, i.e., the crystal shrinks rather than expands on being heated. Negative thermal expansion has been experimentally observed for quartz, cordierite (with beryl structure) [38], zeolite faujasite [39] [Fig. 5.1.13(b)], and other oxides such as ZrW_2O_8 (comprising a network of WO_4 tetrahedra and ZrO_6 octahedra) [40]. Figure 5.1.5 shows a result of a MD simulation that has reproduced the negative thermal expansion for quartz [18].

We can put some other elements in the voids formed by these open structures made of units. A good example is the feldspar family [1, 2], which is indeed the most

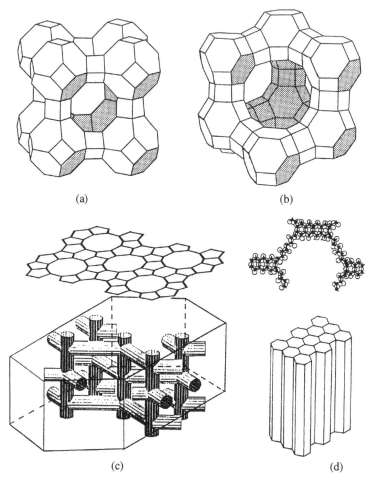

(a)   (b)

(c)   (d)

Figure 5.1.13 Zeolites in which truncated octahedron cages are connected
in (a) a cubic (Linde type A) or (b) diamond (faujasite) network. For
(c) zeolites ZSM-5 and (d) FSM-16, topologies of the structures as well
as cross-sectional atomic structures are shown.

common rock-forming mineral. In these materials, usually expressed as a triangular
diagram with $KAlSi_3O_8$, $NaAlSi_3O_8$, and $CaAl_2Si_2O_8$ on the vertices, metallic ions
are accommodated in the interstices of an array of tetrahedral units, where a fraction of
Si atoms are replaced with Al atoms to maintain the electrical neutrality. For another
class of stoichiometries richer in metallic content (feldspathoid), the arrangements of
the tetrahedra can be isomorphic to silica polymorphs, as in $(Na, K)AlSiO_4$ with a
tridymite framework.

The units of the structure can be layers. In addition to layer-type covalent materials
(such as graphite), MX_2 (M is a transition metal ranging from group IV to group VIII
elements, and X is either S, Se, or Te) is also an important class of layered materials
with many polytypes differing in the stacking of the MX_2 layers. The unit of a layer
can be either an octahedral coordination of Xs around M or a trigonal prismatic one.

Figure 5.1.14 Hollandite structure containing a bundle of hollow channels [43].

Accordingly the d-electron levels are split from the ligand field, which explains a trend that the MX_2 crystal can be metallic with E_F lying in the d_{z^2} band when the structure is prismatic and X is a group V element (as in NbS_2, which is superconducting with $T_C = 7.1$ K). A recent band calculation shows that the nature of the states around E_F comprises the $3d$ component from M and $3p$ component from X with comparable magnitudes [41]. The polytypism corresponds to the way in which the octahedral and the prismatic layers are stacked with a van der Waals gap in between.

The unit can even be one dimensional (1D). In fact a class of transition-metal compounds M_3B (such as V_3Si or Nb_3Sn) has a β-tungsten structure (known as A15 to physicists), which consists of bunches of 1D arrays of M atoms running along three (x, y, z) spatial directions. The Laves phase compounds are another example. Curiously, some of these compounds are superconducting with $T_C > 10$ K [42], called the high-T_C superconductivity before the advent of the cuprate superconductors.

Another class of transition-metal oxides contains a bundle of edge-shared octahedra that leave hollow channels, as in $K_{2-x}Mn_8O_{16}$, in hollandite structures (Fig. 5.1.14). Magnetism in such structures has been investigated [43].

5.1.3.2 High-Pressure – A Unique Tool for Realising Novel Structures

It has long been recognised that ordinary insulators like rare-gas elements such as Xe or molecular crystals such as I_2 or, most recently O_2 [44], can be made metallic if high pressure is applied. The diamond-anvil-cell method has thus turned out to be an indispensable tool for material science.

Although in usual molecular solids the metal-insulator transition can basically be understood in terms of the standard band picture, an entirely different, and generic, example is the high-pressure form of solid hydrogen, i.e., metallic hydrogen. This is a dramatic and generic example because the electrons are so strongly interacting that we have an electron-correlation problem. Some authors suggest that metallic hydrogen, with a light-mass ionic core (=proton), is a high-T_C superconductor [45].

Another interesting example is the most typical semiconductor, silicon. In the high-pressure form with the β-tin structure, silicon becomes metallic, in which the tetrahedral

sp^3 coordination in diamond structure is rearranged into a distorted sixfold coordination; that is, the next-nearest-neighbour distance (dashed lines in Fig. 5.1.1) in the β-tin structure is so close to the nearest-neighbour distance that the coordination may be regarded as sixfold. Metallic silicon is superconducting as well, in which T_C becomes maximum, 8.2 K, at 15 GPa [46]. Other elemental solids such as sulphur have been shown to be superconducting in high pressures (with $T_C = 10$ K at $p = 93$ GPa going up to 17 K at $p = 160$ GPa, which is the highest T_C for elemental solids so far detected) [47].

5.1.3.3 Common Sets of Polymorphs in Different Classes of Materials

The polymorphs described above for silica may not be just an accident for this particular material. For instance, there is an analogy between silica (SiO_2) structures and those of ice (H_2O), one of the most common materials [1]; ordinary ice (denoted as I_h) is similar to silica in high-tridymite form, and ice I_c, a form of ice near $T \simeq 140$ K, is isostructural to cristobalite.

It has also been shown that $GaPO_4$, an isostructural analogue of silica, crystallises into low cristobalite and Cmcm phases at high pressures [48]. An analogue of the Cmcm silica has also been experimentally reported for other ABO_4-type compounds such as $CrVO_4$ [49].

More recently, it was shown that cyanides such as $Cd(CN)_2$, whose structure is similar to that of high cristobalite, can take other structures that are topologically equivalent to high tridymite, stishovite, or beryl, when guest molecules are incorporated [50]. This work has been motivated by prussian blue, $Fe_4^{3+}(Fe^{2+}(CN)_6)_3$, a discovery from the eighteenth century, in which the ability of cyanides to bridge between metal atoms (Fe in prussian blue) into a complex is exploited.

The Pa3 structure, mentioned above as a high-pressure form of silica, is also ubiquitous, appearing in many molecular crystals such as dry ice (solid CO_2) (Fig. 5.1.15), the solid deuterium molecule in the low-temperature phase or the solid hydrogen molecule at high pressures [52], or solid fullerene below $T = 260$ K [53]. Kihara [51] interprets

Figure 5.1.15 Pa3 structure of a molecular crystal of CO_2, dry ice (from Kihara [51]).

the structure as a packing of ellipsoids that have the van der Waals hard cores and the electric quadrupolar interactions.

For covalent materials, a wide range of exotic structures has been examined [54]. Curiously, the structures considered for pure silicon are rather similar to those for silica. Hydrate structures shared between silicon and silica are discussed in Subsection 5.1.5.1 below. A Pa3 structure was also proposed [54] for high pressures as a binary-compound analogue of the β-tin structure for compound semiconductors such as GaAs, which originally has a zinc-blende structure. More recently, another polymorph, Cmcm structure, occurring in silica [Fig. 5.1.2(f)] was proposed as a more stable high-pressure phase of GaAs [55].

The analogies among SiO_2, H_2O, and other materials mentioned in this chapter are compiled in the following table. Interestingly, such an analogy can work both ways. After some clathrasil polymorphs that have no counterparts in hydrates were identified, the polymorphs were in fact shown to exist in hydrates [56].

SiO_2	H_2O	$Cd(CN)_2$	$GaPO_4$	PbO_2	Si	GaAs
Cristobalite	I_c	◯	◯		diamond str	zinc-blende str
Tridymite	I_h	◯				
Keatite	Ice III, IX					
Stishovite		◯				
Pa$\overline{3}$ silica				◯		◯
Cmcm silica		◯				◯
α-PbO_2 silica				◯		
Melanophlogite	Hydrates				$(Na, Ba)_x Si_{46}$	

We have stressed that silica structures can be viewed as an assembly of polyhedra. Intriguing structures that can be viewed as networks of polyhedra are also known to appear in metals or intermetallic compounds as well, with a well-known class being the Laves phases of transition-metal alloys. In fact one Laves phase (known to physicists as C15) comprises tetrahedra that have the same configuration as that of SiO_4 tetrahedra in cristobalite [Fig. 5.1.6(a)].

5.1.4 Pressure-Induced Amorphisation

It was only a decade or so ago when applying pressures was recognised as a way to convert crystals into amorphous states [57]. It was further realised that the amorphisation can occur reversibly, i.e., a crystalline state can be resumed on decompression of the solid.

It may first seem strange that a compression can destroy, rather than help, ordering of atoms. In our view, a class of pressure-induced amorphisation is closely related to the martensitic nature of some classes of the crystal-to-crystal transitions; namely, some of the novel high-pressure forms discussed for silica above have specific crystallographic axes (as exemplified above for C2 silica). If the high-pressure phase is realised in such a way that the phase is a mosaic of microcrystals having different orientations of axes, then the structure is a kind of amorphous state if the dimension of the microcrystals is of atomic scale.

We have stressed that the conversion of fourfold coordination into sixfold is a key process. Even in glasses, four-to-six conversion in the coordination can occur at high pressures as experimentally [58] and theoretically [59] shown for silica glass.

In the case of clathrasils, Tse et al. suggest that the presence of rigid units is important for the reversibility in their study of pressure-induced amorphisation [60].

In a totally different idea for crystal structures, Rivier and Sadoc [61] came up with a unique proposal that atoms are forced to be frustrated in our world of flat 3D space in that even the simplest close packing of spheres, starting from the ideal icosahedron, cannot continue in a periodic fashion in three dimensions, whereas this would be possible if the 3D space were curved fictitiously. So in the ordinary flat space a series of topological defects (disclinations or dislocations) has to occur in which the various ways of introducing them correspond to various crystal structures. The topological defects can occur (1) periodically (which gives a description of Laves phases and A15 structures), (2) in a self-similar manner (resulting in quasicrystals), or (3) in a disordered manner (resulting in amorphous states). It would be interesting if we could extend this view to high-pressure phases for which the pressure may be regarded as controlling the manner in which the atoms are frustrated.

5.1.5   Superstructures

5.1.5.1   Hydrate Structures – Silica and Silicon Clathrates

We have mentioned that melanophlogite is topologically identical to a common (type I) clathrate hydrate structure. Hydrate is a crystal structure comprising a 3D periodic network of H_2O molecules that encapsulate the solute atoms or molecules. Hydrates containing CH_4 molecules, for instance, naturally occur abundantly in Siberia, Canada, and other places. Although the discovery of hydrates dates back to 1810, their crystal structures were explored only in the 1940s by Stackelberg and were determined in the 1950s [62, 63]. We have mentioned that silica and ice have common polymorphs. Thus melanophlogite is a case in which the analogy extends to hydrate structures. A systematic study led to a notion that silica can take a class of clathrate structures, called clathrasils [64].

Si can also take clathrate structures. Indeed, an example $(Na, Ba)_x Si_{46}$ [Fig. 5.1.16(a) [66]] has a Si network that is topologically the same as that of melanophlog-ite $(SiO_2)_{46}$ [Fig. 5.1.2(i) and right panel of Fig. 5.1.16(a)], having a type I clathrate hydrate structure. In the Si clathrate, arrangement of the metallic atoms and Si_{20} cages is identical to that of the solid fullerene Ba_3C_{60} [66], in the A15 phase, which has a space group Pm$\bar{3}$n. Remarkably, $Na_2Ba_6Si_{46}$ is superconducting with $T_C \simeq 2.5$ K. Although this reminds us of the superconductivity in the solid C_{60} doped with

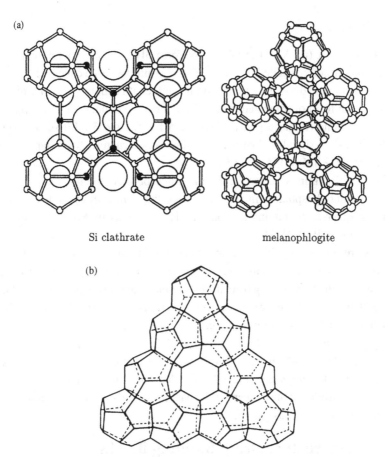

Si clathrate   melanophlogite

Figure 5.1.16 Si clathrate structures: (a) $(Na, Ba)_x Si_{46}$ [65], which has a
type I clathrate hydrate structure displayed along with melanophlogite [11],
(b) type II hydrate structure [1] adopted by $Na_x Si_{136}$.

alkali–metal atoms, an important difference is that in the Si clathrates there is a rigid
network of Si–Si bonds in which metallic atoms are incorporated, whereas there are
only weak van der Waals bonds between the molecules in the solid fullerene.

As seen from the atomic structure, the Si–Si bonding in Si clathrates is primarily
tetrahedral (i.e., sp^3 bonding), unlike in the superconducting elemental Si in its high-
pressure (β-tin) form. From a band-structure calculation for $Na_2 Ba_6 Si_{46}$, the density
of states around the Fermi energy is shown to be enhanced because of the hybridisation
of the Ba orbitals [65].

Si can take another of typical clathrate hydrate structure, type II [63], with a stoichiometry of $Na_x Si_{136}$ [Fig. 5.1.16(b)]. So the analogy between hydrates and clathrasils
is rather extensive. The material is metallic for $x > 10$, but superconductivity has not
been found so far.

A different class of intermetallic compounds of Si, a Zintl phase, can form rather
open structures (such as an α-ThSi$_2$ structure) [67], some of which, like LaSi$_2$, are
superconducting.

5.1.5.2 Zeolites

The ability of SiO_4 (or AlO_4) tetrahedra to form 3D networks can be exploited to construct open structures called zeolites [68] that comprise large cages. The discovery of zeolite as a mineral dates back to the middle eighteenth century [69]. A typical class has cages in the form of truncated octahedra that can be connected in a cubic (Linde type A) or diamond (faujasite) network [Figs. 5.1.13(a) and 5.1.13(b)]. Zeolite from pure silica is called silicalite [70], whose framework structure is similar to that of a synthetic zeolite, ZSM-5 [Fig. 5.1.13(c)].

Beside these aluminosilicates, zeolitelike microporous structures can also be realised in aluminophosphates. In this case, too, the connectivity of the framework structures is diverse, containing a bundle of 1D chains as in $Al_{12}P_{12}O_{48}$ or layered structures as in $Al_3P_4O_{16}$ [71].

Zeolites are interesting also from the viewpoint that they can be made to contain molecules or clusters of atoms as guests in the cages of the zeolite structures [72, 73]. A remarkable finding, intriguing from the electronic structure, is that clusters of alkali–metal atoms in some zeolites exhibit ferromagnetic behaviour [72], whereas the bulk alkali metals are of course paramagnetic.

A class of zeolites such as $Na_nSi_{96-n}Al_nO_{192}$ [Fig. 5.1.13(c)] is another example of the packing of rods, which are hollow in this case. A more recently synthesised example made from kanemite is displayed in Fig. 5.1.13(d). Incidentally, Andersson and O'Keeffe [74] introduced the concept of the packing of rods to aid in the understanding of silicate structures such as garnet.

5.1.5.3 Superstructures from Building Blocks

A way to make zeolitelike structures may be to polymerise cagelike molecules into 3D networks. Fullerene is an ideal cage. A number of authors have proposed zeolitelike structures from fullerene under the name C_{60} zeolite [Fig. 5.1.2(j)] [13] or negative-curvature fullerene [75].

When the atomic network is smoothed into a continuous membrane, these sponge or labyrinth structures define curved surfaces. Some structures, such as C_{60} zeolite, are topologically similar to what mathematicians call minimal surfaces. A surface is called minimal when the mean curvature (sum of the two principal curvatures) vanishes everywhere on the surface. Periodic minimal surfaces are particularly interesting in condensed-matter physics, as they extend over the whole space with a periodic structure, so we may call them a crystal composed of surfaces. MacKay and Terrones have actually classified them group theoretically (just as the ordinary crystals composed of atoms are classified by the space group); they coined the term flexicrystallography, meaning that it has to do with curved surfaces. An attempt has been made to solve Schrödinger's equation on these curved surfaces [77].

Experimentally, fullerene polymers have been synthesised by various methods that include (1) photopolymerisation [78], (2) high-pressure polymerisation [79, 80], and (3) alkali–metal intercalated polymerisation [81].

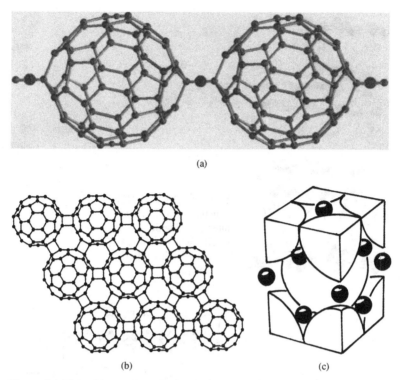

(a)

(b)   (c)

Figure 5.1.17 (a) Linear [82] and (b) 2D [79] fullerene polymers are
schematically shown along with (c) a 3D solid fullerene, M_3C_{60} (after Ref. [83]).
In (c) the solid spheres are alkali–metal atoms, and the large spheres represent
C_{60} molecules.

1D and two-dimensional (2D) polymers have been synthesised with method (1),
and 1D polymers have also been formed with method (3). These 1D and 2D structures
are shown in Fig. 5.1.17 along with a 3D structure (M_3C_{60}) [83]. Electronic band
structure has been calculated for these crystals [82].

Organic molecules, which are also assembled in a 3D fashion, comprise a vast field
in themselves. It is possible to construct network structures with open channels from
organic molecules (such as porphyrin) [84].

For layered structures, intercalation [85] is a way to control the system. This has
been most typically done for graphite or transition-metal dichalcogenides. Both of these
systems can be superconducting. One of the most recent classes of superconductors
is $CdCl_2$-structured compounds of transition metals with nitrogen and halogen, such
as β-HfNCl (Fig. 5.1.18) [86], which, when intercalated with alkali–metal atoms, has
$T_C = 25.5$ K, which is higher than that of Nb_3Ge, an A15 compound.

There have also been attempts [87] at intercalating layer-type silicates, i.e., clay
minerals [Fig. 5.1.19(a)] [88], such as vermiculite. A clay mineral (montmorillonite
here) is also used to intercalate arrays of rods, i.e., the aluminosilicate nanotubes
imogolite [Fig. 5.1.19(b)] [89].

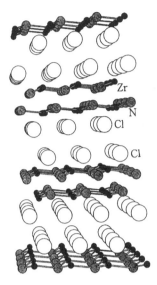

Figure 5.1.18 β-HfNCl structure, which, when intercalated with alkali–metals, becomes superconducting (after Ref. [86]).

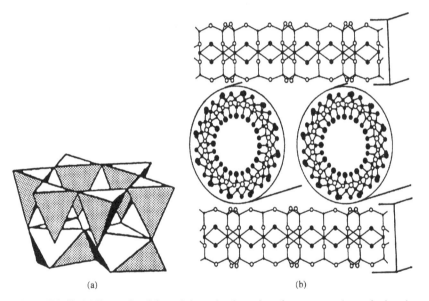

(a)   (b)

Figure 5.1.19 (a) Example of the unit layer in clay mineral structures. A tetrahedron is SiO_4, and an octahedron is an Al or Mg atom surrounded by six oxygen atoms or a hydroxyl group. (b) A clay mineral (montmorillonite here) intercalated with arrays of aluminosilicate nanotubes imogolite [88].

5.1.6   Metal–Insulator Transition – An Example of the Electron-Correlation Effect

Mott's 1949 paper, which proposed what is now called Mott's transition, was a land-mark in solid-state physics because the concept opened up a whole area of electron-correlation problems. Although most of the electronic properties of solids can be understood in terms of the electronic band structure, there is a conspicuous class of materials, typically transition-metal oxides such as NiO, that defies the band-structure description – these materials should be good metals from the band theory, whereas they are, in actual fact, insulators. A most recent example is a class of cuprates such as La_2CuO_4, the host material (i.e., the material before it is doped with other elements) of the high-T_C superconductivity. Again, they should be metals from the band theory whereas they are, in fact, insulators before being doped.

The insulating property cannot be understood when the electron–electron interaction is considered in a mean-field theory or even when the antiferromagnetic spin ordering is considered (because the materials remain insulating above the Néel temperature). Because of the strong Coulomb repulsion among the electrons in d orbitals in the transition elements, electrons have to move in a strongly correlated manner, which is indeed the cause of the insulating behaviour, and this kind of physics is nowadays called electron correlation.

There are many transition-metal oxides that fall into this category. V_2O_3, whose structure is corundum (Fig. 5.1.10), is a textbook example. The material does indeed undergo a metal–insulator transition at $T_C = 150$ K, at which antiferromagnetism emerges at low temperatures, as expected for a Mott's insulator.

Its brother compound, VO_2, is nonmagnetic and less simple. Its crystal structure, rutile [Fig. 5.1.6(b)] at higher temperatures, degrades into monoclinic at lower T, and the insulating electronic structure has been interpreted in terms of either an electron-correlation effect (Mott's insulator) or a one-electron picture (spin-Peierls or band insulator). Recently Wentzcovitch et al. [90] examined this by using constant-pressure MD in which the shape of the unit cell was allowed to deform. They concluded that VO_2 can be understood from band theory.

Perovskite has been an important class of structures, both in geophysics and in condensed-matter physics. The structure [Fig. 5.1.20(a)] comprises a cubic packing of SiO_6 octahedra, which are usually distorted from the ideal cubic symmetry (except in ferroelectric $BaTiO_3$).

In the geophysical context, perovskite is a typical high-pressure phase of silicates such as $MgSiO_3$ in the Earth's mantle for pressure near $p \sim 25$ GPa. Theoretically, when and how the perovskite structure appears in the ABO_3 stoichiometry has been discussed from the time when Goldschmidt introduced the idea of 'tolerance factor' in the 1920s. Recent studies include MD calculations [92] and the energetics of the displacive transition in the perovskite structure [93].

As for the electronic properties, magnetism is the central issue in transition-metal oxides in a perovskite structure. Most recently, half-metallic ferromagnetism (i.e., metallic for one spin component, insulating for the other) has been proposed for manganese perovskite, $La_{1-x}Sr_xMnO_3$ [94]. More recently, Kuwahara et al. have experimentally shown that an interesting class of charge ordering takes place in some transition-metal oxides: This can be controlled by magnetic fields, resulting in what they call giant or colossal magnetoresistance [95].

High-T_C cuprates, discovered by Müller and Bednorz to be superconducting, have layered perovskite structures (Fig. 5.1.21). Immediately after the discovery, Anderson made a seminal proposal that the superconductivity might come from the strong

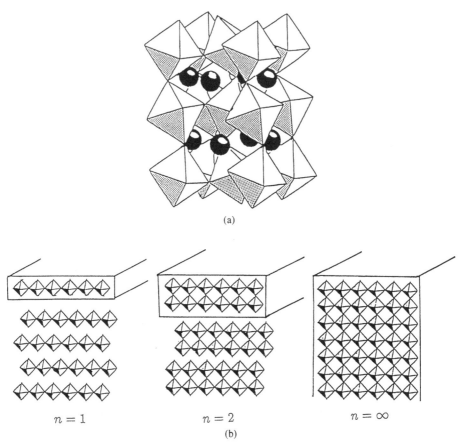

(a)

$n = 1$ $n = 2$ $n = \infty$

(b)

Figure 5.1.20 (a) Perovskite structure AMO_3 with MO_6 octahedra and A atoms (solid spheres), (b) structures of $La_{n+1}Mn_nO_{3n+1}$ that have $n = 1, 2, 3, \ldots$ layers of CuO_2 planes sandwiched between the ionic layers are shown schematically (after Ref. [91]).

electron correlation. Anderson's theoretical approach has deep roots in the physics of the Mott transition, as reviewed in Ref. [96]. Tokura's group [91] has also looked into a series of layered perovskite structures $(La_{n+1}Mn_nO_{3n+1})$ that have $n = 1, 2, 3, \ldots$, layers of CuO_2 planes sandwiched between the ionic layers, as schematically shown in Fig. 5.1.20(b).

5.1.7   Electron-Correlation Engineering in Novel Structures

5.1.7.1 One-to-One Correspondence between the Crystal Structure and the Band Structure

From the condensed-matter physics point of view, a most interesting implication is that diverse crystal structures can offer ample opportunities for novel electronic structures. For one-electron band structures, the band has a one-to-one correspondence with the crystal structure. We can then switch on the electron–electron interactions to look for more fascinating electronic properties.

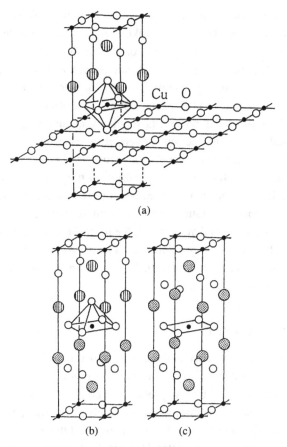

Figure 5.1.21 High-T_C cuprates with layered perovskite
structures. Each plane of CuO_2 consists of a layer of
CuO (a) octahedra (as in La_2CuO_4), (b) pyramids (as in
$YBa_2Cu_3O_{7-x}$), or (c) squares (as in Nd_2CuO_4).

Even on the textbook level, the tight-binding model, in which electrons are assumed
to hop between localised orbitals, is seen to accommodate quite diverse band structures. In the tight-binding band structure in three dimensions with nearest-neighbour
hopping, a bcc lattice has a squared logarithmic divergence at the band centre, whereas
fcc and hcp lattices have logarithmic divergences at the band bottom [97], where the
divergence is an example of the van Hove singularity. If hoppings further than near-
est neighbours are introduced, the divergence persists, with its position shifting away
from the bottom or centre. The peculiar divergence for fcc was in fact an important
motivation when Kanamori constructed an electron-correlation theory of magnetism
for transition metals [98].

Even for 2D lattices, tight-binding band structures are quite diverse: A square
lattice has a logarithmic divergence in the density of states at the band centre, a
honeycomb lattice has a pair of k-linear dispersions that form zero-gap semimetallic
bands (or zero-mass Dirac particle in field-theoretic language), and a Kagomé lattice
has a flat (i.e., dispersionless) band at the bottom.

In more general terms, electronic structures of low-dimensional systems are of great interest [99]. The chapter by Tsuneyuki et al. (Chap. 5.2) in this volume also starts from such a viewpoint. In this section some examples are shown in which novel structures give rise to surprisingly rich electronic properties such as ferromagnetism or superconductivity. It gradually becomes clearer that ferromagnetism or electron-mechanism superconductivity comes from electron correlations. Electron correlation is a term in condensed-matter physics that stands for quantum effects that originate from the correlated motion of electrons governed by electron–electron interactions that cannot be described in terms of mean-field pictures. A historical invention of such a concept arose for the metal–insulator transition in transition-metal oxides as put forward theoretically by Mott, but now we begin to realise that the spectrum of the electron-correlation problem is much wider than hitherto suspected.

An important observation is that, although the mean-field-theoretic band picture is incapable of describing electron-correlation effects, the way in which electron correlation is exerted does depend sensitively on the the starting one-band structure. Some examples are given in the next two subsections.

5.1.7.2 Ferromagnetism

One example is the relation between the crystal structure and ferromagnetism. The problem of itinerant ferromagnetism (i.e., the polarised spin states in metals) has been a long-standing, and one of the toughest, problems in condensed-matter physics. Alignment of spins decreases the electron–electron repulsive interactions (due to Pauli's exclusion principle) but increases the kinetic energy, and the trade-off usually works unfavourably for the ferromagnetic states. Historically, Stoner's criterion claims, from mean-field theory, that ferromagnetism should be obtained when $U D(E_F)$, the repulsive interaction U times the density of states around the Fermi energy, exceeds unity. Ferromagnetism in transition metals (Ni, Co, Fe, etc.) is basically understood along this line, in which the modern implementation of the idea is the spin-density functional formalism. If we pursue the electron-correlation problem, however, the more precisely we treat the electron–electron interaction, the more stringent the criterion for ferromagnetism becomes.

Kanamori's picture for ferromagnetism was mentioned above. We can indeed have a modern look at the link between the crystal structure and magnetism with analytic methods such as the fluctuation exchange (FLEX) approximation. If we survey 3D lattices (fcc, bcc, as well as simple cubic), the fcc, with the divergence in the density of states around the band bottom, is seen to have the strongest tendency towards ferromagnetism [100].

One unique approach is due to Lieb [101], who noted that a special class of crystal structures has the ground state with ferromagnetism (or ferrimagnetism, to be more precise). This magnetism is obviously in a different league from that of Stoner's, as the former extends surprisingly over the Hubbard repulsion $U = $ infinitesimal up to $U = \infty$. The required lattice structures are the bipartite lattice (in which the atoms can be divided into A and B sublattices and the electron hopping in the tight-binding model occurs only between an A site and a B site) that has different numbers of A and B sublattice sites. Another way of expressing the condition is that the one-electron band structures contain flat (dispersionless) band(s). Later, different classes of structures were proposed by Mielke and by Tasaki [102], in which the flat band(s) are located at the bottom, on which the spins align ferromagnetically. This magnetism has been shown [103, 104] to be robust in that the spin stiffness (the energy required for twisting the aligned spins) is finite regardless of $U = $ infinitesimal or $U = \infty$, or the ferromagnetism survives a degradation of the flatness of the bands. We can also relax the condition of the on-site (Hubbard) repulsion to near-neighbour repulsions, which can even assist the ferromagnetism [105].

A next question is obviously how to prepare these structures systematically. Shima and Aoki have shown, from group theory, that a class of superstructures, which may be viewed as a periodic array (or network) of clusters, does indeed have the required flat bands [106] (Fig. 5.1.22). These ideas can be extended to 3D crystals. Fujita et al.

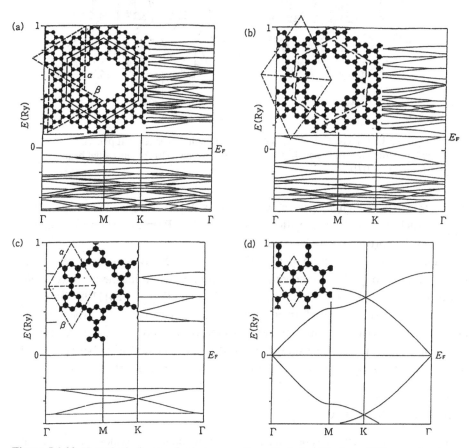

Figure 5.1.22 Hypothetical examples in a group-theoretical classification of 2D superstructures in which classes (c) and (d) have flat bands, as shown (after Ref. [106]). Dashed triangles indicate unit structures.

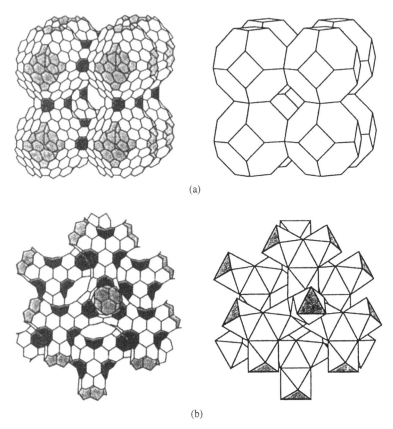

(a)

(b)

Figure 5.1.23 Hypothetical examples of 3D graphitic sponges (left panels: atomic structures; right panels: their topology) in which (b) contains a flat band (from Ref. [107]).

have considered zeolitelike open carbon structures (Fig. 5.1.23) made from graphitic sheets (which they call Pearcene after the architect Peter Pearce) to show that some of them [Fig. 5.1.23(b)] contain flat bands [107].

Another proposal came from quite a different avenue. Many cuprates were synthesised in search of high-T_C superconductors and in one case ferromagnetism was found. The material is La_2BaCuO_5, which has a set of 1D chains (Fig. 5.1.24) [108] for which a band calculation has been performed [109]. Tasaki [110] interprets this to be a realisation of the flat-band magnetism, although further theoretical investigation is needed for this identification. Yet another example for the flat-band magnetism has been proposed for an array of atoms (or atomic quantum wire) fabricated on a surface [111].

5.1.7.3 Superconductivity

Among the vast variety of superconductors, superconducting oxides are of special interest in the present context. Among a number of oxide superconductors [112], materials with perovskite structures such as $SrTiO_{3-\delta}$ or $Ba_{1-x}K_xBiO_3$ are well known,

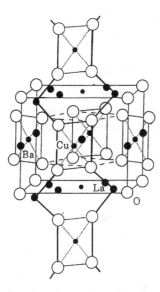

Figure 5.1.24 Structure of a cuprate, La_2BaCuO_5, that has a set of 1D chains (after Ref. [108]).

although spinel-structured superconductors such as $Li_{1+x}Ti_{2-x}O_4$ also exist. The high-T_C cuprates, typically La_2CuO_4, belong to layered perovskite structures, which have long been known as K_2NiF_4 structures. More precisely, each plane of CuO_2 comprises CuO octahedra (as in La_2CuO_4), pyramids (as in $YBa_2Cu_3O_{7-x}$), or squares (as in Nd_2CuO_4), and the layer is sandwiched between ionic layers [Fig. 5.1.21]. There is also a body of experimental work for fabricating structures that have $n = 1, 2, 3, \ldots$ layers of CuO_2 planes sandwiched between the ionic layers.

Intercalation (such as iodine intercalation into $Bi_2Sr_2CaCu_2O_y$) has been attempted to control the layer distance in these layered perovskites [113, 114]. Surprisingly, a recent paper [115] reports that T_C does not change appreciably, even when the interlayer distance is expanded to as large as 45 Å.

As for the mechanism of the superconductivity, a consensus has still not been reached. However, the following picture is emerging. The host material, such as un-doped La_2CuO_4, is an insulator purely because of the electron–electron repulsive interactions, whereas the usual band calculation predicts a metal. In other words, the system is a Mott's insulator. When the system is doped with holes or electrons, either by doping other elements or by controlling the oxygen deficiency, the system becomes metallic and superconducting below T_C, which exceeds 100 K for some cuprates. The metal has a well-defined Fermi surface, as detected from angle-resolved photoemission spectroscopy. The shape of the Fermi surface differs from one cuprate to another.

A most interesting point is whether the superconductivity arises from the electron–electron repulsive interactions. There is a body of theoretical work on this point, in which we usually start from a model that incorporates the copper $3d$ orbitals and the oxygen $2p$ orbitals (a $d–p$ model) or simpler models derived from that. Some quantum Monte Carlo studies, a powerful method for treating the electron-correlation effect numerically, have been performed; the conclusion by Kuroki and Aoki is that the 2D $d-p$ model has indeed a sign to superconduct [116], which continues to be the case with a simpler one-band Hubbard model in two-dimensions, a generic model [117].

From the point of view of the correspondence between structures and electronic properties, the concept of fermiology in correlated electron systems in general and the relation of Cooper pair symmetry with the fermiology in particular emerged from the above works [118]. Namely, the important process for electron correlation (the pair-hopping processes for superconductivity) is a sensitive function of the shape of the Fermi surface (i.e., whether the surface is close to a van Hove singularity or a special point in the Brillouin zone, etc.). So here again we have an opportunity for electron-correlation engineering. Physically, the superconductivity from repulsive interactions is mediated by spin fluctuations (which are usually antiferromagnetic), in which the hopping of Cooper pairs is a key process. An important point is that Cooper pairing is anisotropic (d wave), for which the pair-hopping interaction works as an attraction. The spin structure (spin–spin correlation function) is of crucial interest in this context. Experimentally the spin structure has been measured from neutron scattering, with an interesting asymmetry in the doping dependence over electron- and hole-doped regimes, which can be interpreted in terms of a FLEX approach [119]. Also, the superconducting phase has to compete with magnetic ones, and there is a numerical suggestion that the superconductivity will give way to antiferromagnetism when the doping level is too close to the Mott transition [120].

Another avenue that emerged from the high-T_C cuprates is that of the ladder systems. In a class of cuprates, typically $Sr_{n-1}Cu_nO_{2n-1}$, there is a layer that comprises ladder-shaped oxide chains (Fig. 5.1.25) [121]. A spin-off of a theory for high-T_C

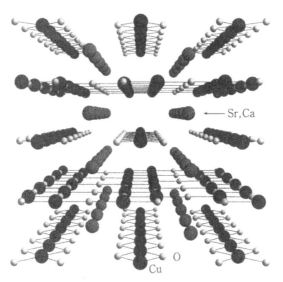

Figure 5.1.25 Example of a ladder cuprate, $Ca_{14}Cu_{24}O_{42-x}$, that contains a layer comprising ladders.

cuprates predicted the occurrence of electron-correlation-originated superconductivity in ladder structures that have an even number of legs. Superconductivity has in fact been detected for a two-legged compound [122]. Cooper pairing has also been detected from a quantum Monte Carlo study [123], which is found to be the case even for an odd number of legs [124, 125]. Recently it was suggested that the ladder compound should be considered as a 2D system, but the fascination towards various crystal structures in cuprates continues.

Acknowledgments

It is an utmost pleasure to acknowledge Yoshito Matsui for introducing me to the fascination of mineralogy and for collaboration on a number of papers. I also thank him for numerous discussions and correspondences on classical music, which belongs, after all, to the same branch of human activity as the natural science from the Greek era. I also thank him for a critical reading of the manuscript. I also acknowledge Shinji Tsuneyuki for invaluable collaborations and discussions for a number of years and also for help in preparing this manuscript. I also extend my thanks to Volker Heine, Renata Wentzcovitch, Nick Rivier, Alan MacKay, Ron Cohen, Yoshinori Tokura, Mitsutaka Fujita, Yasuo Nozue, and Toschitake Iwamoto for discussions on crystal structures at some time or other in these past several years. Collaborations with Kazuhiko Kuroki on superconductivity, Koichi Kusakabe on ferromagnetism, Nobuyuki Shima on group theory for superstructures, and Takashi Kimura and Ryotaro Arita on ladders have also been appreciated.

References

[1] A. F. Wells, *Structural Inorganic Chemistry*, 5th ed. (Clarendon, Oxford, U.K., 1984).

[2] C. Klein and C. S. Hurlbut, Jr., *Manual of Mineralogy*, 21st ed. (Wiley, New York, 1977).

[3] B. K. Vainshtein, V. M. Fridkin, and V. L. Indenbom, *Structure of Crystals*, 2nd ed. (Springer, New York, 1995).

[4] P. W. Bridgeman, Z. Kristallogr. **67**, 363 (1928); L. Pauling, Z. Kristallogr. **69**, 35 (1928).

[5] S. Froyen and M. Cohen, Phys. Rev. B **28**, 3258 (1983); S. B. Zhang and M. Cohen, Phys. Rev. B **35**, 7604 (1987).

[6] C. Z. Wang and K. M. Ho, in *Advances in Chemical Physics*, I. Prigogine and S. A. Rice, eds. (Wiley, New York, 1996), Vol, XCIII, p. 651.

[7] S. Tsuneyuki, M. Tsukada, H. Aoki, and Y. Matsui, in *Dynamic Processes of Material Transport and Transformation in the Earth's Interior*, F. Marumo, ed. (Terra Scientific, Tokyo, 1990), p. 1.

[8] S. Tsuneyuki, H. Aoki, and Y. Matsui, in *Computer Aided Innovation of New Materials*, M. Doyama et al., eds. (Elsevier, New York, 1991), p. 381.

[9] S. Tsuneyuki, M. Tsukada, H. Aoki, and Y. Matsui, Phys. Rev. Lett. **61**, 869 (1988).

[10] S. Tsuneyuki, Y. Matsui, H. Aoki, and M. Tsukada, Nature (London) **339**, 209 (1989).

[11] H. Aoki and S. Tsuneyuki, Nature (London) **340**, 193 (1989).

[12] M. T. Dove et al., Ferroelectrics **136**, 33 (1992); A. P. Giddy et al., Acta Crystallogr. A **49**, 697 (1993); K. D. Hammonds et al., Am. Mineral. **81**, 1057 (1996).

[13] M. O'Keeffe, Nature (London) **352**, 674 (1991); M. O'Keeffe, G. B. Adams, and O. F. Sankey, Phys. Rev. Lett. **68**, 2325 (1992); D. Vanderbilt and H. Tersoff, Phys. Rev. Lett. **68**, 511 (1992).

[14] S. Tsuneyuki, H. Aoki, M. Tsukada, and Y. Matsui, in *Proceedings of the 20th International Conference on Physics of Semiconductors*, E. M. Anastassakis and J. D. Joannopoulos, eds. (World Scientific, Singapore, 1990), p. 2221.

[15] Approaches beyond the pairwise potentials have been sought for semiconductors and metals by, e.g., A. E. Carlsson in *Solid State Physics*, H. Ehrenreich and D. Turnbull, eds. (Academic, New York, 1990), Vol. 43, p. 1.

[16] The density for C2 (Cmcm) silica is quoted for $p = 50$ (15) GPa, as they cannot be quenched at ambient pressure, whereas the other values are for ambient pressure. This is why C2 silica appears to be denser than Pa$\bar{3}$ silica.

[17] Band structures of polymorphs of silica have been obtained with the first-principles method by F. Liu et al., Phys. Rev. B **49**, 12528 (1994).

[18] S. Tsuneyuki, H. Aoki, M. Tsukada, and Y. Matsui, Phys. Rev. Lett. **64**, 776 (1990).

[19] Y. Matsui and K. Kawamura, in *High-Pressure Research in Mineral Physics*, M. H. Manghnani and Y. Syono, eds. (Terra Scientific, Tokyo, 1987), p. 305; Y. Matsui and M. Matsui, in *Structural and Magnetic Phase Transitions in Minerals*, S. Ghose, J. M. D. Coey, and E. Salje, eds. (Springer-Verlag, New York, 1988), p. 129.

[20] K. T. Park, K. Terakura, and Y. Matsui, Nature (London) **336**, 670 (1988); K. Terakura, K. T. Park, and Y. Matsui in *Dynamic Processes of Material Transport and Transformation in the Earth's Interior*, F. Marumo, eds. (Terra Scientific, Tokyo, 1990), p. 23.

[21] R. J. Hemley et al., Eos Trans. Am. Geophys. Union **66**, 357 (1985); R. J. Hemley, in *High-Pressure Research in Mineral Physics*, M. H. Manghnani and Y. Syono, eds. (Terra Scientific, Tokyo, 1987), p. 347.

[22] Y. Tsuchida and T. Yagi, Nature (London) **340**, 217 (1989).

[23] R. E. Cohen, in *High-Pressure Research: Application to Earth and Planetary Sciences*, Y. Syono and M. H. Manghnani, eds. (Terra Scientific, Tokyo, 1992).

[24] Y. Matsui and S. Tsuneyuki, in *High-Pressure Research: Application to Earth and Planetary Sciences*, Y. Syono and M. H. Manghnani, eds. (Terra Scientific, Tokyo, 1992), p. 433.

[25] J. S. Tse et al., Phys. Rev. Lett. **69**, 3647 (1992).

[26] B. Kamb, Science **148**, 232 (1965).

[27] H. Gies, Z. Kristallogr. **164**, 247 (1983).

[28] M. Hemmati and C. Austin Angell, Chap. 6.1 of this volume.

[29] S. Tsuneyuki and Y. Matsui, Phys. Rev. Lett. **74**, 3197 (1995).

[30] J. Haines, J. M. Léger and O. Schulte, J. Phys. Condens. Matter **8**, 1631 (1996).

[31] P. W. Anderson, Phys. Rev. **102**, 1008 (1956).

[32] N. Binggeli and J. R. Chelikowsky, Nature (London) **353**, 344 (1991).

[33] S. Mardix, Phys. Rev. B **33**, 8677 (1986).

[34] A. K. Sharma et al., J. Phys. Condens. Matter **8**, 5801 (1996).

[35] The nucleation of diamond in graphite was studied by W. R. L. Lambrecht et al., Nature (London) **364**, 607 (1993).

[36] M. J. Rutter and V. Heine, J. Phys. Condens. Matter **9**, 2009 (1997).

[37] P. Sollich, V. Heine, and M. Dove, J. Phys. Condens. Matter **6**, 3171 (1994) and references therein.

[38] R. M. Smart and F. P. Glasser, Sci. Ceram. **9**, 256 (1977).

[39] J. W. Couves et al., J. Phys. Condens. Matter **5**, L329 (1993).

[40] T. A. Mary et al., Science **272**, 90 (1996).

[41] N. Suzuki, T. Yamasaki, and K. Motizuki, J. Phys. C **21**, 6133 (1988).

[42] S. V. Vonsovsky, Y. A. Izyumov, and E. Z. Kurmaev, *Superconductivity of Transition Metals* (Springer, New York, 1982).

[43] H. Sato et al., J. Alloys Compd. **262–263**, 443 (1997).

[44] K. Shimizu et al., Nature (London) **393**, 767 (1998).

[45] N. W. Ashcroft, in *From Quantum Mechanics to Technology* (Springer, New York, 1996), p. 2; B. Edwards and N. W. Ashcroft, Nature (London) **388**, 625 (1997).

[46] K. J. Chang et al., Phys. Rev. Lett. **54**, 2375 (1985).

[47] V. V. Struzhkin et al., Nature (London) **390**, 382 (1997).

[48] J. L. Robeson et al., Phys. Rev. Lett. **73**, 1644 (1994).

[49] S. M. Sharma and S. K. Sikka, Phys. Rev. Lett. **74**, 3301 (1995).

[50] Iwamoto et al., J. Chem. Soc. Dalton Trans. **1997**, 4127 (1997).

[51] T. Kihara, Acta Crystallogr. **16**, 1119 (1963); **21**, 877 (1966); Acta Crystallogr. A **26**, 315 (1970); **31**, 718 (1975).

[52] H. Mao and R. J Hemley, Rev. Mod. Phys. **66**, 671 (1994).

[53] W. I. F. David et al., Nature (London) **353**, 147 (1991).

[54] J. Crain, G. J. Ackland, and S. J. Clark, Rep. Prog. Phys. **58**, 705 (1995).

[55] A. Mujica and R. J. Needs, J. Phys. Condens. Matter **8**, L237 (1996).

[56] J. A. Ripmeester et al., Nature (London) **325**, 135 (1987).

[57] O. Mishima, L. D. Calvert, and E. Whalley, Nature (London) **310**, 393 (1984); O. Mishima, *ibid*, **384**, 546 (1996).

[58] C. Meade et al., Phys. Rev. Lett. **69**, 1387 (1992).

[59] W. Jin et al., Phys. Rev. Lett. **71**, 3146 (1993).

[60] J. S. Tse et al., Nature (London) **369**, 724 (1994).

[61] N. Rivier and J. F. Sadoc, Europhys. Lett. **7**, 523 (1988).

[62] L. Pauling and R. E. Marsh, Proc. Nat. Acad. Sci. **38**, 1112 (1952).

[63] E. D. Sloan, Jr., *Clathrate Hydrates of Natural Gases* (Marcel Dekker, New York, 1990).

[64] F. Liebau et al., Zeolites **6**, 373 (1986).

[65] S. Saito and A. Oshiyama, Phys. Rev. B **51**, 2628 (1995).

[66] A. R. Kortan et al., Phys. Rev. B **47**, 13070 (1993); S. Yamanaka et al., Fullerene Sci. Tech. **3**, 21 (1995).

[67] H. Nakano and S. Yamanaka, J. Solid State Chem. **108**, 260 (1994) and references therein.

[68] See, e.g., E.G. Derouane et al., eds., *Zeolite Microporous Solids – Synthesis, Structure, and Reactivity* (Kluwer, Dordrecht, The Netherlands, 1992); H. van Bekkum, E.M. Flanigen, and J.C. Jansen, eds., *Introduction to Zeolite Science and Practice* (Elsevier, Amsterdam, 1991).

[69] A. F. Cronstedt, Akad. Handl. Stockholm **18**, 120 (1756).

[70] See, e.g., J. Kärger and D. M. Ruthven, *Diffusion in Zeolites and Other Microporous Solids* (Wiley, New York, 1992), Chap. 14 and references therein.

[71] See, e.g., J. Yu et al., Chem. Mater. **10**, 1208 (1998).

[72] Y. Nozue et al., Mater. Sci. Eng. A **217/218**, 123 (1996); Y. Nozue et al., in *Materials and Measurements in Molecular Electronics*, K. Kajimura and S. Kuroda, eds. (Springer-Verlag, New York, 1996), p. 151.

[73] P. A. Anderson et al., in *Magnetic Ultrathin Films, Multilayers and Surfaces* (Materials Research Society, Pittsburgh, PA, 1995), p. 9.

[74] S. Andersson and M. O'Keeffe, Nature (London) **267**, 605 (1977).

[75] A. L. MacKay and H. Terrones, Nature (London) **352**, 762 (1991); T. Lenovsky et al., Nature (London) **355**, 333 (1992). This has been generalized to graphite sponges by M. Fujita et al. (see Ref. [107] below).

[76] A. L. MacKay and H. Terrones, in *Growth Patterns in Physical Sciences and Biology*, J. M. Garcia-Ruiz et al., eds. (Plenum, New York, 1993) p. 315; A. MacKay, Curr. Sci. **69**, 151 (1995).

[77] H. Aoki, M. Koshino, H. Morise, D. Takeda, and K. Kuroki, submitted.

[78] A. M. Rao et al., Science **259**, 955 (1993).

[79] Y. Iwasa et al., Science **264**, 1570 (1994).

[80] M. Nunez-Regueiro et al., Phys. Rev. Lett. **74**, 278 (1995).

[81] O. Chauvet et al., Phys. Rev. Lett. **72**, 2721 (1994).

[82] T. Ogitsu et al., Phys. Rev. B **58**, 13925 (1998).

[83] D. W. Murphy et al., J. Phys. Chem. Solids **53**, 1321 (1992).

[84] B. F. Abrahams et al., Nature (London) **369**, 727 (1994).

[85] See, e.g., K. Nakao and S. A. Solin, eds., *Graphite Intercalation Compounds* (Elsevier Sequoia, Lausanne, Switzerland, 1985); W. Müller-Warmuth and R. Schöllhorn, eds., *Progress in Intercalation Research* (Kluwer, Dordrecht, The Netherlands, 1994).

[86] S. Yamanaka et al., Nature (London) **392**, 580 (1998).

[87] N. Wada, D. R. Hines, and S. P. Ahrenkiel, Phys. Rev. B **41**, 12895 (1990).

[88] See, e.g., R. E. Grim, *Clay Mineralogy* (McGraw-Hill, New York, 1968); G.W. Brindley and G. Brown, *Crystal Structures of Clay Minerals and Their X-Ray Identification* (Mineralogical Society, London, 1980).

[89] I. D. Johnson, T. A. Werpy, and T. J. Pinnavaia, J. Am. Chem. Soc. **110**, 8545 (1988).

[90] R. M. Wentzcovitch, W. W. Schulz, and P. B. Allen, Phys. Rev. Lett. **72**, 3389 (1994).

[91] T. Kimura et al., Mater. Res. Soc. Symp. Proc. **494**, 347 (1998).

[92] R. M. Wentzcovitch, J. L. Martins, and G. D. Price, Phys. Rev. Lett. **70**, 3947 (1993).

[93] L. Stixrude and R. E. Cohen, Nature (London) **364**, 613 (1993).

[94] J. H. Park et al., Nature (London) **392**, 794 (1998).

[95] Y. Tokura (ed.) *Colossal Magnetoresistive Oxides* (Gordon and Breach, London, 2000).

[96] P. W. Anderson, in *Frontiers and Borderlines in Many-Particle Physics*, R. A. Broglia and J. R. Schrieffer, eds. (North-Holland, Amsterdam, 1988), p. 1; P. W. Anderson, *A Carrier in Theoretical Physics* (World Scientific, Tokyo, 1994).

[97] T. Morita and T. Horiguchi, J. Math. Phys. **12**, 986 (1971).

[98] J. Kanamori, Prog. Theor. Phys. **30**, 275 (1963).

[99] H. Aoki, M. Tsukada, M. Schlüter, and F. Lévy, eds., *New Horizons in Low-Dimensional Electron Systems* (Kluwer, Dordrecht, The Netherlands, 1992).

[100] R. Arita, S. Onoda, K. Kuroki, and H. Aoki, J. Phys. Soc. Jpn. **69**, 785 (2000).

[101] E. H. Lieb, Phys. Rev. Lett. **62**, 1201 (1989).

[102] A. Mielke, J. Phys. A **24**, L73 (1991); H. Tasaki, Phys. Rev. Lett. **69**, 1608 (1992).

[103] K. Kusakabe and H. Aoki, Phys. Rev. Lett. **72**, 144 (1994).

[104] K. Kusakabe and H. Aoki, Physica B **194-196**, 215 (1994).

[105] R. Arita et al., Phys. Rev. B **57**, 10609 (1998).

[106] N. Shima and H. Aoki, Phys. Rev. Lett. **71**, 4389 (1993).

[107] M. Fujita et al., Phys. Rev. B **51**, 13778 (1995); in *Proceedings of the 22nd International Conference on the Physics of Semiconductors*, D. J. Lockwood, eds. (World Scientific, Tokyo, 1995) p. 2069.

[108] H. Masuda et al., Phys. Rev. B **43**, 7871 (1991).

[109] V. Eyert, K. H. Höck, and P. S. Riseborough, Europhys. Lett. **31**, 385 (1995).

[110] H. Tasaki, Progr. Theoret. Phys. **99**, 489 (1998).

[111] R. Arita et al., Phys. Rev. B **57**, R6854 (1998).

[112] See, e.g., R. Micnas, J. Ranninger, and S. Robaszkiewicz, Rev. Mod. Phys. **62**, 113 (1990).

[113] X. D. Xiang et al., Nature (London) **348**, 145 (1990).

[114] A. Fujiwara et al., Phys. Rev. B **52**, 15598 (1995).

[115] J. H. Choy, S. J. Kwon, and G. S. Park, Science **280**, 1589 (1998).

[116] K. Kuroki and H. Aoki, Phys. Rev. Lett. **76**, 4400 (1996); *ibid* **78**, 161 (1997).

[117] K. Kuroki and H. Aoki, Phys. Rev. B **56**, R14287 (1997).

[118] K. Kuroki and H. Aoki, J. Phys. Soc. Jpn. **67**, 1533 (1998).

[119] K. Kuroki, R. Arita and H. Aoki Phys. Rev. B **60**, 9850 (1999).

[120] K. Kuroki and H. Aoki, Phys. Rev. B
       60, 3060 (1999); R. Arita, K. Kuroki,
       and H. Aoki, Phys. Rev. B **60**, 14585
       (1999).

[121] Z. Hiroi et al., J. Solid State Chem.
       95, 230 (1991).

[122] M. Uehara et al., J. Phys. Soc. Jpn.
       65, 2764 (1997).

[123] K. Kuroki, T. Kimura, and H. Aoki,
       Phys. Rev. B **54**, R15 641 (1996).

[124] T. Kimura, K. Kuroki, and H. Aoki,
       Phys. Rev. B **54**, R9608
       (1996).

[125] T. Kimura, K. Kuroki, and H. Aoki, J.
       Phys. Soc. Jpn. **66**, 1599 (1997); **67**,
       1377 (1998).

Chapter 5.2

Theoretical Search for New Materials – Low-Temperature Compression of Graphitic Layered Materials

S. Tsuneyuki, Y. Tateyama, T. Ogitsu, and K. Kusakabe

The synthesis of new materials by compressing graphitic layered materials at low temperatures is proposed on the basis of first-principles molecular-dynamics simulations. It is predicted that a class of BCN heterodiamonds are derived from graphite/h-BN superlattices. It is also suggested that compression of a graphite intercalation compound results in an exotic diamondlike material. Differences in layer stacking, intercalants, and stage structures of intercalants in the starting layered materials will result in fruitful variations of the resultant materials.

5.2.1 Introduction

Predicting the structures and properties of materials from a knowledge of their chemical composition has been a long-standing problem of materials science [1]. Thanks to the recent development of computational science approaches, it is often not as difficult to predict stable structure and elctronic properties of even unknown materials theoretically, if we know or assume a rough arrangement of the constituent atoms in the material. It is much more difficult, however, to know how to make it and how stable it is compared with other unknown structures beyond our imagination, as we need overall knowledge of the potential-energy surface in the multidimensional configuration space to do so. An exceptionally simple and hopeful situation for such theoretical designs of materials can be found in the low-temperature compression of crystals.

Some crystals compressed at low (room) temperature undergo structural transformation without atomic diffusion (martensitic transformation), resulting in metastable

H. Aoki et al. (eds), *Physics Meets Mineralogy* © 2000 Cambridge University Press.

structures inaccessible at thermal equilibrium or by rapid quenching of high-tempera-ture/pressure phases.

Typical examples can be found in pressure-induced amorphization of crystals. Low quartz, a polymorph of framework silica (SiO_2), undergoes amorphization at 20–30 GPa at room temperature [2]. Although it might be expected that pressure-amorphized materials are uniform because of the lack of the diffusion processes needed for phase separation or recrystallization, it has been found that the pressure-amorphized quartz is elastically anisotropic and retains the memory of its original crystallographic orien-tation [3]. More surprising is the case of $AlPO_4$ berlinite, which becomes amorphous above 18 GPa at 300 K and transforms back into a single crystal with the same orienta-tion as that of the starting crystal when the pressure is released [4]. Amorphous Fe_2SiO_4, synthesized by compression of fayalite, is reported to exhibit a Néel transition at a tem-perature identical to that observed in the crystalline form [5], suggesting that the struc-tural order in the crystalline phase is somewhat recovered. Many molecular crystals are also known to be amorphized by pressure reversibly. These experimental results clearly indicate distinctive characteristics of the pressure-amorphized materials compared with normal amorphous phases obtained by rapid quenching of their liquid phases.

We can also find some examples that novel crystalline phases are reached by room-temperature compression. Low-cristobalite, which is another polymorph of SiO_2, un-dergoes structural transformation into an unidentified crystal structure at ~ 10 GPa [6]. It has turned out that the low-cristobalite phase of $GaPO_4$ transforms into a *Cmcm* phase accompanied by tetrahedral-to-octahedral oxygen-coordination change around Ga atoms [7]. The hexagonal diamond is obtained by room-temperature compression of graphite [8], although the cubic diamond is much more abundant in nature. The transformation occurs martensitically accompanied by a change in the chemical bonds from sp^2 to sp^3, retaining the direction of its crystalline axes [8]. A molecular crystal, SnI_4, is reported to undergo reversible transition to an fcc crystal [9]. The fcc phase is considered to be a mixed crystal unattainable under thermal equilibrium because of phase separation [10].

The examples mentioned above suggest that we could somehow design and obtain novel materials by compressing crystals at room (low) temperature. The key point is that the system is trapped at a local-potential minimum close to the starting low-pressure structure in the configuration space, as the temperature is much lower than the activation energy needed for diffusion or phase separation into thermodynamically stable structures. Because the atomic displacement is quite local, it is relatively easy to consider all the possible rearrangement of atoms in theoretical investigation.

In this context, we study the possibility of synthesizing novel diamondlike materials from graphitic layered material by compression at low temperatures. There are several reasons why we focus on the graphitic layered materials. First, sp^2 and sp^3 chemical bonds are energetically competitive in these materials, so that a drastic structural change is expected at a reasonably low pressure. Second, transition between the two types of chemical bonds usually requires appreciable activation energy at ambient pressure. This is important if we want to quench the diamondlike structure. Third, graphitic layered materials show variations in layer stacking, intercalants, or stage structure of the intercalants, so that we can consider a variety of resultant materials. Below we review two examples of our trials on this line [11, 12].

Throughout this work the constant-pressure first-principles molecular-dynamics (CP-FPMD) method [13, 14] is used to optimize the structures under pressure. Calculations of electronic states are based on the density functional formalism within the local-density approximation (LDA). Norm-conserving soft pseudopotentials [15] are used with a plane-wave basis set for expanding the electronic wave functions and a correction to the energy functional [16] is introduced to retain a constant-energy cutoff for the basis, irrespective of change in the unit-cell volume.

5.2.2 BCN Heterodiamond

BCN heterodiamonds have attracted significant interest as candidates for superhard materials or wide-bandgap semiconductors with high electron mobility and high thermal conductivity. Since they are ternary compounds with a large number of structural freedoms, they are expected to show a variety of properties more than those obtained from binary semiconductors if their atomic composition and/or arrangement could be controlled.

Experimental efforts have been concentrated on high-temperature compression of graphitic BC_2N obtained by chemical-vapor deposition, yet there are only a few reports on their synthesis [17]. This is probably because BCN compounds have a strong tendency toward phase separation into diamond and c-BN at high pressure. Also, in theoretical investigation, other than pioneering works on (110) superlattices of diamond and cubic BN [18], there have been few reports on the BCN heterodiamonds to our knowledge. Recently we searched for energetically favorable structures of BC_2N heterodiamonds and the synthesis paths by using the CP-FPMD method. From the result, we propose a high-pressure synthesis of a class of BCN heterodiamonds by utilizing diffusionless structural transformation at low temperatures [11]. To explore the possibility of the application to electronic devices, we have studied the electronic states of such probable structures by using first-principles band calculations [19].

The total-energy calculations of several possible structures have revealed that a bond-counting rule holds in the B—C—N system. The rule is that more stable structures have more C—C and B—N bonds but neither B—B nor N—N bonds. This is consistent with the tendency toward phase separation. Thus the BN/C_2 (111) superlattice, as shown in Fig. 5.2.1(b), which is called β-BC_2N in Ref. [11], is more stable than the (110) superlattice proposed previously [18] and energetically the most favorable among atomically mixed structures of the BC_2N heterodiamond.

As for the synthesis of this superlattice, it might be difficult to make a well-crystallized sample by ordinary epitaxial growth methods, because island growth is expected to be predominant because of the bond-counting rule. On the other hand, if we can prepare a proper starting material, the desired sample could be made by

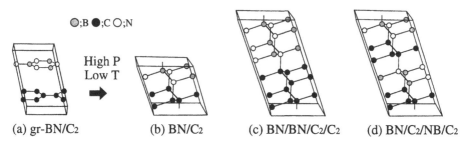

Figure 5.2.1 Energetically favorable BC$_2$N heterodiamond (111) superlattices investigated in this work. They are synthesized by the compression of graphitic superlattices at low temperatures.

high-pressure synthesis from layered structures at a temperature low enough to suppress atomic diffusion. From this viewpoint, we propose compression of a graphitic superlattice (gr-BN/C$_2$), as shown in Fig. 5.2.1(a), which consists of alternate stacking of graphite and hexagonal BN monolayers at low temperatures [11]. Because gr-BN/C$_2$ is energetically very stable and free from orientational disorder as well as intralayer defects, it is more promising for a well-crystallized sample. The calculated pressure dependence of the enthalpies indicates that gr-BN/C$_2$ can transform into the BN/C$_2$ (111) superlattice with compression up to ~16 GPa, suggesting that this synthesis path will be experimentally feasible [11].

The results on the structural stability of BC$_2$N imply that the synthesis of BN/C$_2$ heterodiamond (111) superlattices with different stacking orders of diamond and cubic BN will also be probable by means of compression of graphitic structures. The structures shown in Figs. 5.2.1(b) and 1(c) are those of 1 + 1 (BN/C$_2$) and 2 + 2 (BN/BN/C$_2$/C$_2$) superlattices, respectively. The structure shown in Fig. 5.2.1(d) is also a 1 + 1 superlattice, although BN orientation in alternating BN layers is reversed (BN/C$_2$/NB/C$_2$). This type of structure might appear locally as a stacking fault.

The band structures calculated with the optimzed structures of Figs. 5.2.1(b)– 5.2.1(d) are shown in Fig. 5.2.2. All of the band structures can be roughly explained as a folding structure of diamond or cubic BN, as known for (110) superlattices [18], and the bandgap is indirect in all the cases. However, there are differences in the band-edge states. The conduction-band minima (CBMs) of BN/C$_2$ and BN/BN/C$_2$/C$_2$ are located at the D point, corresponding to the X point in the fcc first Brillouin zone, at which the CBM of cubic BN is located. On the other hand, the CBM of BN/C$_2$/NB/C$_2$ is located at the 0.7 D point on the Γ–D line, which is similar to the case of diamond. The LDA bandgap is also different: those of BN/C$_2$ and BN/BN/C$_2$/C$_2$ are 3.6 and 3.5 eV, respectively, whereas that of BN/C$_2$/NB/C$_2$ is 2.8 eV. At the valence-band maximum (VBM), a difference is also found in the splitting energy from triply degenerated p-like states into a p_z-like single state in the [111] direction and doubly degenerated states perpendicular to it. The spatial distributions of the CBM and the VBM wave functions also show dramatic differences. The CBM states of BN/C$_2$ and BN/BN/C$_2$/C$_2$ are spread throughout the entire crystal, whereas that of BN/C$_2$/NB/C$_2$ is localized in a C layer, which is sandwiched between the two N layers.

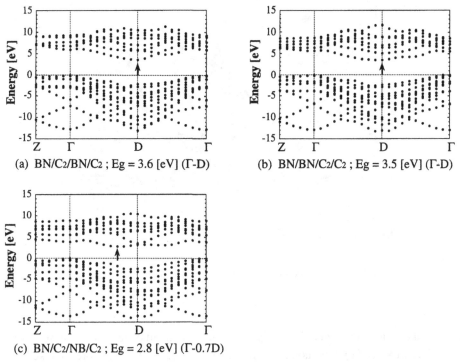

(a) BN/C$_2$/BN/C$_2$; Eg = 3.6 [eV] (Γ-D)

(b) BN/BN/C$_2$/C$_2$; Eg = 3.5 [eV] (Γ-D)

(c) BN/C$_2$/NB/C$_2$; Eg = 2.8 [eV] (Γ-0.7D)

Figure 5.2.2 Band structures of (a) BN/C$_2$/BN/C$_2$, (b) BN/BN/C$_2$/C$_2$, and (c) BN/C$_2$/NB/C$_2$ in a rhombohedral cell (space group $R3m$). The Z–Γ, Γ–D, and D–Γ lines correspond to (ΓL/4)–Γ, Γ–X, and L–Γ in the fcc first Brillouin zone, respectively. Each arrow indicates the conduction-band minimum in each structure. The bandgaps (E_g) are calculated within LDA.

Besides these superlattices, we have studied several types of stacking sequences to find systematic changes in their bandgap and stability [20]. Unfortunately all the structures investigated are found to have indirect bandgaps so that application of BCN heterodiamond to optoelectronic devices might be difficult. Nevertheless it is good news that the bandgap engineering seems to make sense in this system.

5.2.3 Li-Encapsulated Diamond

Cohesion of C atoms in graphite occurs in two ways: in a single graphite layer, C atoms are tightly connected with each other by means of sp^2 covalent bonds, and interaction between the layers is attributed to weak van der Waals force. The weak interlayer coupling allows various elements or molecules to be intercalated between the graphite layers accompanied by cationic or anionic ionization of the intercalants. Graphite intercalation compound (GIC) is a generic term for the compounds obtained in this way.

Because the graphite structure is well retained in the monolayer of GICs, we expect that a GIC might undergo structural transformation into a diamondlike structure if the

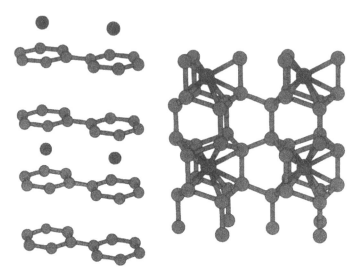

Figure 5.2.3 Atomic configuration of $LiC_{12}(G)$ (left) and $LiC_{12}(D)$ (right).

intercalant is small enough to be encapsulated in a diamond cage, as is the case of Li impurity introduced into diamond by ion implantation [21, 22]. Therefore we take Li-GICs as starting materials and investigate the stability of Li-doped diamonds, $LiC_6(D)$ and $LiC_{12}(D)$, expected from stage 1 and stage 2 Li-GICs, respectively [12].

Stage 1 Li-GIC [$LiC_6(G)$] has A–α–A–α $\cdots$ stacking, in which the α layer is a $\sqrt{3} \times \sqrt{3}$ triangular lattice of Li atoms located at the center of facing hexagons in adjacent graphite layers. In case of stage 2 Li-GIC [$LiC_{12}(G)$], the stacking structure is reported to be A–α–A–A–α–A $\cdots$. The diamond structure naively expected from these graphite-layer stackings is hexagonal diamond with the same c-axis orientation as the GICs. The Li atoms are expected to reside in the interstitial site, keeping the triangular lattice in the ab plane. The structures of $LiC_{12}(G)$ and hypothetical $LiC_{12}(D)$ are illustrated in Fig. 5.2.3.

Starting from the hypothetical diamondlike structure with the same lattice constant as pure diamond, we performed structural optimization without symmetry restriction by means of the CP-FPMD method. The unit cell for calculation contains 12 C atoms and 1 or 2 Li atoms.

We have found that $LiC_6(D)$ is unstable and spontaneously collapses into a GIC phase, whereas $LiC_{12}(D)$ is quenchable at ambient pressure at least at the low-temperature limit, if it is once formed. The C—C bond length in $LiC_{12}(D)$ is elongated to1.60–1.63 A in the diamond cage encapsulating a Li atom, whereas it remains 1.52–1.54 A elsewhere. The system contains an odd number of electrons in the unit cell and is metallic.

Because of the large lattice distortion, $LiC_{12}(D)$ is less stable than the GIC at low pressure, whereas it could be stabilized more than GIC at ultrahigh pressure, as is the

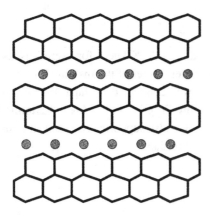

Figure 5.2.4 Schematic view of a possible high-pressure phase of a higher-stage GIC.

case of pure diamond. Therefore this exotic material would probably be obtained by compression if any phase separation leading to carbide or Li segregation is prohibited at low temperature.

It is likely that the diamond cage cannot be preserved if the intercalants are larger than Li. Even so, in case of higher-stage GIC, we imagine that thin diamond layers are formed and sandwich intercalant layers (Fig. 5.2.4). Then the dangling bonds would be partly passivated by the electrons transferred from the intercalants. Another possibility is that a three-dimensional network of $sp2$ covalent bonds is formed just like a high-pressure phase of silicide, or it may be that some C dimers appear to form carbide locally. In any case, low-temperature compression of GICs would result in novel structural transformations and new high-pressure phases worthy of detailed investigation.

Finally we mention that the pseudopotential for Li adopted here is too soft for quantitative study of the Li-encapsulated diamond with an extraordinarily small Li–C distance. Reexamination of the energetics and a detailed study of the electronic properties are now under way with a revised pseudopotential.

5.2.4   **Conclusion**

We have proposed a method of synthesizing exotic diamondlike materials from graphitic layered materials, gr-BC$_2$N and Li-GIC. The compression should be carried out at low temperatures under nonequilibrium conditions so that atomic diffusion leading to phase separation is prohibited, and the low temperature tends to lower the transformation probability and degrade crystallinity of the sample. Thus it is important to search for the best temperature condition for each synthesis.

Differences in layer stacking, intercalants, and stage structures of intercalants in the starting materials result in fruitful variations of the resultant materials. More generally, low-temperature compression of crystals is a promising method for designing and obtaining new materials inaccessible at thermal equilibrium.

Acknowledgments

We thank S. Itoh, T. M. Briere, H. Ogawa, and A. Kikuchi for useful discussions. The calculation was done at the Supercomputer Center of the Institute for Solid State Physics, University of Tokyo. This work was conducted as Japan Society for the Promotion of Science Research for the Future Program in the Area of Atomic-Scale Surface and Interface Dynamics. This work was also supported by Grant-in-Aids from the Ministry of Education, Science and Culture, Japan.

References

[1] J. Maddox, Nature (London) **335**, 201 (1988); M. L. Cohen, Nature (London) **338**, 291–292 (1989); F. C. Hawthorne, Nature (London) **345**, 297 (1990).

[2] R. J. Hemley, A. P. Jephcoat, H. K. Mao, L. C. Ming, and M. H. Manghnani, Nature (London) **334**, 53 (1988).

[3] L. E. McNeil and M. Grimsditch, Phys. Rev. Lett. **68**, 83 (1992).

[4] M. B. Kruger and R. Jeanloz, Science **249**, 647 (1990).

[5] M. B. Kruger, R. Jeanloz, M. P. Pasternak, R. D. Taylor, B. S. Snyder, A. M. Stacy, and S. R. Bohlen, Science **255**, 703 (1992).

[6] Y. Tsuchida and T. Yagi, Nature (London) **347**, 267 (1990).

[7] J. L. Robeson, R. R. Winters, and W. S. Hammack, Phys. Rev. Lett. **73**, 1644 (1994).

[8] T. Yagi, W. Utsumi, M. Yamakata, T. Kikegawa, and O. Shimomura, Phys. Rev. B **46**, 6031 (1992).

[9] N. Hamaya, K. Sato, K. Usui-Watanabe, K. Fuchizaki, Y. Fujii, and Y. Ohishi, Phys. Rev. Lett. **58**, 796 (1997).

[10] S. Tsuneyuki and S. Ohta, in *Proceedings of the Third Symposium on Atomic-scale Surface and Interface Dynamics*, Fukuoka (Japan Society for the Promotion of Science, Tokyo, 1999).

[11] Y. Tateyama, T. Ogitsu, K. Kusakabe, S. Tsuneyuki, and S. Itoh, Phys. Rev. B **55**, 10161R (1997).

[12] S. Tsuneyuki, T. Ogitsu, Y. Tateyama, K. Kusakabe, and A. Kikuchi, in *Advances in High Pressure Research in Condensed Matter*, S. K. Sikka et al., eds. (National Institute of Science Communication, New Delhi, India, 1997), p. 104.

[13] R. M. Wentzcovitch, J. L. Martins, and G. D. Price, Phys. Rev. Lett. **70**, 3947 (1993).

[14] Y. Tateyama, T. Ogitsu, K. Kusakabe, and S. Tsuneyuki, Phys. Rev. B **54**, 14994 (1996).

[15] N. Troullier and J. L. Martins, Phys. Rev. B **42**, 1993 (1991).

[16] M. Bernasconi, G. L. Chiarotti, P. Focher, S. Scandolo, E. Tosatti, and M. Parrinello, J. Phys. Chem. Solids **56**, 501 (1995).

[17] A. R. Badzian, Mater. Res. Bull. **16**, 1385 (1981); Y. Kakudate et al., in *Proceedings of the 3rd IUMRS International Conference on Advanced Materials*, (Elsevier Science, Amsterdam, 1994) p. 1447; S. Nakano

et al., Chem. Mater. **6**, 2246 (1994); E. Knittle et al., Phys. Rev. B **51**, 12149 (1995).

[18] W. E. Pickett, Phys. Rev. B **38**, 1316 (1988); W. L. R. Lambrecht and B. Segall, Phys. Rev. B **40**, 9909 (1990); W. L. R. Lambrecht and B. Segall, Phys. Rev. B **47**, 9289 (1993).

[19] Y. Tateyama, K. Kusakabe, T. Ogitsu, and S. Tsuneyuki, in *Proceedings of the International Conference on Silicon Carbide, III-Nitrides and Related Materials – 1997* (Trans Tech Publications LTD, Switzerland; Stockholm, Sweden, 1997).

[20] Y. Tateyama, Ph.D. dissertation, The University of Tokyo (1998).

[21] G. Braunstein and R. Kalish, Appl. Phys. Lett. **38**, 416 (1981).

[22] S. A. Kajihara, A. Antonelli, J. Bernholc, and R. Car, Phys. Rev. Lett. **66**, 2010 (1991).

Chapter 5.3

H...H Interactions and Order–Disorder at High Pressure in Layered Hydroxides and Dense Hydrous Phases

J. B. Parise, H. Kagi, J. S. Loveday, R. J. Nelmes, and W. G. Marshall

Observations of order–disorder phenomena at high pressure in hydrous phases are reinterpreted with the results of Rietveld analysis and neutron-diffraction data. The reported partial amorphization of the hydrogen sublattice in β-Co(OD)$_2$ at 11.2 GPa was not confirmed in powder-diffraction data collected with the Paris–Edinburgh cell to 15.5 GPa. The diffraction data, and perhaps the spectroscopic data on which the observations of amorphization are based, are consistent with an increase in the H...H repulsion with pressure. The structural consequences of competition between H...H repulsion and H-bond (O—H...O) formation is observed in the M(OH)$_2$ compounds in general. It is also observed in the dense high-pressure phases recovered from high-pressure synthetic experiments.

5.3.1 Introduction

The hydrogen bond (X—H...Y) is one of the most studied bond geometries in the mineralogical, biological, and solid-state organic chemical communities [1, 2]. For nonmineral and mineral structures alike, the published literature, consisting mainly of crystal-structure determinations at ambient pressure, provides a means to study the bond as donor (X) and acceptor (Y) vary over a variety of structures and chemistries [3, 4]. The secondary environment, however, is important in considering the effects of structure on H-bond geometry [5]; in many cases gross changes in this environment from one structure type to the next make it difficult to separate the effects of the relatively weak H bonding from the steric effects because of the framework making up the

H. Aoki et al. (eds), *Physics Meets Mineralogy* © 2000 Cambridge University Press.

remainder of the structure [5]. Isochemical studies, in which interatomic distances are varied for a fixed chemistry or for an isostructural class of materials, by use of pressure for example, have been carried out in an attempt to overcome these ambiguities [6, 7].

The richness of phenomena associated with the interactions of the hydroxide ion in the solid state has been a topic of great interest of late [8, 9]. Interest in the mineralogical community has been focused on the stabilization of H in novel hydrous [10–18] and nominally anhydrous phases [10, 19–26] recovered from synthetic experiments carried out at high pressures and temperatures. The amount of H_2O, its effects on phase relations [27], crystal chemistry, elasticity, strength, rheology [28–31], and the dehydration reactions of these high-pressure phases [32–34] have a direct bearing on a number of earth processes. These include the amount of H_2O stored in the Earth's mantle [35–41], recycling and continent growth [33, 34, 42], the storage of stress in high-pressure hydrous phases [30, 31, 43], and the origin of deep-focus earthquakes [42, 44].

5.3.1.1 Dense Hydrous Phases Recovered in Quench Experiments

The structures of several novel dense hydrous phases have been determined [12–14, 16–18] and the H-bond geometry discussed. With the exception of phase A [45] (Fig. 5.3.1) these materials were studied with x-ray techniques, which is not as precise as neutron diffraction for the study of H (D) [48]. Although imprecise, the positions of the H atoms in these structures suggest that the H...H and H...O interactions share common features. Disorder of the H positions has been noted in several cases [18, 46] and symmetry reduction, on at least the local level, has been reported from spectroscopic studies that utilize nonlinear optical measurements [46] or magnetic-resonance (NMR) spectroscopy [47]. The H...H distances in many of these phases (Fig. 5.3.1) are short, and may be responsible for the partially occupied H-sites in the case of fully hydrated topaz [46] and the reduction in local symmetry noted in Shy-B [47]. The distance between deuterium positions in the case of phase A (Fig. 5.3.1) is larger than twice the van der Waals radius for H, and preliminary NMR data are consistent with the crystallographic symmetry [45]; there is little evidence for the disorder noted in the other phases (Fig. 5.3.1). Finally, many of the high-pressure hydrous phases possess similar H-bond geometries with more than one H (D) involved in bonding to the same O (Fig. 5.3.1). Given the difficulties in synthesizing large quantities of these materials for high-pressure neutron experiments, the difficulties inherent in deconvolving the effects of H bonding on these highly covalent and complex materials, and the desirability of studying the relative contributions of H-bonding and H...H repulsion (Fig. 5.3.1) in high-pressure phases, suitable analogs were sought for investigations of these phenomena.

5.3.1.2 $M(OD)_2$ Hydroxides

The CdI_2-related layered hydroxides are suitable for the study the effects of H bonding (Figs. 5.3.1 and 5.3.2). The fixed parts of the framework, the MO_6 octahedra, are

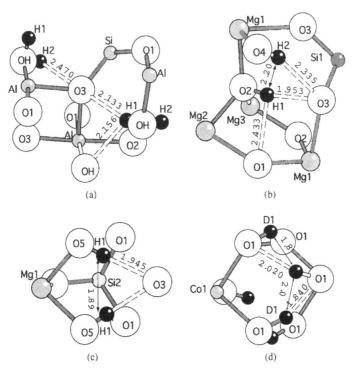

Figure 5.3.1 H-bonding environment in (a) hydrous topaz, determined from single-crystal x-ray data [46], (b) phase A [17] determined from powder-neutron-diffraction data [45], (c) Shy-B determined from single-crystal x-ray data [12], and (d) β-Co(OD)$_2$ determined from powder-neutron–powder-diffraction data collected at 9.5 GPa. In the case of β-Co(OD)$_2$ only one of the threefold disordered sites is shown [6, 7]. Dashed lines indicate possible H bonds, and the double arrows indicate the short H...H distances in each structure. In those cases in which these distances are less than the sum of the van der Waals radii [Shy-B, topaz-OH, and β-Co(OD)$_2$] there is evidence of disorder [6, 47]. For phases A, in which this distance is close to but not less than twice the van der Waals radius, the hydrogens appear more ordered [45].

structurally distinct from that part of the structure containing H (Fig. 5.3.2). Several transition-metal (Mn, Fe, Co, Ni, Cd) and alkaline-earth (Mg, Ca) hydroxides crystallize in this structure. They are easily synthesized in either H or D forms. Little real difference is expected in the behavior of the H or the D forms [49]; although there are reports of very large differences in this behavior [50, 51] from powder-diffraction studies, a comparison of powder and single-crystal neutron refinements reveals a consistent picture [50, 52]. The geometry of the H bond in these hydroxides (Figs. 5.3.1 and 5.3.2) share characteristics in common with the dense hydrous phases, including short O...H(D) and H(D)...H(D) distances at high pressures [6, 7, 50, 51]. Further, with the exception of Ca(OH)$_2$ [53–55], the CdI$_2$ structure is stable for the majority of these compositions to high pressures, ensuring that the H-bond geometry is maintained in the gross sense over a wide range of pressures.

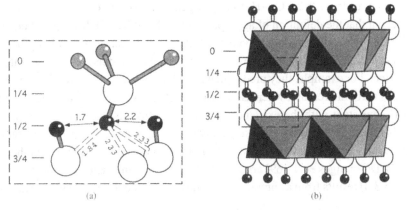

Figure 5.3.2 (a) H bond geometry in (b) the layered structure of the CdI$_2$-related β-Co(OD)$_2$. The (001) layers of CoO and H alternate along the c axis and are located close to the z fractional coordinates indicated. Only one of three symmetry-equivalent H sites is shown in (a). These sites are displaced from the ideal position $2d$ at ($\frac{1}{3}$, $\frac{2}{3}$, z) for space group $P\bar{3}m1$ shown in (b) to $6i$ at ($\frac{1}{3} \pm \delta$, $\frac{2}{3} \pm 2\delta$, z). This results in a decrease in one of the H...O distances, favoring H bonding and changes in the H...H distances (see Fig. 5.3.3). The value of δ and its sign is dependent on pressure, which affects the H...H distance (Fig. 5.3.4).

For these reasons the CdI$_2$-structured hydroxides are used as models for H bonding in their own right and as analogs for H-bonding interactions in the dense hydrous phases (Fig. 5.3.1) [6, 7, 50, 51, 56–64]. The use of pressure as a means to probe the O—H...O interaction isochemically is complementary to studies that extract the geometry of this moiety from published structural data [3]. Such studies also reduce some of the ambiguity introduced by differences in steric effects when systems with very different chemistries and structures are compared [5].

During some of the high-pressure investigations of the M(OH)$_2$ hydroxides there have been indications of subtle phase transitions [62, 65, 66]. The mechanism for these transitions is attributed to the increase in H-bond strength at pressure [50, 67] or on cooling [66]. For example Duffy et al. [62] have interpreted the growth of new Raman peaks in terms of the disordering of the H atom from its threefold position to the sixfold site, first reported from neutron-diffraction data [50]. Earlier studies had suggested that pressure enhances H bonding [67]. Indeed, subsequent single-crystal neutron-diffraction studies [52] noted structural features, such as displacement of H from the ideal threefold position (Fig. 5.3.2) consistent with increased H bonding even at room pressure. There was little evidence for an increase in the O—H(D) bond length as a function of pressure in the layered hydroxides [6, 7] or other compounds with H bonding [8, 9], in contradiction to the interpretation of spectroscopic results [67]. Many of these effects are under investigation and, although a consensus has yet to emerge, will undoubtedly benefit from the insights of theoretical studies.

One interaction not emphasized is the role of H...H contacts in the high-pressure behavior of the CdI$_2$-related hydroxides. This is despite several indications that this interaction competes with the O—H. . .O hydrogen bond even at room pressure. Jeffrey [1] points out that the definition of appropriate O. . .O and H. . .O distances indicative of H bonding is a long-standing problem [2]. Attempts to rationalize this in the hydroxides and the hydrates include the use of bond-valence models [6, 7, 68], but these are imperfect. For example, they predict that no H bonding exists in any of the layered hydroxides at room pressure, whereas accurate single-crystal studies indicate a geometry of H consistent with the existence of this interaction [52, 69]. Megaw [66, 70] argued that the split-site model, subsequently observed in neutron-diffraction studies several decades later [50, 52, 69], results from the tendency for O to form tetrahedral sp^3 bonds. Desgranges et al. [69] argue that, in addition to this tendency, the threefold site (Fig. 5.3.3) results from a compromise in the covalent bonding of H with O and the

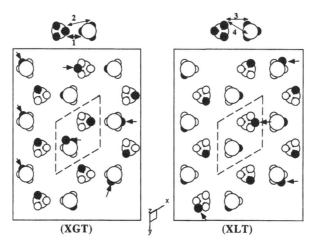

Figure 5.3.3 Idealized models for the H. . .H interactions in the layered hydroxides derived from the published neutron-diffraction investigations [6, 50, 52, 69], viewed along the c axis, for structures related to CdI_2. In this case H is displaced from the ideal position $2d$ at $(\frac{1}{3}, \frac{2}{3}, z)$ to $6i$ at $(\frac{1}{3} \pm \delta, \frac{2}{3} \pm 2\delta, z)$ by (a) $\delta > 0$ (XGT) and (b) $\delta < 0$ (XLT) and $|\delta| = 0.06$ for β-Co(OD)$_2$ at ~10 GPa [6]. Small circles represent three possible positions for H atoms on each O (large open circles) for two opposing layers of the hydroxyl unit (Fig. 5.3.2). The small filled circles represent sites filly occupied by H(D) so as to minimize the short H. . .H contacts. Distances of less than the sum of the van der Waals radii are noted in the two models shown at the top: (1) 1.3, (2) 1.8, (3) 1.7, (4) 1.8 Å. To avoid these unfavorable contacts, defects (arrowed) of one scheme, with δ either less than or greater than zero, can be introduced into the ideal scheme. This would be modeled by an increased preference of anisotropic displacement models at lower average H. . .H distances, as occur at high pressures. Different environments for H are introduced in this fashion and are expected to merge at the highest pressures into a temporal and spatial mixture at the highest pressures where H. . .H contacts of less than the sum of the van der Waals radii, ~2 Å, are more numerous.

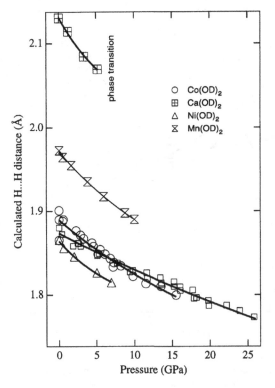

Figure 5.3.4 Calculation of idealized H...H distances in the layered hydroxides by use of the variation in unit-cell size with pressure (Fig. 5.3.5) reported in the literature [6, 7, 62, 64]. All H are presumed to lie on ideal 2d positions [Fig. 5.3.2(b)], with the metal and oxygen atoms at the positions indicated in Fig. 5.3.2, and so the shortest H...H are under estimated by ~0.05 Å.

Coulombic interaction of the electropositive H and the electronegative O atom of the adjacent layer (Fig. 5.3.2). There is an additional consideration, however (Fig. 5.3.3), involving the interactions between H atoms in different layers. The distance between these atoms at room pressure is close the sum of their Van der Waals radii, and this decreases with increased pressure (Figs. 5.3.1–5.3.4).

5.3.1.3 β-Co(OD)$_2$

The behavior of β-Co(OD)$_2$ at pressure has attracted interest of late as it appears to be anomalous with respect to increasing pressure. Reports of partial (sublattice) amorphization [65] are intriguing and suggest deconstruction of one part of the structure while the remainder remains crystalline. These observations were made with a combination of IR and Raman spectroscopies, which clearly show the broadening and the disappearance, respectively, of peaks attributed to the O—H stretching frequency. The relative constancy of the x-ray-diffraction pattern suggests that only the H sublattice is affected and that amorphization occurs only in this structural unit (Figs. 5.3.2 and 5.3.3). Neutron diffraction is a structural probe most appropriate for the investigation of the long-range ordering state of H(D) in crystalline and amorphous solids. Recent advances in high-pressure devices pioneered by the Paris-Edinburgh (P-E) team [71] now allow exploration beyond the nominal 10 GPa of our first study [6] of β-Co(OD)$_2$ and into the pressure regime in which sublattice

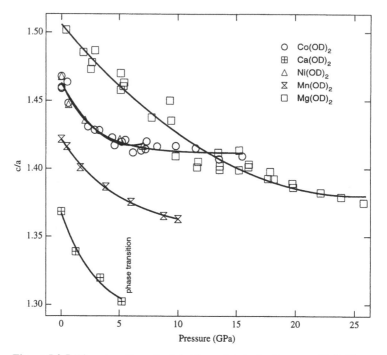

Figure 5.3.5 The c/a ratios calculated from data in the literature [6, 7, 62, 64]. For all materials in the CdI$_2$ structure over the pressure ranges studies, the ratio decreases rapidly with increasing pressure [and indeed appears to pass through a minimum for Mg(OH)$_2$ above 10 GPa [62]]. Note that for Co(OD)$_2$ the ratio appears to reach a constant value. For portlandite the range of data is restricted by a low-pressure phase transition [54].

amorphization is reported to occur. We hope to provide some confirmation of the phenomenon and insight into the mechanism of what would be a remarkable structural reconstruction. We found no such confirmation [72] but instead found evidence of the important role of H...H repulsion in determining disorder in these systems at high pressure.

5.3.2   Experimental Details

5.3.2.1 Sample Preparation

The elimination of incoherent scattering from H [48] makes the use of deuterated samples for neutron scattering desirable. There are commonly only very small differences in the strength of the OD$^-$...O hydrogen bonds compared with that of OH$^-$...O [58]. The sample of β-Co(OD)$_2$ was synthesized as previously described [6]; it was not recrystallized because the compound is not stable over 100 °C and forms a well-crystallized material close to room temperature.

5.3.2.2 Calculation of Powder-Diffraction Data

Neutron-powder-diffraction patterns were calculated for several structural models to simulate the sublattice amorphization in β-Co(OD)$_2$. These models included (1) the removal of D altogether from the structure, (2) displacement of a fully occupied D site from its expected position to increase the O–D distance, (3) decreasing the occupancy of D in the structure as well as displacing it from the ideal position (Fig. 5.3.2), (4) increasing the ratio of the U_{11}/U_{33} anisotropic displacement parameters [6, 66], and (5) simulation of a sheet scattering from D situated at various values of z between the CoO sheets (Fig. 5.3.2). Several variations and combinations of these models were tested, and in all cases there were dramatic differences in the neutron-diffraction peak intensities compared with those previously observed at room pressure [6]; not an unexpected result given that D contributes $\sim$50% of the total elastic neutron scattering from Co(OD)$_2$: $\sim$4% for x-radiation.

5.3.2.3 Neutron Diffraction

Data were collected in a standard P-E cell [71] fitted with tungsten carbide anvils for pressures to 10 GPa [6]. For higher pressures a double-toroid cell made from sintered diamond was used. Before the collection of data suitable for Rietveld refinement [73, 74], an equation of state was determined for β-Co(OD)$_2$, with NaCl as an internal standard. These data were collected for $\sim$3 h at average proton currents of 175 μA. Fresh samples either free of NaCl or including minimal amounts (5 mol.%) to act as an internal calibrant of intensity were used for Rietveld structure analysis. The data were taken at 300 K and between 0 and 15.5 GPa on the PEARL beamline of the U.K. pulsed source ISIS at the Rutherford–Appleton Laboratory. Data were collected in a detector bank at $2\theta = 90°$ for 12–18 h. Some data were taken with the cell turned 90° to the incident beam to ensure that the powder sample was randomized with little preferred orientation. Care was taken to correct the data for the absorption of the cell and the sintered diamonds [75].

5.3.2.4 Structure Refinement

The refinement of crystal structure was initiated with the previous ambient-pressure model [6] with Co at (0, 0, 0), and O and D at $(\frac{1}{3}, \frac{2}{3}, z)$ (Fig. 5.3.2). Because of interference from the diamond anvils, both midfocus (MF) and full-focus data sets were used [6]. The results from both sets of data were consistent, but the MF data were used at the highest pressure to maximize the sample scattering contributions to the pattern (Fig. 5.3.6). Two classes of models were refined: (1) a model assuming isotropic displacement parameters and D situated at $(\frac{1}{3} \pm \delta, \frac{2}{3} \pm 2\delta, z)$ with $\delta = 0$, $\delta > 0$ (XGT), and $\delta < 0$ (XLT) [61]; (2) an anisotropic model (ANISO) with D situated in the ideal position ($\delta = 0$) [52, 66, 69, 70].

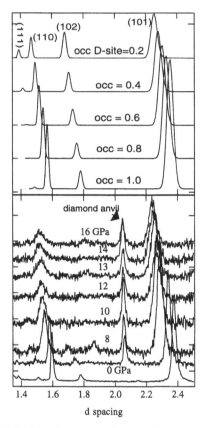

Figure 5.3.6 Calculated (top) and observed (bottom) neutron-powder-diffraction patterns for β-Co(OD)$_2$ over the range of interplanar spacing indicated. The calculated patterns are for split-site positions of the deuteron (Figs. 5.3.2 and 5.3.3) with occupancy decreasing from bottom to top. The cell parameters, atomic positions, and displacement parameters used to calculate the pattern were the same for all models; only the occupancy of the D site was varied, and each pattern is arbitrarily offset to simulate the decrease in lattice parameter expected at high pressure. As the deuteron occupancy decreases or as the displacement parameter is increased to simulate amorphization [65], the (102) and the (111) peaks increase while the (101) and (110) peaks weaken in relative intensity. No such systematics are observed in the real neutron-diffraction data collected at the ISIS spallation neutron source in a P-E cell at the pressures indicated (bottom set of curves).

5.3.3   Results

Before presentation of the results of Rietveld refinement, a great deal can be ascertained from inspection of the observed and the calculated neutron-powder-diffraction patterns (Fig. 5.3.6). First, there was no evidence for overall amorphization of the sample (Fig. 5.3.6), consistent with the result reported from x-ray studies in the diamond-anvil cell [65]. Although considerable broadening of the pattern is observed, which limits the resolution and the peak-to-background discrimination, the ratio of scattering from NaCl and the hydroxide is constant within experimental error, suggesting no overall amorphization of β-Co(OD)$_2$ beyond that experienced by NaCl to 16 GPa. The broadening of the pattern is presumed to arise from deviatoric stress introduced into the sample at high pressures [76, 77].

The ratio of peak intensities in the pattern is also constant (Fig. 5.3.6). At the highest pressures (>10 GPa), deviatoric stress [76, 77] and limitations on the amount of precious beam time allocated to the project restrict the resolution obtainable to d spacings >0.7 Å. The data obtained at the highest pressure, however, clearly show (1) a constant ratio and (2) no appearance of new peaks, as would be predicted from any of the models discussed above (Fig. 5.3.6). Indeed, the changes observed in the neutron-powder-diffraction patterns are consistent with a continuation of the displacement of D from the ideal $2d$ site [6, 50].

Preliminary refinements confirmed that the scattering from the D sites indicated full occupancy by D to 16 GPa for all models refined. This is further evidence that the structure of β-Co(OD)$_2$ above 10 GPa is consistent with the split-site model reported below this pressure [6]. Full structure refinements of either the ANISO, XGT, or XLT models (see above) suggest that, as pressure increases and the interlayer spacing decreases (Fig. 5.3.2), the average positions occupied by D move farther from the threefold inversion axis. They do so in such a manner, however, as to maintain an O—D bond distance of ~1 Å, as expected for a strong O—D bond. There is no evidence in either the raw data (Fig. 5.3.6) or in the refinements of a loss of the diffraction signature from the D; the loss of diffraction is a common criterion for the onset of amorphization. Instead the pattern (Fig. 5.3.6) and Rietveld refinement are consistent with a deuteron in close proximity to the O (Figs. 5.3.2 and 5.3.3).

To distinguish between the two split-site models (Fig. 5.3.3), which contain sites separated by between 0.2 Å at the lowest pressures and ~0.5 Å at the highest, diffraction data to d spacings of ~0.5 Å would be ideal [6]. Realistically, these data, with reasonable counting statistics, were available only for the pressures below 12 GPa. Comparisons of the fits to the data suggest that at all pressures there is a preference for the XGT and ANISO models over the XLT model. This result is in line with refinements of models for the structures of the Mg, Ca, Fe, and Mn hydroxides [7, 50, 52, 69] at ambient pressure. A reexamination of the structure of Ni(OD)$_2$, on the other hand, showed a slight preference for an XLT ordering scheme (Fig. 5.3.3) at ambient pressures and a preference for the ANISO model at pressures between 2 and 7 GPa. These results are preliminary and need to be interpreted with caution, however. Although of sufficient quality to distinguish between sublattice amorphization and order–disorder phenomena [6, 7, 50], the data were not sufficient to unambiguously distinguish among XGT, XLT, and ANISO models, as was possible from single-crystal neutron-diffraction data collected at ambient pressure [52, 69]. Also, at the highest pressures there is a lack of discrimination among the three models.

5.3.4 **Discussion**

The neutron-diffraction results and observations that the OH stretching bands of β-Co(OH)$_2$ broaden with increasing pressure in both the IR and the Raman experiments need to be reconciled. Some broadening of IR and Raman bands with increasing pressure is due to deviatoric stresses within the diamond-anvil cell; as Nguyen et al. [65] point out, however, the band broadening they observe is too large to be due to differences in pressure within the sample alone. It is also unlikely that this degree of broadening is caused by an increase in H-bond strength and lengthening of the O—H bond; at ~11 GPa the H. . .O distance is estimated to be ~2.2 Å, a distance not associated with usually strong H bonds [1, 2, 66, 70]. Decomposition of the structure does not occur (Fig. 5.3.6), and so an abrupt change in the ordering scheme of the H atoms remains the only obvious possibility to explain (1) the increased IR bandwidth with pressure and (2) the abrupt increase in width at ~11 GPa reported [65]. Nguyen et al. [65] pointed out that the rapid increase in the IR bandwidth was associated with the flattening of the c/a ratio (Fig. 5.3.5) above 11 GPa. This flattening is not observed for the other deuteroxides or hydroxides, except perhaps for Ni(OD)$_2$ (Fig. 5.3.6), but it is observed in deuterated Co(OD)$_2$ at approximately the same pressure; this is further strong evidence that the H and the D materials display very similar behavior.

Inspection of the decrease in the H. . .H contacts (Figs. 5.3.3 and 5.3.4) suggests a plausible mechanism for the behavior of the hydroxides at high pressure. Except for the Mn and the Ca hydroxides, the calculated H. . .H contacts in these materials (Fig. 5.3.4) are less than the sum of the van der Waals radii and less than the shortest distances observed in the high-pressure hydrous phases synthesized at above 10 GPa (Fig. 5.3.1). These calculations were carried out with the assumption that the H atom is located on the threefold inversion axis; however, it is clear that H. . .H repulsion will compel the H to move off the ideal site to minimize unfavorable contacts. That the H does so in such a manner as to maximize H bonding in the case of Ca, Mn, Mg, and Co suggests that, despite argument to the contrary [68], the H. . .O interactions provide sufficient stabilization in these cases to overcome the H. . .H repulsive contacts and to promote H bonding between layers [6, 50, 52, 69]. It appears that the H. . .H contact is short enough in the case of Ni(OD)$_2$, however, that the XLT ordering scheme is favored because it minimizes H. . .H contacts.

Neither the XLT nor the XGT model (Fig. 5.3.3) will be favored as the average H. . .H distance decreases, however, and this is observed in the Rietveld refined structure models. For β-Co(OD)$_2$ in particular, the IR observations and structural data are consistent with a increasing admixture of the two displacement models (Fig. 5.3.3) and eventual lack of discrimination between them so that the environment of the H(D) site is no longer discrete (Fig. 5.3.2). This situation would explain the breadth of the IR band [65] and would be consistent with the increased anisotropy observed in the displacement parameters for all the hydroxides at higher pressures [6]. In crystallographic terms this situation does not fit the usual understanding of the term amorphous because the hydrogens (deuterons) are located close to the oxygen and within a distance range expected for a strong O—D bond. A change over from one ordering scheme to another could also explain the observed phase transition reported earlier in $Mg(OH)_2$ by Duffy et al. [62]. Clearly these interactions are important in determining the H bonding in these materials, and they require further careful study for deriving a consistent picture for H. . .H and O. . .H interactions.

5.3.5 Conclusion and Future Work

The relative importance of the H. . .H repulsion and H. . .O attraction has implications for hydrous minerals in general and for high-pressure hydrous phases in particular. For example, the dehydration temperatures for Cd, Co, and Ni [78] are quite different (lower) from those of Mg and Ca hydroxides and is pressure dependent. The role of H. . .H contacts in destabilizing the transition-metal hydroxides requires further study as does the investigation of these contacts in the high-pressure hydrous phases (Fig. 5.3.1). Studies in these later phases, with high-pressure neutron-diffraction techniques, will be difficult given the structural complexity in the dense hydrous phases [12, 13, 15–18] and the nominally anhydrous phases [20–24, 26]. The implications for deep-earth studies, particularly hydration–dehydration reactions in the Earth's mantle [19, 33, 41, 79], are sufficiently interesting to warrant this effort. This work has begun, and we are in the process of studying the higher-symmetry hydrous phases (A) [45] and D at high pressures [Fig. 5.3.1(b)]. Eventually heating will be required for relieving the effects of deviatoric stress and allow collection of data to sufficiently low values of d spacing to distinguish between possible models for H-atom distribution within their crystal structures at the pressure and temperature relevant to the Earth. Finally, theoretical calculations, consistent with the crystal structures derived from neutron-diffraction studies, are required for determining the relative contributions of O—H, H. . .O, H. . .H, M—O, and M. . .H (Fig. 5.3.2) interactions and bond strengths on the relative stability of model structures. Given the complexity of the dense hydrous phases, further experimental and theoretical [80] work is required for providing data on the simple layered hydroxides. As the spectroscopic and diffraction data alluded to here demonstrate, these materials provide excellent tests for the relative contributions of interatomic forces for stabilizing and destabilizing crystal structures containing the O—H. . .O moiety. A recent theoretical investigation [80] confirms the importance of H. . .H repulsion [72].

Acknowledgments

This work is supported by the U.S. National Science Foundation through grant EAR-9909145. We acknowledge the help and advice of S. Klotz and J. M. Besson of Université P et M Curie, S. Peter and H. D. Lutz of Siegen, and D. Klug of NRC,

Ottawa. We acknowledge the full access to the P-E facilities and techniques at ISIS, the ISIS facility itself, PEARL beamline scientists, Duncan Francis and R. Smith of ISIS for help during the setup of these experiments; JSL and WGM acknowledge the support of the U.K. Engineering and Physical Sciences Research Council.

References

[1] G. A. Jeffrey, *An Introduction to Hydrogen Bonding* (Oxford U. Press, Oxford, U.K., 1997).

[2] W. C. Hamilton and J. A. Ibers, *Hydrogen Bonding in Solids* (Benjamin, New York, 1968).

[3] F. C. Hawthorne, Acta Crystallogr. B **50**, 481 (1994).

[4] F. H. Allen et al., Acta Crystallogr. B **52**, 734 (1996).

[5] F. A. Cotton and G. Wilkinson, *Advanced Inorganic Chemistry: A Comprehensive Text* (Wiley-Interscience, New York, 1980).

[6] J. B. Parise et al., Phys. Chem. Mineral. **25**, 130 (1998).

[7] J. B. Parise et al., Rev. High Press. Sci. Technol. **7**, 211 (1998).

[8] W. G. Marshall et al., Rev. High Press. Sci. Technol. **7**, 565 (1998).

[9] R. J. Nelmes et al., Phys. Rev. Lett. **71**, 1192 (1993).

[10] Y. Kudoh, T. Inoue, and H. Arashi, Phys. Chem. Miner. **23**, 461 (1996).

[11] Y. Kudoh et al., Phys. Chem. Miner. **22**, 295 (1995).

[12] R. E. Pacalo and J. B. Parise, Am. Mineral. **77**, 681 (1992).

[13] Y. Kudoh et al., Phys. Chem. Miner. **19**, 357 (1993).

[14] L. W. Finger, R. M. Hazen, and C. T. Prewitt, Am. Mineral. **76**, 1 (1991).

[15] L. W. Finger and C. T. Prewitt, Geophys. Res. Lett. **16**, 1395 (1989).

[16] L. W. Finger, R. M. Hazen, J. Ko et al., Nature (London) **341**, 140 (1989).

[17] H. Horiuchi et al., Am. Mineral. **64**, 593 (1979).

[18] H. Yang, C. T. Prewitt, and D. J. Frost, Am. Mineral. **82**, 651 (1997).

[19] G. R. Rossman and J. R. Smyth, Am. Mineral. **75**, 775 (1990).

[20] J. R. Smyth, Am. Mineral. **72**, 1051 (1987).

[21] J. R. Smyth, D. R. Bell, and G. R. Rossman, Nature (London) **351**, 732 (1991).

[22] J. R. Smyth, Am. Mineral. **79**, 1021 (1994).

[23] S. D. Jacobsen, J. R. Smyth, and T. Kawamoto, Eos Trans. Am. Geophys. Union **77**, F662 (1996).

[24] J. R. Smyth and T. Kawamoto, Earth Planet. Sci. Lett. **146**, E9 (1997).

[25] H. Kagi et al., Japan Earth Planet. Sci., Joint Meet. Abstract G42-09, 506 (1997).

[26] J. R. Smyth et al., Am. Mineral. **82**, 270 (1997).

[27] T. Inoue, Phys. Earth Planet. Inter. **85**, 237 (1994).

[28] T. Inoue et al., Earth Planet. Sci. Lett. **160**, 107 (1998).

[29] R. E. G. Pacalo and D. J. Weidner, Phys. Chem. Miner. **23**, 520 (1996).

[30] J. Chen et al., Eos Trans. Am. Geophys. Union **77**, F716 (1996).

[31] J. Chen et al., Geophys. Res. Lett., **25**, 575 (1998).

[32] A. R. Pawley and B. J. Wood, Contrib. Mineral. Petrol. **124**, 90 (1996).

[33] R. P. Rapp and E. B. Watson, J. Petrol. **36**, 891 (1995).

[34] R. P. Rapp, J. Geophys. Res. **100**, 15601 (1995).

[35] M. Akaogi and S. Akimoto, J. Geophys. Res. **85**, 6944 (1980).

[36] M. Akaogi and S. Akimoto, Phys. Chem. Mineral. **13**, 161 (1986).

[37] S. Akimoto and M. Akaogi, Phys. Earth Planet. Inter. **23**, 268 (1980).

[38] S. Akimoto and M. Akaogi, *Possible Hydrous Magnesian Silicates in the Mantle Transition Zone* (Terra Scientific, Tokyo, 1984).

[39] D. R. Bell and G. R. Rossman, Science **255**, 1391 (1992).

[40] P. C. Burnley and A. Navrotsky, Am. Mineral. **81**, 317 (1995).

[41] A. B. Thompson, Nature (London) **358**, 295 (1992).

[42] C. Meade and R. Jeanloz, Science **252**, 68 (1991).

[43] D. J. Weidner, Y. Wang, and M. T. Vaughan, Science **266**, 419 (1994).

[44] P. G. Silver et al., Science **268**, 69 (1995).

[45] H. Kagi et al., Phys. Chem. Mineral. **27**, 225 (2000).

[46] P. A. Northrup, K. Leinenweber, and J. B. Parise, Am. Mineral. **79**, 401 (1994).

[47] B. L. Phillips et al., Phys. Chem. Miner. **24**, 179–190 (1997).

[48] G. E. Bacon, *Neutron Diffraction* (Oxford U. Press, Oxford, 1962).

[49] K. Beckenkamp and H. D. Lutz, J. Mol. Struct. **270**, 393 (1992).

[50] J. B. Parise et al., Am. Mineral. **79**, 193 (1994).

[51] M. Catti, G. Ferraris, and A. Pavese, Phys. Chem. Miner. **22**, 200 (1995).

[52] L. Desgranges, G. Calvarin, and G. Chevrier, Acta Crystallogr. B **52**, 82 (1996).

[53] K. Leinenweber et al., J. Solid State Chem. **132**, 267 (1997).

[54] M. Kunz et al., High Pressure Res. **14**, 311 (1996).

[55] S. Ekbundit et al., J. Solid State Chem. **126**, 300 (1997).

[56] X. Xia, D. J. Weidner, and H. Zhao, Am. Mineral. **83**, 68 (1997).

[57] A. Pavese et al., Phys. Chem. Miner. **24**, 85 (1997).

[58] H. D. Lutz, H. Möller, and M. Schmidt, J. Mol. Struct. **328**, 121 (1994).

[59] M. Kunz et al., Eos Trans. Am. Geophys. Union, Session T42E-1 (1994).

[60] D. M. Sherman, Am. Mineral. **76**, 1769 (1991).

[61] S. A. T. Redfern and B. J. Wood, Am. Mineral. **77**, 1129 (1992).

[62] T. S. Duffy et al., Am. Mineral. **80**, 222 (1995).

[63] T. S. Duffy and T. J. Ahrens, J. Geophys. Res. **14**, 14319 (1991).

[64] Y. Fei and H. K. Mao, J. Geophys. Res. **98**, 11875 (1993).

[65] J. H. Nguyen, M. B. Kruger, and R. Jeanloz, Phys. Rev. Lett. **49**, 1936 (1997).

[66] H. D. Megaw, Rev. Mod. Phys. **30**, 96 (1958).

[67] M. B. Kruger, Q. Williams, and R. Jeanloz, J. Chem. Phys. **91**, 5910 (1989).

[68] H. D. Lutz et al., J. Mol. Struct. **351**, 205 (1995).

[69] L. Desgranges et al., Acta Crystallogr. B **49**, 812 (1993).

[70] H. D. Megaw, *Crystal Structures: A Working Approach* (Sanders, Philadelphia, PA, 1973).

[71] R. J. Nelmes et al., Trans. Am. Crystallogr. Assoc. **29**, 19 (1993).

[72] J. B. Parise et al., Phys. Rev. Lett. **83**, 328 (1999).

[73] A. C. Larson and R. B. Von Dreele, Report AUR-86-748 (Los Alamos National Laboratory 1986).

[74] H. M. Rietveld, J. Appl. Crystallogr. **2**, 65 (1969).

[75] R. M. Wilson et al., Nucl. Instrum. Methods Phys. Res. A **354**, 145 (1995).

[76] D. J. Weidner, Y. Wang, and M. T. Vaughan, Geophys. Res. Lett. **21**, 753 (1994).

[77] Y. Meng, D. J. Weidner, and Y. Fei, Geophys. Res. Lett. **20**,1147 (1993).

[78] J. B. Parise et al., Photon Factory Activity Rep. **14**, 458 (1996).

[79] T. J. Ahrens, Nature (London) **342**, 122 (1989).

[80] S. Raugei, P. L. Silvestrelli, and M. Parrinello, Phys. Rev. Lett. **83**, 2222 (1999).

Part VI

Melts and Crystal–Melt Interactions

Chapter 6.1

Comparison of Pair-Potential Models for the Simulation of Liquid SiO$_2$: Thermodynamic, Angular-Distribution, and Diffusional Properties

M. Hemmati and C. A. Angell

We extend a recent comparison of the abilities of different pair-potential models for SiO$_2$ to reproduce the IR spectrum of the glassy state (a short-time dynamic property) to other properties. The comparison is extended to include static (thermodynamic and structural) and long-time dynamic (diffusivity) properties. As with the IR spectrum, none of the existing potentials succeeds in correctly locating the temperature of the density maximum $T_{\rho(\max)}$. Although the modified Matsui potential, which best represented the IR spectrum, also performs well in describing $T_{\rho(\max)}$ and the intertetrahedral bond angle, it seriously overestimates the diffusivity and hence incorrectly represents the shape of the potential far from the minimum. The best performance across the range of properties examined is obtained with the van Beest–Kramer–van Santen potential. Some evidence for a higher-order phase transition below the density maximum is given.

6.1.1   Introduction

It is now two decades since it was found (unexpectedly to many) that key properties of liquid and glassy SiO$_2$ could be reproduced semiquantitatively by molecular-dynamics (MD) computer simulations in which the Si and the O components are treated as simple ions with full formal charges [1]. The internal energy at 300 K was obtained to within 3%, and the qualitatively unique features of (1) high liquid compressibility contrasting with low liquid expansivity, (2) anomalous increase of diffusivity with pressure, and (3) large pressure-induced glass densification were also demonstrated. The same simple two term potentials (now called TRIM, for transferable rigid-ion model, potentials

H. Aoki et al. (eds), *Physics Meets Mineralogy* © 2000 Cambridge University Press.

because only one parameter, σ, is used to distinguish between different cations in oxide glass formers) also predicted correctly that it would not be possible to retain Si in six coordination on decompression from the six-coordinated state, regardless of whether the compression was carried out in the liquid or the glassy state [1, 2]. These simple transferable potentials, of the form

$$U(r) = z_i z_j / r + A_{ij} \exp[(\sigma_i + \sigma_j - r)/\rho], \qquad (6.1.1)$$

have been very useful for establishing qualitative trends in the behavior of liquid and glassy silicates across a wide range of multicomponent systems [3–5], and for obtaining insight into structure–property relations in nonoxide glasses as well [6], although they are too simple to give quantitative predictions.

Professor Matsui was one of the first to recognize the possibilities of such simulation studies for solving geophysical problems, and his 1980 study of liquid MgSiO$_3$ [7] did much to awaken the geoscience community to what has now become a major field of activity [8]. He has also been very instrumental in the development of more accurate potential functions for use in these simulations, although, as will be seen, the provision of a quantitatively valid pair-potential model for SiO$_2$ is still an unfulfilled need.

A good example of the qualitative success but quantitative failure of the simple transferable potentials is provided by the unusual behavior of the density with varying temperature to which we devote considerable attention in this chapter. Maxima in density with varying temperature are very rare among liquids. The famous case of water at $4\,°\mathrm{C}$ is often regarded as unique. However, it is also encountered in liquid SiO$_2$, not far above T_g, at 1820 K [9]. The existence of a density maximum in SiO$_2$ is qualitatively verified with the TRIM potential. However, the temperature at which the maximum occurs, 7–9000 K [10–12], is far higher than that observed in the laboratory material. Another failure of the simple model concerns the crystal state. The prediction of distinct polymorphs existing at different pressures and temperatures is an essential requirement of a good set of pair potentials – and is one that is not met by TRIM potentials. Accordingly, a variety of more accurate, but of course more complicated, potentials have been developed, of which the potential developed by Matsui and co-workers, the so-called Tsuneyuki potential [13], is a leading example.

Pair potentials that have been developed for the accurate evaluation of crystal-phase properties of SiO$_2$ have proven useful for the study of certain liquid- and glassy-state properties of this important substance, although they may be unsuitable for others. For instance, the Tsuneyuki potential, which was developed to describe the crystals [13] and was applied with apparent success to the calculation of glassy-state properties by Della Valle and Andersen [14], proves unsatisfactory for the calculation of the liquid-state diffusivities, which are some 3 orders of magnitude higher than expected from the measured viscosities (vide infra). In this latter respect, the Tsuneyuki potential is considerably less satisfactory than the simple TRIM potential. Indeed, there are some properties that none of the pair potentials yet developed appear capable of reproducing satisfactorily. We recently [15] documented the failure of all available pair potentials to reproduce the separation of the principal peaks of the IR spectrum. We now document the failure of all available pair potentials to reproduce accurately the temperature of the density maximum, a liquid-state anomaly that is known from aqueous studies to be closely related to the anomalous pressure coefficient of the fluidity; hence it is of major geophysical importance [16, 17].

In the case of the IR spectrum, the addition of the many-body (i.e., non-pairwise-additive) interaction, arising from the polarizability of the oxide ion, was found necessary to obtain a correct description [18]. A variety of three-body (non-pairwise-additive) potentials have been developed [19–23] to improve the modeling of properties such as the average bond angle Si—O—Si, which in turn seems to be related to the IR band separation problem. Whether or not calculations with three-body potentials are more economic than the algorithms by which oxide-ion polarization is currently [18] being accounted for is not clear, but the latter approach is simpler and more satisfactory in principle. Ultimately it will be necessary to use the more fundamental Car–Parrinello approach [24] in which the forces acting on the atoms are evaluated at each step by high-order quantum-mechanical calculations, and the accelerations, hence the trajectories, are evaluated by the normal classical mechanical methods. The first applications of this approach to the simulation of SiO_2 have recently been reported [25]. Unfortunately, the computer time consumed in such simulations is so great that at this time they may be applied to only a very limited range of problems.

The last-mentioned ab initio calculations are for some purposes not helpful. There are many problems of fundamental as well as geophysical interest for which the ability to follow the system evolution over very many (preferably of the order of 10^6) particle oscillations is essential. For such problems, which are on the border between class A and class B simulation problems [26], the use of pairwise-additive potentials is mandatory, given present computing speeds and algorithms. By way of example, a general problem for fragile supercooling liquids is that of establishing the physics of the bifurcation of the relaxation function into α and (slow)β processes. This is found to occur at a temperature that is very near the mode-coupling-theory critical temperature and at characteristic times that are some 3 orders of magnitude beyond the range of present computing capabilities. For the particular case of SiO_2 there is a more immediate problem. It is the problem of understanding the discrepancy (vide infra) between the diffusivity and the thermodynamic behavior of the best pair potentials (namely a strongly non-Arrhenius diffusivity coupled with – and expected from Ref. [27] – a considerable excess heat capacity [5, 28]) and that which is observed in the laboratory (an Arrhenius diffusivity and a very small excess heat capacity). We need to know whether the discrepancy has the same structural/geometrical origin as the anomalies seen in the simulation of H_2O by use of the best pair potentials. In the H_2O case there is a crossover from fragile liquid to strong liquid behavior during cooling. At ambient pressure such a transition might occur by means of a continuous transition [29–31] or (as in the case of liquid Si [32, 33]) by a first-order transition.

Thus there is a real need to evaluate the different pair potentials proposed for SiO_2 in order to determine how best to invest the considerable resources of computer time that will be necessary to resolve the above types of problem. In this chapter we therefore extend our previous comparative study of the IR spectrum, a short-time dynamic property, to the study of static (thermodynamic) anomalies on the one hand and long-time dynamic properties on the other. In our earlier paper [15], we stressed how insensitive the gross structural characteristics of the glass (and liquid), such as the radial distribution functions (RDFs), are to differences in the pair potentials that were found to produce large differences in the IR spectrum. Here we will compare a different structural property, the Si—O—Si bond-angle distribution, which will be seen to be rather more sensitive than the RDF. However, it is the diffusivity, a long-time dynamic property, that proves, not surprisingly [15, 34], to be the most sensitive of all the properties studied to these differences in potential function.

6.1.2 Procedures

In the previous paper [15] we provided, in tabular form, the parameters of the pair potentials being compared. We cast all potentials in the same form to help identify where the essential differences lie. We do not repeat the tabulation here but note the origins of the potentials being compared. They are, in order of their appearance in the literature, (1) the TRIM potential, (2) the Kubicki–Lasaga potential [35], (3) the Tsuneyuki potential [9], (4) the van Beest–Kramer–van Santen (BKS) potential [36], and (5) the modified Matsui potential [15, 37]. There have been minor variations of the TRIM potential, of which we have used the version studied in detail by Poole et al [12]. There are also different implementations of the BKS potential (of which we have used the original version without the truncation adopted by Vollmayr et al. [28]). The modified Matsui potential, which gave the best representation of the IR spectrum [15], is a potential adapted by us from the potential devised by Matsui [37] for the simulation of a binary silicate. It is closely related to the Tsuneyuki potential but differs in the short-range attractive component and is more stable at high temperatures against the collapse that occurs easily in the Tsuneyuki (and also BKS) potential because of the vanishing of repulsive forces between particles on close approach [38].

The emphasis in this chapter, except for the bond-angle distribution, is on the behavior of the simulated system in the ergodic state, i.e., in states that are liquid on the computational time scale. Runs were carried out by stepwise cooling of a 450-ion system with periodic boundary conditions at fixed density. The density chosen was 2.20 g/cm³, that of the laboratory material that is almost independent of temperature up to 2000 K [9].

Starting from high temperatures at which equilibration is very rapid, the samples were cooled stepwise to the target temperature, equilibrated under constant-temperature conditions for an appropriate period, and then run at constant energy to collect data for subsequent comparisons. The statistical averages are taken from a period long enough for the average particle to diffuse at least 2.0 particle diameters as this is the surest way of ensuring ergodicity, for the reasons given below. The cooling was then continued stepwise, with results being recorded from constant-energy runs made after equilibration, until ergodicity could no longer be restored in an acceptable time (maximum run length 200 ps, except for the BKS and the modified Matsui potentials, for which the lowest temperature runs were extended to 300 and 400 ps, respectively).

Breaking of ergodicity can be detected in different ways:

1. In general, the temperature of a nonergodic system will not hold steady under constant-energy conditions because of the continuing conversion of potential to kinetic energy (or vice versa if the system is reequilibrating after a temperature up-jump).

2. In the $P-T$ isochores used to establish the position in temperature of the density maximum, loss of ergodicity is indicated by pressures that, in the negative expansivity regime below the density maximum, do not rise as much as expected from the ergodic data (however, see the unexpected development highlighted in the next section). This corresponds to the different degrees of expansion of the volume below the density maximum found for different cooling rates by Vollmayr et al. [28].

3. In diffusivity plots with data acquired with our descending temperature protocol, loss of ergodicity is indicated by diffusivity values that fall above the trend established by equilibrated samples. As pointed out long ago [34] and emphasized again recently [15, 39], it is the relaxational processes of amorphous systems, rather than spectroscopic or structural features, that are the most sensitive to the thermodynamic state and hence to the ergodicity or otherwise of the system. This important point is further substantiated by the data presented in this chapter.

Broken ergodicity is always associated with mean-square particle displacements substantially smaller than the value mentioned above, 2.0 d, where d is the diameter of the particle. The criterion of (root-mean-square) distance diffused >2 particle diameters is based on the observation that structural relaxation times are, in the Stokes–Einstein approximation as well as in experiment, approximately the times needed for the average particle to diffuse $\sim d/3$ [40]. Equilibration, which we take to mean the decay of a perturbation to 2% of its initial value, requires the elapse of some 5–10 relaxation times, depending on the exponentiality or otherwise of the relaxation process [41]. For processes observed on the time scales available to MD, departures from exponentiality are usually quite small.

Before leaving this section we note that, for the four-coordinated SiO_2 of this study, a strong finite-size effect has been detected in the dynamic properties, such that small samples relax more slowly than larger samples, an effect that persists until sample sizes exceed 8000 particles[42]. (This interesting effect, which we believe is a manifestation of the harmonicity of strong liquid structures, is expected to affect equally all the models studied here, although this needs to be established.)

6.1.3   Results

We present first the static functions (thermodynamic and structural) and then pass to the dynamic properties.

6.1.3.1 Heat Capacity and Glass Transition

All models show well-defined glass transitions (as in, e.g., Ref. [10]) at which the constant-volume heat capacity departs from its classical harmonic oscillator value of 24.4 J/mol K and rises by over 50% to values between 36 and 41.2 J/mol K, depending on the model. In this respect, all models are apparently at variance with experiment because C_p in the laboratory experiment, although also reaching classical excitation in the glassy state, shows very little change at T_g. The observed change, is only ∼10% [43, 44] at 1200 °C, where the relaxation time reaches 100 s. C_v, which is the quantity obtained in the present simulations, must show even less (as $C_v < C_p$), except at the density maximum where the expansivity α is zero and $C_v < C_p$. Of course the experimental values are known at only lower temperatures, and it is quite possible that, if the measurements could be made, the laboratory substance would develop an increasing specific heat at higher temperatures (as BeF$_2$, a silica analog, tends to do).

Of the models studied, the BKS potential shows the largest ΔC_v and the TRIM potential shows the smallest. Both Matsui and TRIM potentials show a rapid decrease in C_v at higher temperatures. The T_g value for the BKS potential, indicated by the change of slope in the enthalpy–temperature relation, agrees with the temperature at which the anomalous low-T expansion freezes, according to the study of Vollmayr et al., namely 3650 K [see Fig. 6.1.1(b)].

6.1.3.2 Density Maximum and Possible Higher-Order Transition

Turning now to the matter of the density maximum, we note that this occurs in the study by Vollmayr et al. at ∼4500 K [28]. This is confirmed in the present study by the presence of an extremum at this temperature in the isochore P versus T seen in Fig. 6.1.1. As discussed many times before, the extremum in P versus T corresponds to a zero in the ratio of expansivity to compressibillity, α/κ_T, which can occur only if α passes through zero at a density maximum (or minimum). According to Fig. 6.1.1, a density maximum occurs for each of the potentials studied, but in every case the temperature of the maximum occurs far above the temperature of maximum density observed in the laboratory substance, 1820 K [9]. Indeed the laboratory phenomenon occurs at such a high viscosity that it would not be observable in a computer simulation made with an exact potential.

For some of the lowest temperature points in Fig. 6.1.1 (temperatures indicated by filled symbols), the data did not satisfy the criterion for ergodicity and hence do not represent the equilibrium behavior. The breaks in the P–T plots therefore generally represent the breaking of ergodicity (however, see below).

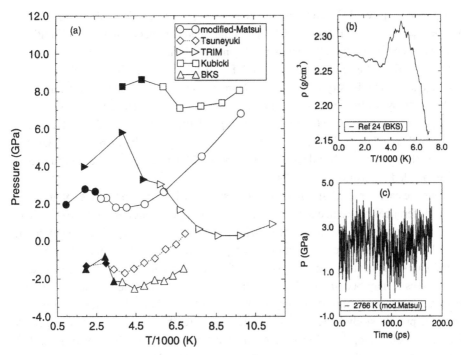

Figure 6.1.1 Pressure–temperature isochores at density 2.20 g/cm³ for SiO₂ liquid and glass in the different pair-potential representations indicated in the inset. A minimum pressure (or maximum tension) in the isochore indicates the existence of a density maximum. Filled symbols indicate nonergodic (glassy) states. Part (b) shows the isobaric variation of density, in which the density maximum is seen directly, from Vollmayr et al., with the BKS potential [28]. Part (c) shows the pressure fluctuations during passage of 180 ps of a constant-energy run at average temperature 2766 K and indicates the difficulty of obtaining precise isochore data at temperatures below the density maximum.

The temperature at which the density maximum occurs, $T_{\rho(max)}$, in the different models is highest in the case of the TRIM potential, in which it is found at ~9000 K, and lowest in the cases of the Tsuneyuki and the modified Matsui potentials, in which it falls at ~4000 K. In the much-studied BKS case, $T_{\rho(max)}$ is somewhat higher, ~4500 K, but the pressure at the density maximum is quite negative. If the density were lowered so that the maximum density would be reached at ambient pressure, then $T_{\rho(max)}$ would presumably fall at a somewhat lower temperature (but then the density would not be the experimental value. Note that this problem may be resolved by truncating the short-range interaction [28]. The truncation evidently not only economizes computer time but also leads to improved agreement with experiment).

For the modified Matsui potential, which is the most diffusive case (see Subsection 6.1.3.4 below), something more interesting may have happened. The point at 2750 K that failed to increase in pressure as expected was obtained from a run in which the ergodicity requirement was satisfied. This case was subjected to extended calculations to ensure that the values obtained represented equilibrium states. The fact that this extended study failed to change the value of P at the lowest T suggests that some sort of thermodynamic transition that arrests the anomalous expansion may have occurred in this case. This would be consistent with the presence of a compressibility maximum at $T < T_{\rho(\text{max})} = 6000$ K for the TRIM potential, seen with very lengthy equilibrations [12]. The density at which this isothermal compressibility maximum occurs, happens to be the same, 2.2 g/cm^3, as that of all the Fig. 6.1.1 isochores. In this light, the near coincidence of the density versus temperature plots for the two slowest cooling rates in the study by Vollmayr et al. [28] may be interpreted as evidence that the observed behavior is the equilibrium (cooling-rate-independent) behavior (see Fig. 6.1.1 inset). This plot then indicates the presence of a rather sharp expansivity extremum at 3500 K for the BKS potential. Our data for the modified Matsui potential would indicate a more rounded "transition" at ∼2750 K.

Thus the case for a transition in the liquid state of SiO₂, which might explain some of the differences between experiment and simulation, is strengthened. Indeed, a thermodynamic argument, due to Sastry et al. [45], shows that below any density maximum must lie a compressibility maximum. A compressibility maximum, like a heat capacity maximum, is usually regarded as a criterion for a (higher order) transition, and in some cases it may become a true phase transition i.e., a thermodynamic singularity. Unfortunately, we cannot draw any strong conclusions about this interesting matter from the present study because of the increasing magnitude of fluctuations (expected from increasing compressibility), which must be averaged over as the temperatures fall increasingly below $T_{\rho(\text{max})}$. The situation is worsened by the fact that the fluctuations, which are structural in origin, have the time scale of the structural relaxation that is lengthening rapidly in the domain of interest. The fluctuations, which have short-time vibrational and long-time structural components, are illustrated in Fig. 6.1.1(c) which makes clear the problem of obtaining precise data in this domain. Further discussion of this interesting phenomenon will be deferred to future articles.

6.1.3.3 Bond-Angle Distribution

As mentioned above, the presence of a density maximum in the liquid is thought to be linked to temperature-dependent variations in the intertetrahedral angle distribution. According to the study by Vollmayr et al. [28], this angle is increasing with decreasing temperature at the lower temperatures as it was found to be largest in the glasses formed during the slowest cooling. We have compared this angle at 300 K for the different potential models cooled stepwise at the same effective rate 2×10^{14} K/s. Figure 6.1.2 shows that the overestimate of the most probable intertetrahedral angle, which is a weakness of the TRIM potential, is largely rectified in the cases of the BKS and modified Matsui potentials. Differences in the widths of the distributions are discernible but are not readily characterized within the present data noise.

6.1.3.4 Diffusivities

Finally, Figs. 6.1.3 and 6.1.4 show the behavior of the diffusivities. Figure 6.1.3 shows the individual mean-square displacement plots, from whose long time slope the diffusivities are obtained, using the case of the modified Matsui potential. With

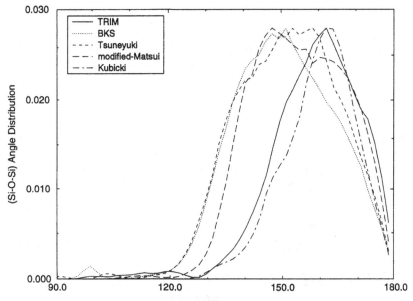

Figure 6.1.2 The SiO_2 glass intertetrahedral angle distribution for the different pair potentials listed in the inset. The effective cooling rate, 2×10^{14} K/s, was the same in each case. Experimental estimates (summarized in Ref. 15) range from 144° to 153°.

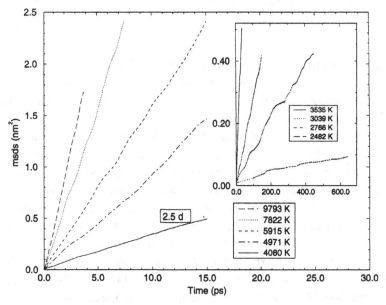

Figure 6.1.3 Mean-square displacement of oxygen species as a function of time for the modified Matsui potential. The inset shows the two lowest temperatures studied. The displacement at 2482 K is insufficient to give a valid measure of the ergodic diffusivity. The displacement at 2766 K is marginally sufficient.

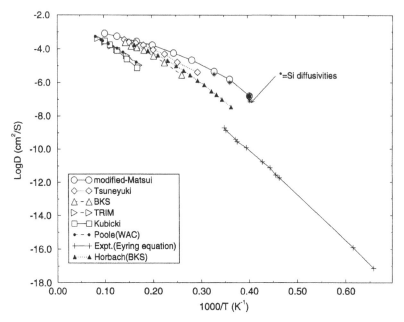

Figure 6.1.4 Diffusivities of oxygen in liquid SiO$_2$ in the different pair-potential models compared with high-temperature experimental values obtained from measured viscosities by means of the Eyring equation. Filled triangles are recent data from Horbach et al. [50] obtained with the BKS potential with truncated short-range potential. Si diffusivities are only included in one case, since they are always similar, in the temperature range of this study.

one exception, data points are not included in Fig. 6.1.4 unless the mean-square displacement during the constant-energy run is greater than 2.0 oxide-ion diameters. The exception is the lowest temperature point (2482 K) of Fig. 6.1.3 for the modified Matsui potential for reasons that are given below.

Figure 6.1.4 then shows the collection of diffusivity Arrhenius plots for the different potentials and includes experimental data based on use of the Eyring equation [46] to convert the viscosity data measured in the temperature range 1500–2800 K to oxygen diffusivities. The most recent justification for this conversion formula is found in the work of Poe et al. [17]. The jump distance used is the same as that in Refs. [10], [17], and [46], viz., the oxide-ion diameter, 0.28 nm. At the temperatures of simulation the diffusivities of Si and O are very similar, so the Si values are omitted in most cases.

At temperatures near the experimental T_g it seems that the oxide-ion diffusivity is much the larger, i.e., a decoupling of the two occurs [47] as is also seen in high-pressure simulations [3]. It should also be remembered, in comparing these diffusivities with experiment, that dynamic properties of small SiO$_2$ systems at normal densities are subject to finite-size effects that slow down the response functions relative to those of large systems, >8000 ions [42]. The effect amounts to some 500% in D for 450-ion systems and can be expected to be similar for all the potentials of this study as it is a function of the same harmonicity that allows the other liquid-state anomalies to show up above thermal smearing effects. Because the effect can be linked to the more harmonic behavior of the strong liquids, it is likely to be more pronounced at the lower temperatures. This might account for the very low values of D obtained for the modified Matsui potential at the lowest temperature at which, because of (Fig. 6.1.3, inset), an overestimate would have been expected.

In Fig. 6.1.4 we see, at the lowest common temperature, 6000 K, a spread of oxygen diffusivities for the different pair potentials that approaches 2 orders of magnitude. The modified Matsui potential gives quite the highest diffusivity, whereas TRIM and Kubicki potentials group together as the least diffusive. There are two slightly different versions of the TRIM potential result. That marked Poole et al. was obtained with the original Woodcock–Angell–Cheeseman potentials (see Ref. [1], Table 2, and appended notes), and the other was obtained with the parameters listed in Ref. [4].

6.1.4   **Discussion**

The comparison of results given in Figs. 6.1.1–6.1.4 is quite instructive. It shows that although the modified Matsui potential gave the best representation of the IR spectrum, and also yields the temperature of maximum density closest to the experimental value, it yields a diffusivity behavior that is quite incompatible with the experimental behavior. The simulated temperature range extends to 2750 K. This overlaps the experimental viscosity range, which extends to 2800 K. We find a discrepancy of some 4 orders of magnitude at 2800 K, and it will be increased even further by the finite-size effect. Thus we find that, although the time scale for oscillation of the particles in their potential minima is relatively well represented by the modified Matsui potential, the probability of their escaping from the potential minimum and thus diffusing is grossly overestimated. This implies that although the shape of the potential near its minimum is quite accurate, the curvature at distances more removed from the minimum is not well represented. In the latter respect the original TRIM potential, the Kubicki potential, and the BKS potential are all superior.

The TRIM potential diffusivity data, which show a moderate curvature, appear to be on a natural course to connect to the experimental data. The BKS data, although at first sight appearing to be too diffusive to be compatible with the Arrhenius experimental data, show considerable curvature, and if this accelerates, as observed for TRIM SiO_2 at higher densities [49], the data could conceivably join up with the experiments.

Recent extensions of the Ref. [28] studies, by Horbach et al. [50], in which extremely long runs were used, however, show a return to Arrhenius behavior at the lowest temperatures, presumably after passage through a thermodynamic transition of the sort we have discussed above. Such behavior would imply the existence of an interesting crossover in the liquid state from fragile to strong behavior as a function of temperature alone, as seems to be the case for H_2O according to simulation studies by Tanaka [51] and Sciortino et al. [29] and according to evidence from experimental data from Angell and co-workers [52, 53]. The apparent equilibrium cessation of the anomalous expansion observed in Fig. 6.1.1 for the modified Matsui potential would presumably be thermodynamic evidence that such a transition can occur. Previously, evidence for a (continuous) strong liquid to fragile liquid transition in SiO$_2$ has been provided only by behavior under compression [54, 55].

The possibility that a higher-order transition occurs within the liquid state of SiO$_2$ is a matter of great interest as the existence of such a transition would link the behavior of silica into the phenomenology of H$_2$O even more strongly than has been suggested to date [12, 49]. In laboratory SiO$_2$, however, it would seem to be possible only as a sub-T_g phenomenon, as the density maximum, below which the transition would have to occur, lies at 1800 K. Furthermore, confusion arises from the fact that laboratory SiO$_2$ is already a strong liquid at higher temperatures. This problem is certainly one that deserves further study [56]. We point out that the imminence of a compressibility maximum near T_g is susceptible to a convenient laboratory study with ambient-temperature x-ray-scattering determinations of density fluctuations that are frozen on quenching samples that have been equilibrated at different temperatures near and below the normal (10 K/min) glass transition.

As far as the best of the currently available pair potentials for SiO$_2$ simulations is concerned, the BKS potential seems to give a good representation of all aspects of the laboratory material's behavior [28, 50, 55], although it evidently needs further adjustment to yield a lower temperature for the maximum density (which will necessarily be pushed below the ergodicity curtain in the case of an accurate potential) and a wider splitting of the IR spectrum.

6.1.5   Conclusions

Our results highlight problems with, as well as strengths of, the pair potentials that have been proposed for the simulation of liquid SiO$_2$. It has been shown that pair potentials that will reproduce the behavior of silicalike systems are not only of importance to geophysics but are also of interest in the overall development of liquid-state phenomenology. The existence of waterlike anomalies and the associated possible existence of second critical points [30, 31] and even first-order polyamorphic transitions [57] in these quasi-ionic systems suggest that the further investigation of such potentials will be rewarding. Concerning the simulation of laboratory SiO$_2$, however, it must be concluded on the basis of the present study and the earlier vibrational spectroscopy simulations that a really adequate pairwise additive model has not yet been developed.

Acknowledgment

This work has been carried out under U.S. National Science Foundation Solid State Chemistry Grant DMR 9614531.

References

[1] L. V. Woodcock, C. A. Angell, and P. A. Cheeseman, J. Chem. Phys. **65**, 1565 (1976).

[2] This question was also examined, with the same results, for the analog weakened network system BeF_2 [B. Boulard, C. A. Angell, J. Kieffer, and C. C. Phifer, J. Non-Cryst. Solids **140**, 350 (1992)].

[3] C. A. Angell, P. A. Cheeseman, and S. Tamaddon, Bull. Mineral. **1–2**, 87 (1983); Science **218**, 885 (1982).

[4] C. A. Angell, P. A. Cheeseman, and R. R. Kadiyala, Chem. Geol. **62**, 85 (1987).

[5] C. A. Angell, C. Scamehorn, C. C. Phifer, R. R. Kadiyala, and P. A. Cheeseman, Phys. Chem. Miner. **15**, 221 (1988).

[6] J. Lucas, C. A. Angell, and S. Tamaddon, Mater. Res. Bull, **19**, 945 (1984)

[7] Y. Matsui, and K. Kawamura, Nature (London) **285**, 648 (1980). Note: A somewhat earlier MD study (P.A. Cheeseman, Ph.D. dissertation, Purdue University,1980) on the crystal dynamics of $MgSiO_3$ and its geochemical significance was never published in the open literature (after an initial rejection).

[8] P. H. Poole, P. F. McMillan, and G. H. Wolf, in *Structure, Dynamics and Properties of Silicate Melts*, J. F. Stebbins, P. F McMillan, and D. Dingwell, eds., Vol. 32 of Reviews in Mineralogy Series (Mineralogical Society of America, Washington, D.C. 1996), p. 563.

[9] R. W. Douglas and J. O. Isard, J. Soc. Glass Technol. **35**, 206 (1951); R. Bruckner, J. Non-Cryst. Solids **5**, 281 (1971).

[10] C. A. Angell, P. A. Cheeseman, and C. C. Phifer, Mater. Res. Soc. Symp. Proc. **63**, 85 (1986).

[11] D. R. Perchak and J. M. O'Reilly, J. Non-Cryst. Solids **167**, 211 (1994).

[12] P. H. Poole, M. Hemmati, and C. A. Angell, Phys. Rev. Lett. **79**, 2281 (1997)

[13] S. Tsuneyuki, M. Tsukada, M. Aoki, and Y. Matsui, Phys. Rev. Lett. **61**, 869 (1988).

[14] R. G. Della Valle and H. C. Andersen, J. Chem. Phys. **97**, 2682 (1992).

[15] M. Hemmati and C. A. Angell, J. Non.-Cryst. Solids **217**, 236 (1997).

[16] C. Angell, P. A. Cheeseman, and S. Tamaddon, Science **218**, 885 (1982).

[17] B. T. Poe, P. F. McMillan, D. C. Rubie, S. Chakraborty, J. Yarger, and J. Diefenbacher, Science **276**, 1245 (1997).

[18] M. Wilson, P. A. Madden, M. Hemmati, and C. A. Angell, Phys. Rev. Lett. **77**, 4023 (1996).

[19] S. H. Garofalini, J. Chem. Phys. **76**, 3189 (1982); Feuston and Garofalini (1988).

[20] P. A. Vashishta, R. K. Kalia, J. P. Rino and I. Ebbsjo, Phys. Rev. B **41**, 12197 (1990).

[21] B. Vessal, M. Amini, D. Fincham, and C. R. A. Catlow, Philos. Mag. B **60**, 753 (1989).

[22] D. C. Anderson, J. Kieffer, and S. Klarsfield, J. Chem. Phys. **98**, 8978 (1993).

[23] M. J. Sanders M. Leslie, C. R. A. Catlow, J. Chem. Soc. Chem. Commun. 1271 (1984).

[24] R. Car and M. Parrinello, Phys. Rev. Lett. **55**, 2471 (1985).

[25] A. A. Demkov, J. Ortega, O. F. Sankey, and M. P. Grumbach, Phys. Rev. B **52**, 1618 (1995): J. Sartheim, A. Pasquarello, and R. Car, Phys. Rev. Lett. **74**, 4682 (1995).

[26] C. A. Angell, Comput. Mater. Sci. **4**, 285 (1995).

[27] G. Adam and J. H. Gibbs, J. Chem. Phys. **43**, 139 (1965).

[28] K. Vollmayr, W. Kob, and K. Binder, Phys. Rev. B **54**, 15808 (1996)

[29] F. Sciortino, P. Gallo, P. Tartaglia, and S.-H. Chen, Phys. Rev. E **54**, 6331 (1996)

[30] P. H. Poole, F. Sciortino, U. Essmann, and H. E. Stanley (1992) Nature (London) **360**, 324; (1993); Phys. Rev. E **48**, 3799; Phys. Rev. E **48**, 4605 (1993).

[31] S. T. Harrington, R. Zhang, P. H. Poole, F. Sciortino, and H. E. Stanley, Phys. Rev. Lett. **78**, 2409 (1997).

[32] H. Lüdke and U. Landman, Phys. Rev. B **37**, 4656 (1988): **40**, 1164 (1989).

[33] C. A. Angell, S. Borick, and M. Grabow, J. Non-Cryst. Solids **205–207**, 463 (1996): C. A. Angell, J. Shao, and M. Grabow, in *Non-Equilibrium in Supercooled Fluids, Glasses and Amorphous Materials*, Eds. M. Giordano, D. Leporini and M. P. Tosi (World Scientific, Singapore, 1996), pp. 50–57.

[34] J. Wong and C. A. Angell, *Glass: Structure by Spectroscopy* (Marcel Dekker, New York, 1976), Chap. 11.

[35] J. D. Kubicki and A. C. Lasaga, Am. Mineral. **73**, 941 (1988).

[36] B. W. H. van Beest, G. J. Kramer, and R. A. van Santen, Phys. Rev. Lett. **64**, 1955 (1990).

[37] Y. Matsui, Phys. Chem. Miner. **169**, 234 (1988).

[38] The problem of high-temperature instability arises in the BKS and Tsuneyuki potentials from the way repulsion vanishes at small r values because of the parameters in the short-range van der Waals attractive terms. Such small r values are not seen in the range of temperatures originally studied with these potentials, but are obtained in higher-temperature studies, with disastrous consequences. To avoid the problem we follow Vollmayr et al. [28] in replacing the original potential function with a harmonic potential when r_{ij} is smaller than a critical distance [for BKS $r_{c(Si-O)} = 1.1936$ Å, $r_{c(O-O)} = 1.439$ Å, and for Tsuneyuki, $r_{c(Si-O)} = 1.280$ Å, $r_{c(O-O)} = 1.510$ Å].

[39] W. Kob and J.-L. Barrat, Phys. Rev. Lett. **78**, 4581 (1997)

[40] C. A. Angell, J. Non-Cryst. Solids **131–133**, 13 (1991).

[41] C. A. Angell and L. M. Torell, J. Chem. Phys. **78**, 937 (1983); S. A. Brawer, *Relaxation in Viscous Liquids* (American Ceramic Society, Columbus, OH, 1985).

[42] J. Horbach, W. Kob, K. Binder, and C. A. Angell, Phys. Rev. B **54**, 5897 (1996).

[43] P. Richet, Y. Bottinga, L. Denelieu, J. P. Petitet, and C. Tequi, Geochim. Cosmochim. Acta **46**, 2639 (1982).

[44] A. Navrotsky, Rev. Mineral. **29**, 309 (1994).

[45] S. Sastry, P. G. Debenedetti, F. Sciortino, and H. E. Stanley, Phys. Rev. E **53**, 6144 (1996).

[46] H. Yinnon and A. R. Cooper, Phys. Chem. Glasses **21**, 204 (1982).

[47] J. C. Mikkelsen, Jr., Mater. Res. Soc. Symp. Proc. **59**, 19 (1986); G. Brebec, R. Sebuin, C. Sella,

J. Bevenot, and J. C. Martin, Acta Metallogr. **28**, 327 (1980).

[48] J. Kieffer and C. A. Angell, J. Chem. Phys. **90**, 4982 (1989).

[49] C. A. Angell, P. H. Poole, and J. Shao, Nuovo Cimento **16D**, 993 (1994).

[50] Horbach, W. Kob, and K. Binder, Phil. Mag. B **77**, 297 (1998).

[51] H. Tanaka, Nature (London) **380**, 328 (1996).

[52] C. A. Angell, J. Phys. Chem. **97**, 6339, (1993).

[53] K. Ito, C. A. Angell, and C. T. Moynihan, Nature **398**, 492 (1999).

[54] J. Shao and C. A. Angell, in *Proceedings of the XVIIth International Congress on Glass* (1995), p. 311.

[55] B. Guillot and Y. Guissanni, Phys. Rev. Lett. **78**, 2401 (1997).

[56] Considerable progress has recently been made in this respect – see C. A. Angell, R. D. Brassel, M. Hemmati, E. J. Sare, and J. C. Tucker, Phys. Chem. Chem. Phys., **2**, 1559, 2000.

[57] P. H. Poole, T. Grande, C. A. Angell, and P. F. McMillan, Science **275**, 322 (1997).

Chapter 6.2

Transport Properties of Silicate Melts at High Pressure

Brent T. Poe and David C. Rubie

Diffusivity measurements in silicate liquids at high pressures have now been obtained up to 15 GPa and 2800 K by use of a multianvil apparatus. Most striking from these investigations is that oxygen and silicon self-diffusivities in $Na_2Si_4O_9$ liquid increase continuously as a function of pressure from 2.5 to 15 GPa. According to the Eyring relation, this would suggest that the viscosity of the liquid decreases over the same pressure range. However, as the composition of the liquid becomes more polymerized, approaching metaluminous albite, diffusivities reach a maximum at pressures below 10 GPa and decrease with further increases in pressure. These results demonstrate the importance of considering compositional parameters, such as degree of polymerization and $Al/Al + Si$ ratio, as well as pressure, when we are modeling magmatic processes in the Earth's deep interior.

6.2.1 Introduction

Magmatic processes within the Earth are controlled to a large extent by transport properties of silicate liquids. For example, both rates of magma ascent and rates of crystal settling during fractionation processes are controlled by the viscosity of silicate melts. In addition, ionic diffusion in silicate liquids controls rates of magma mixing, homogenization, and chemical equilibration and is therefore an important controlling parameter at all stages of magma evolution, from initial partial melting at depth to eruption or emplacement at or near the Earth's surface. Because a knowledge of the transport properties of silicate melts is essential for quantifying magmatic processes in the Earth, the viscosities of such melts have been studied extensively as

H. Aoki et al. (eds), *Physics Meets Mineralogy* © 2000 Cambridge University Press.

functions of temperature and chemical composition at 1 bar with a range of experimental techniques [1]. Although the results of such investigations can be applied to understanding magmatic processes at shallow depths in the Earth's crust, it is essential to take into account the effect of pressure on viscosity and diffusion when considering processes at depths greater than a few kilometers.

The effects of pressure on the transport properties of silicate liquids are particularly important for understanding the early history of the Earth. It is widely believed that the early Earth was involved in a collision with one or more planetary bodies roughly the size of Mars. The result of such a collision would have been extensive melting, leading to the Earth's being covered by a deep (e.g., 1000 km) magma ocean [2]. It is also likely that magma oceans formed on the Moon and perhaps on other planets of the solar system. The crystallization of a deep magma ocean is therefore likely to have determined the early chemistry and mineralogical structure of planetary interiors. Some models for magma ocean crystallization involve differentiation by simple fractional crystallization (involving crystal settling) and lead to an unlikely mineralogical and geochemical structure for the early interior of the Earth [3]. However, Tonks and Melosh [4] proposed that such models are not realistic because convection in a deep planetary-scale magma ocean would be very different from that in small magma bodies. They argued that convection would be so vigorous that even large crystals would be held in suspension, and they compared the process to aeolian transport in a deep planetary atmosphere. Their calculations were based on melt viscosities estimated from data obtained at 1 bar and moderate temperatures. To make such models more realistic, the effect of high pressure on viscosity must be considered (note that the pressure at a depth of 1000 km is of the order of 40 GPa). Furthermore, there are indications that the activation energy for viscous flow of silicate melts may decrease significantly with increasing pressure, at least for polymerized melts [5, 6]. In this case, the application of 1-bar activation enthalpies for estimating melt viscosities at high pressures (e.g., 20–40 GPa) by extrapolation from low temperatures (e.g., 1200 °C) to high temperatures (e.g., 2000–3000 °C) could lead to large errors.

A further motivation for studying transport properties at high pressures is that considerable information can be obtained concerning melt structure and the effects that pressure have on this property. This is especially the case when spectroscopic data are also available, for example for glasses quenched from melts at high pressures [7, 8].

It was first predicted with molecular-dynamics simulations of polymerized silicate liquids that diffusivities of network-forming ions pass through a maximum with increasing pressure [9]. Computer simulations such as these are valuable for predicting high-pressure behavior, but they often require unrealistically high temperatures (6000 K in the study by Angell et al. [9]) for determining diffusivities within reasonable computational times. Because of experimental difficulties, there have been only a few laboratory studies of viscosity and diffusion in silicate melts at elevated pressures (see below). Furthermore, the maximum pressures of most studies have been limited to ~2–3 GPa, which corresponds to depths of <100 km in the Earth. Consequently a systematic understanding of the effect of high pressure on viscosity and other transport properties is currently lacking.

6.2.2   Previous Experimental Studies

Determinations of the viscosity of silicate liquids at high pressures have been made previously by falling-sphere experiments [7, 10–12]. This method involves a sample containing a small high-density sphere (typically platinum) placed near the top of the capsule. At high pressure, the temperature is rapidly increased to superliquidus conditions, whereupon the sphere begins to descend because of the force of gravity. If the experiment is thermally quenched before the sphere reaches the capsule bottom, the rate of descent can be determined, allowing viscosity to be calculated by means of Stoke's Law:

$$\eta = \frac{2gr^2 \Delta\rho}{9v(1 + 3.3r/h_c)},$$

where η is viscosity, v is sphere velocity, g is acceleration that is due to gravity, r is sphere radius, h_c is container height, and $\Delta\rho$ is the density difference (sphere versus liquid). In cases in which the radius of the falling sphere is significant compared with the radius of the container, a wall correction factor is often used [13]. In addition to the difficulty of trapping the sphere midway in its descent by quenching at the appropriate time, one must assume that the sphere began its descent at the time at which the run temperature was reached and that the rate of descent was constant while it was descending. Otherwise, several experiments must be performed at the same P and T conditions, with varying time to construct time–distance relationships that give the sphere velocity [12]. Such experiments are feasible in a piston-cylinder apparatus (up to 3 GPa) because of the possibility of using a relatively long capsule (e.g., 5–10 mm).

It is generally accepted that conventional mechanisms for mass transport result in slower diffusion and higher viscosities as pressure is increased. However, studies up to 3 GPa show that polymerized silicate liquids display the reverse behavior and become more fluid with increasing pressure, as predicted by molecular-dynamics simulations [9]. Typically, viscosities can decrease by a factor of 3–10 when pressure is raised from 1 atm to 2 GPa. For example, Kushiro [10] found that the viscosity of jadeite liquid ($NaAlSi_2O_6$) decreased with increasing pressure by nearly a factor of 10 from 1 atm to 2.4 GPa at 1350 °C, whereas the viscosity of $Na_2Si_3O_7$ also decreased with pressure but by less than a factor of 3 from 1 atm to 2.0 GPa at 1175 °C. This study also illustrates the important role played by Al in the dynamics of strongly polymerized silicate liquids at high pressures (see below). Kushiro [11] later determined the viscosity of albite liquid up to 2 GPa and found that it also decreased with increasing pressure. Several other highly polymerized (NBO/T < 1, where NBO/T stands for the average number of non-bridging oxygens per tetrahedral cation) silicate liquids have also been found to show the same anomalous behavior at elevated pressures [7, 14–17]. This anomalous behavior appears to be related to the degree of polymerization of the melt. This was demonstrated by Brearley et al. [18], who measured viscosities along the albite–diopside join up to 2.5 GPa in a piston cylinder and found that as the melt became increasingly diopside rich, the pressure dependence of viscosity became less negative and eventually became positive near the diopside composition (NBO/T $= 2$). Scarfe et al. [12] also showed that the viscosities of some relatively depolymerized (NBO/T > 1) silicate liquids increase as a function of pressure up to 2 GPa at constant temperature.

Diffusion studies have also been performed at high pressures in a piston-cylinder apparatus, and these have resulted in similar conclusions. Shimizu and Kushiro [5] found that for jadeite ($NaAlSi_2O_6$) liquid, oxygen diffusivities increased as a function of pressure between 1 atm and 2 GPa at 1400 °C, whereas for diopside liquid, diffusivities decreased with pressure. Shimizu and Kushiro [5] also demonstrated that for a number of silicate melts that are polymerized, the inverse correlation between oxygen diffusivity and viscosity is well approximated by the Eyring relation:

$$D = k_B T / \eta \lambda,$$

where D is the oxygen diffusion coefficient, η is the viscosity, k_B is Boltzmann's constant, T is absolute temperature, and the diffusive length or jump distance λ is the diameter of an oxygen ion, 2.8 Å. The fact that the correlation works best with the short jump length suggests that the diffusing species linked to the mechanism for viscous flow is the O^{2-} anion. However, the more depolymerized melts, such as diopside, deviate from the correlation, indicating that larger ion clusters might be the diffusing units. Lesher et al. [19] determined oxygen and silicon diffusivities up to 2 GPa in a basaltic liquid with NBO/T $= 1$ and showed that jump distances of the order of twice the diameter of the O^{2-} ion give better fits to the Eyring relation when compared with viscosities calculated for the liquid by the method of Shaw [20]. Rubie et al. [6] determined oxygen diffusivities for a polymerized silicate liquid ($Na_2Si_4O_9$) up to 10 GPa, and Poe et al. [21] determined both silicon and oxygen diffusivities for $Na_2Si_4O_9$, extending the pressure range to 15 GPa. In the latter study, the effects of changing degree of polymerization and Al/(Al + Si) ratio were also investigated by the inclusion of two aluminosilicate compositions.

Very little viscosity data for silicate liquids exist at pressures greater than 3 GPa, i.e., at conditions beyond those that can be attained in a piston-cylinder apparatus. Although pressures in the range 3–25 GPa and temperatures up to 3000 K can be produced routinely in the multianvil apparatus [22], sample sizes are much smaller than in piston-cylinder experiments, with capsule lengths ranging from 1 to 4 mm, depending on pressure. This small sample size, combined with high liquidus temperatures at high pressures, makes the falling-sphere technique much more difficult and its results subject to much greater uncertainties than in lower-pressure piston-cylinder experiments. One possibility for making viscosity measurements at high pressure by the falling-sphere method is to measure the rate of descent of the sphere by use of an in situ technique. For example, Kanzaki et al. [23] determined the viscosity and density of albite liquid in a multianvil apparatus by monitoring the velocity of the falling sphere by using synchrotron radiation. The difference between the x-ray absorption coefficients for Pt and silicate liquid means that the falling sphere can be imaged by x rays so that its position can be monitored in real time. However, the viscosities determined in this preliminary study were much lower than expected, possibly because the melts were contaminated with volatiles derived from the pressure medium [23]. Further technical developments are required before this method can be considered as reliable. Another possible approach is to monitor the falling sphere by electro-detection, a method that has so far been used up to 3 GPa on Fe–S melts [24].

An alternative to measuring viscosity directly at very high pressures is to determine oxygen diffusivities experimentally and to calculate viscosities by use of the Eyring relation. Such experiments have now been carried out at pressures up to 15 GPa, five times greater than that of the highest-pressure viscosity measurements for similar samples [6, 21]. In this study, we have determined both silicon and oxygen self-diffusivities in highly polymerized silicate melts up to 15 GPa and 2800 K. By using the Eyring relation, which has been demonstrated to be valid for such melts, we are therefore able to predict viscosities for these liquids at pressure and temperature conditions far beyond the range currently possible for the falling-sphere method.

6.2.3   Experimental Methods

Diffusivity experiments have been performed up to 15 GPa with a 1000-tonne split-cylinder multianvil apparatus. In experiments up to 10 GPa, an 18M octahedral assembly (18-mm edge-length MgO octahedron) was utilized into which a capsule 2.0 mm in diameter and 3.5 mm long can be inserted. Experiments above 10 GPa required a 14M assembly (14-mm edge-length octahedron) with capsule dimensions of 1.6 mm in diameter and 2.7 mm long. Sample assemblies contained cylindrical $LaCrO_3$ furnaces, capable of producing temperatures near 3000 K, with relatively small temperature gradients along the length of the sample because of the stepped wall thickness (Fig. 6.2.1). Temperature was measured with WRe_3/WRe_{25} thermocouples (0.25-mm-diameter wires) located adjacent to the sample capsule (Fig. 6.2.1) with no correction for the effect of pressure. Re capsules were loaded with finely ground glasses as starting materials. One half of the capsule was filled with a glass enriched in either ^{18}O or both ^{18}O and ^{30}Si. The remaining half of the capsule was filled with a glass containing O and Si isotopes in their natural abundances. A polished piston used to pack the powders into the capsule provided a sharp initial interface perpendicular to the axis of the capsule between the two glasses. Pressure was generated with a 6–8-type multianvil arrangement in which eight 32-mm tungsten carbide cubes, with truncated corners, compressed the octahedral assembly [22, 25]. Truncation edge lengths on the tungsten carbide cubes were 11 mm for the 18M assembly and 8 mm for the 14M assembly. Sample pressure was estimated from calibrations of pressure as a function of hydraulic oil pressure (i.e., applied force) based on high-temperature phase-equilibria experiments on silicates [6]. In each experiment, the hydraulic oil pressure was first increased to the desired value over several hours before the sample was heated. Heating was performed with a three-step procedure: (1) manual heating to 100 °C, (2) ramping to 800 °C at 1000 °C/min, and (3) ramping to the final run temperature at 2500 °C/min with no overstep. Rapid heating minimizes the amount of diffusion that occurs before

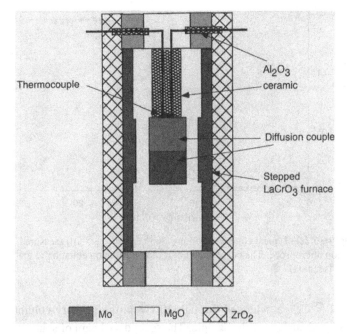

Figure 6.2.1 Cross section of the cylindrical heater + sample
assembly that is contained in the MgO octahedral pressure cell in
multianvil high-pressure experiments. The sample (diffusion couple)
is contained in a Re capsule that is initially 3.5 mm long in
experiments up to 10 GPa (18M assembly) and 2.8 mm long in
higher-pressure experiments (14M assembly). The isotopically
enriched glass is located in the lower half of the capsule because its
slightly greater density stabilizes the sample and helps to prevent
convection. The ZrO_2 sleeve provides thermal insulation.

the desired temperature is reached. Samples were isobarically quenched after a given
length of time of up to 240 s. Quench rates were estimated to be of the order of 400
to 500 °C/s. Following quench, samples were slowly decompressed over 10–15 h.
Recovered samples were transparent glasses with no evidence of crystallization. Test
experiments in which a thin layer of graphite powder was placed at the interface between the isotopically enriched and isotopically normal halves of the sample showed
that any convection that occurred during experiments could not be detected by optical
microscopy. The samples were sectioned lengthwise along the cylindrical axis of the
capsule and polished for ion microprobe analysis. Selected samples were analyzed
for water content by Fourier-transform IR spectroscopy and showed amounts ranging
from 10 to 50 parts in 10^6 by weight.

6.2.4   Ion-Microprobe Probe Analysis

Samples were analyzed for concentrations of ^{16}O, ^{18}O, ^{28}Si, and ^{30}Si with a Cameca
IMS-4f ion microprobe at the University of Edinburgh, U.K. [26]. The primary beam

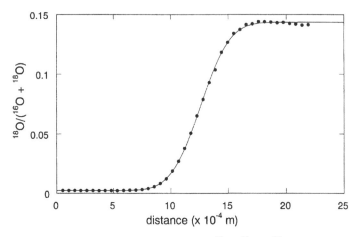

Figure 6.2.2 Typical concentration profile [$^{18}O/(^{18}O + {}^{16}O)$] measured by ion microprobe. The curve shows a fit of the diffusion equation to the data (see text).

was 1–3 nA of $^{133}Cs^+$ accelerated through 10 kV in the primary column onto the Au-coated polished surface of the sample. The secondary beam of sputtered negative ions was accelerated through 4.5 kV into the mass analyzer. The sample surface was illuminated during analysis by a normal-incidence low-energy electron flood gun.

Concentration profiles were measured along the entire length of the samples, perpendicular to the diffusion couple interface. The primary beam was defocused to give a spot diameter of 30–60 μm of which only the central area (8-μm diameter) was sampled. At each analysis point, the sample was sputtered for 2 min before counts were collected for typically 100-s duration. This avoided sampling contaminants, such as ^{16}O derived from absorbed H_2O, that were often present on the sample surface. The errors associated with such measurements have been discussed in detail by Rubie et al. [6]. Two parallel profiles were analyzed on selected samples, and the consistency of these indicates that convection did not occur during the experiments.

After ion-microprobe analysis, the position of each analysis point was measured to an accuracy of ± 5 μm by an optical microscope fitted with a vernier X–Y stage. A typical concentration profile is shown in Fig. 6.2.2.

6.2.5   Results and Discussion

In most cases, diffusivities were obtained by fitting of the concentration profiles with an expression for diffusion between two semi-infinite bodies:

$$C(x) = (C1 + C2)/2 + (C1 - C2)/2^*\mathrm{erf}[(x - Xo)/D],$$

where concentration of the isotope (C) varies with position along the diffusion profile (x), $C1$ and $C2$ are the isotopic concentration limits, Xo is the position of the diffusion

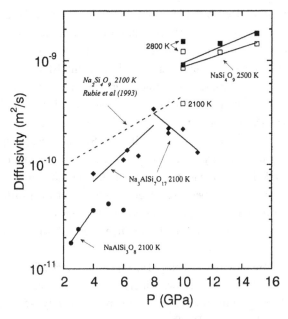

Figure 6.2.3 Oxygen (filled symbols) and silicon (open symbols) self-diffusivities determined in the pressure range 2–15 GPa for the compositions $NaAlSi_3O_8$, $Na_3AlSi_7O_{17}$, and $Na_2Si_4O_9$. The trend of earlier data [6] for $NaSi_4O_9$ is shown for comparison.

couple interface, and D is the diffusion coefficient. In cases of relatively fast diffusion (e.g., $Na_2Si_4O_9$ above 10 GPa), an expression for diffusion in finite media that accounts for boundary interactions was required [27].

Self-diffusion coefficients for silicon and oxygen in $Na_2Si_4O_9$ liquid and for oxygen in $Na_3AlSi_7O_{17}$ and $NaAlSi_3O_8$ liquids are plotted as functions of pressure in Fig. 6.2.3. For $Na_2Si_4O_9$ in which the melt is slightly depolymerized (NBO/T = 0.5), we have extended the experimental pressure range from 10 to 15 GPa. To ensure that all experiments were conducted at superliquidus conditions, experimental run temperatures were as high as 2800 K. Oxygen self-diffusivities continue to increase with increasing pressure from 10 to 15 GPa at 2500 K, which is consistent with the lower-pressure results of Rubie et al. [6]. We find no evidence for a diffusivity maximum, which suggests that, if the Eyring relation continues to be valid at pressures above 10 GPa, the viscosity of $Na_2Si_4O_9$ melt decreases with increasing pressure over the pressure range of 1 atm to 15 GPa. We note that Angell et al. [9] predicted an oxygen diffusivity maximum for sodium trisilicate liquid ($NaSi_3O_7$) above 20 GPa in their molecular-dynamics study. A more recent MD study of $Na_2Si_4O_9$ liquid by Diefenbacher et al. [28] showed an oxygen diffusivity maximum near 15 GPa at a simulation temperature of 6000 K. Silicon self-diffusivities for $Na_2Si_4O_9$ also increase with pressure and are very close to, but slightly lower than, the oxygen diffusivities. Lesher et al. [19] made similar observations for their more depolymerized basaltic liquid and indicated that the diffusion of the two elements is cooperative, although translation distances (λ) greater than the diameter of a single O^{2-} ion were required for fitting the Eyring relation. The pressure dependence for oxygen diffusion in $Na_2Si_4O_9$ from 10 to 15 GPa is in good agreement with the lower-pressure study, giving an activation volume of -2.8 cm^3/mol compared with -3.3 cm^3/mol between 2.5 and 10 GPa at 2100 K determined by Rubie et al. [6]. Although viscosity data for $Na_2Si_4O_9$ liquid are not available at high pressures, Dickinson et al. [7] determined $K_2Si_4O_9$ viscosity up to 2.4 GPa at 1200 °C, revealing a decrease by a factor of ~10. This suggests that the Eyring relation is valid at the alkali tetrasilicate composition although the pressure dependences of the two are not equivalent, because of either the effect of a different network-modifying cation or the fact that the $K_2Si_4O_9$ viscosity measurements were conducted at pressure and temperature conditions lower than those of our $Na_2Si_4O_9$ diffusion experiments.

For the slightly more polymerized $Na_3AlSi_7O_{17}$ liquid (NBO/T = 0.25) at 2100 K, oxygen diffusivities initially increase with increasing pressure and then pass through a maximum near 8 GPa. Activation volumes are approximately -5.4 cm^3/mol below 8 GPa and 4.7 cm^3/mol above 8 GPa. This represents the first experimental observation of such a diffusivity maximum as a function of pressure in a polymerized silicate liquid, confirming the behavior predicted by Angell et al. [9] for liquids such as jadeite ($NaAlSi_2O_6$) and $NaSi_3O_7$.

For the fully polymerized albite ($NaAlSi_3O_8$) liquid (NBO/T = 0), oxygen self-diffusivities initially increase with pressure at 2100 K and also pass through a maximum that is located at a slightly lower pressure near 5 GPa. Because viscosity data for albite liquid are available up to 2 GPa [11], the diffusivities determined for albite liquid allow us to test the validity of the Eyring relation. In Fig. 6.2.4, we compare viscosities calculated from our experimentally determined diffusivities by using the Eyring relation with viscosities determined by Kushiro [11] with the falling-sphere technique in a piston-cylinder apparatus. Because the two studies were not conducted at the same temperature, we used an activation energy of 400 kJ/mol for the viscosity of albite at 1 atm [29] to adjust the viscosities from Kushiro [11] (1350 °C) to the same temperature at which the diffusivities were determined (2100 °C). In addition to absolute viscosities, which are in close agreement where the two sets of data nearly overlap, the pressure dependencies also match well until at higher pressure, where the diffusion data suggest that viscosities pass through a minimum near 5 GPa. Recent viscosity measurements of albite liquid confirm that a local minimum exists at 5 GPa, but viscosity again decreases with pressure between 6 and 7 GPa [30].

Based on the results of this study and previous investigations, the viscous flow behavior of silicate melts and its pressure dependence are strongly correlated with the degree of polymerization, i.e., the proportion of bridging to nonbridging oxygens. Highly polymerized, Al-bearing melts that have an anionic network composed mostly

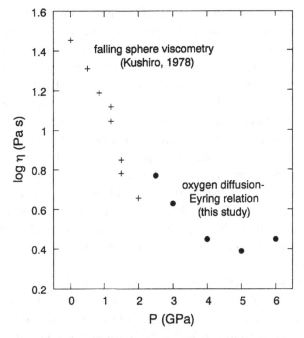

Figure 6.2.4 Viscosities determined for $NaAlSi_3O_8$ liquid by falling-sphere viscometry up to 2 GPa[11] (Kushiro et al. [14]) compared with viscosities estimated in this study up to 15 GPa from oxygen diffusion data by the Eyring relation.

of bridging oxygens linking Si^{4+} and Al^{3+} tetrahedral cations behave anomalously in that their viscosities initially decrease with increasing pressure. The viscosities of these melts pass through a minimum at pressures below 10 GPa and increase again with higher pressures. The viscosities of more depolymerized Al-free melts also decrease with pressure and therefore show similar anomolous behavior, but a viscosity minimum is not apparent up to pressures of 15 GPa.

The anomalous effect of pressure on transport properties of polymerized silicate liquids can be explained by structural changes that occur at high pressure. Waff [31] first suggested that Al in silicate melts transforms from a tetrahedrally coordinated cation to an octahedrally coordinated cation at high pressure. Since then, spectroscopic studies of glasses quenched from superliquidus conditions at high pressure have provided an enormous amount of insight into the structural properties of silicate liquids at high pressure. Evidence for a pressure-induced increase in coordination of a network-forming cation was first made by ^{29}Si MAS nuclear magnetic resonance spectroscopy. Stebbins and McMillan [32] observed both five- and six-coordinated Si in a $K_2Si_4O_9$ glass quenched from 1.9 GPa. Xue et al. [8] found that increasing amounts of both five- and six-coordinated silicon could be quenched into glasses with increasing pressure. Stebbins [33] demonstrated that, even at 1 atm, small amounts of ^VSi could be quenched from melt to glass and that the amount was quench rate dependent.

Angell et al. [9] noted from their simulations that a five-coordinated silicon acts as a transient intermediate molecular species in the oxygen diffusion mechanism. Thus, although not required, increasing pressure favors the formation of high-coordinate silicon and consequently enhances the diffusion process of network forming ions. Formation of high-coordinate Si species is most favored in highly polymerized silicate liquids, such as $K_2Si_4O_9$ (NBO/T = 0.5), in which the melt retains an open three-dimensional network structure but has some available nonbridging oxygens. Normally associated with network-modifying cations, these nonbridging oxygens can attack a neighboring tetrahedral silicate unit to temporarily form a five-coordinate species. Then, either the reverse process occurs such that the original nonbridging oxygen is reformed or one of the other four Si—O bonds is broken to form a new nonbridging oxygen, completing the oxygen diffusion step. Whereas the energy required for breaking a normal tetrahedral Si—O bond is quite high (a few hundred kilojoules per mole), the energy needed to form the five-coordinate silicate species is much lower ($\sim$30 kJ/mol) [33].

The evidence for Al coordination changes in high-pressure melts is even more dramatic. Yarger et al. [34] found that the proportions of ^{V}Al and ^{VI}Al in a partially depolymerized sodium aluminosilicate ($Na_3AlSi_7O_{17}$) glass reach nearly 50% of the total Al when quenched from 8 and 12 GPa. However, when placed in the context of the total network-forming cations (i.e., Al + Si), the results of Yarger et al. [34] for this composition are comparable with those from $Na_2Si_4O_9$ glasses [8], as shown in Fig. 6.2.5. Over a similar range of pressure, glasses quenched from melts contain similar amounts of high coordinate network-forming cations, the concentrations of which increase with increasing pressure. However, for the aluminosilicate glasses, the concentration of five-coordinate Al passes through a maximum near 8 GPa, whereas the abundance of all the other high-coordinate species continue to increase up to the maximum pressure at which the melts could still be quenched to glasses. The amount of ^{V}Al in the aluminosilicate glass at 8 GPa, which corresponds to $\sim$4% of the total number of network-forming cations in the glass, is similar to the amount of ^{V}Si in the Al-free silicate glass at 8 GPa. With increasing pressure, additional ^{V}Si is formed in the silicate but the amount of ^{V}Al decreases in the aluminosilicate. This difference is likely to account for the diffusivity maximum at 8 GPa for $Na_3AlSi_7O_{17}$ liquid and the continued increase in diffusivities for $Na_2Si_4O_9$ liquid observed in this study. ^{VI}Si and ^{VI}Al abundances increase with pressure in each of the glasses over the entire pressure range, suggesting that these octahedrally coordinated cations are not likely to play a role in the oxygen transport mechanism. Only the formation of the five-coordinate Al or Si cation is likely to influence oxygen diffusion, and hence viscosity, in silicate melts at elevated pressures.

In this study we have focused on melt compositions that are highly polymerized in order to explore the nature of the anomalous transport behavior at high pressure. In addition, there are significantly more structural data on glasses quenched from high pressures in this compositional range, thus enabling transport properties and melt

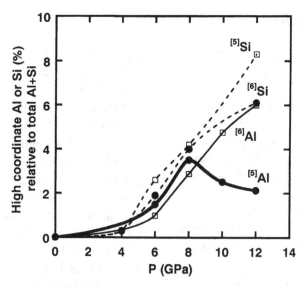

Figure 6.2.5 Concentrations of five- and six-coordinated Al
in $NaAlSi_7O_{17}$ glasses (data from Yarger et al. [34]) and five-
and six-coordinated Si in $Na_2Si_4O_9$ glasses (data from Xue
et al. [8]) quenched from melts as a function of pressure. The
curves (solid curves for Al and dashed curves for Si) are
intended to show only the general trends.

structure to be related. More depolymerized silicate liquids, composed of both monovalent and divalent cations, are expected to be more relevant to magmatic processes in
the deep interior of the Earth and are therefore worthy of future experimental study.
Such melts may not behave as those studied here nor may the Eyring relation be valid.
An additional problem is that depolymerized melts do not quench to glasses easily
at high pressure so that structural information is more difficult to obtain. According
to preliminary studies, the viscosities of depolymerized melts increase with pressure.
The viscosity of $CaMgSi_2O_6$ melt, for example, increases by 0.5 log units between 1
bar and 1.5 GPa at 1600 °C [18]. Such results suggest that viscosity might increase
by orders of magnitude in a deep magma ocean and therefore have a large effect on
fractionation processes at depth.

Acknowledgments

We gratefully acknowledge J. Craven, who performed the ion microprobe analyses at the University of Edinburgh, U.K., using a facility that is supported by the
Natural Environment Research Council. We thank S. Chakraborty, J. Diefenbacher,
P. F. McMillan, and J. Yarger who contributed to this study through discussions and
preparation of isotopically enriched glasses.

References

[1] D. B. Dingwell, in *Mineral Physics and Crystallography, A Handbook of Physical Constants*, T. Ahrens, ed. (American Geophysical Union, Washington, D.C., 1995), p. 209.

[2] H. J. Melosh, in *Origin of the Earth*, H. J. Newsom and J. H. Jones, eds. (Oxford U. Press, New York, 1990), p. 69.

[3] A. E. Ringwood, in *Origin of the Earth*, H. J. Newsom and J. H. Jones, eds. (Oxford U. Press, New York, 1990), p. 101.

[4] W. B. Tonks and H. J. Melosh, in *Origin of the Earth*, H. J. Newsom and J. H. Jones, eds. (Oxford U. Press, New York, 1990), pp. 151–174.

[5] N. Shimizu and I. Kushiro, Geochim. Cosmochim. Acta **48**, 1295 (1984).

[6] D. C. Rubie, C. R. Ross II, M. R. Carroll, and S. C. Elphick, Am. Mineral. **78**, 574 (1993).

[7] J. E Dickinson Jr, C. M. Scarfe, and P. F. McMillan J. Geophys. Res. **95**, 15675 (1990).

[8] X. Xue, J. F. Stebbins, M. Kanzaki, P. F. McMillan, and B. Poe, Am. Mineral. **76**, 8 (1991).

[9] C. A. Angell, P. A. Cheeseman, and S. Tamaddon, Science **218**, 885 (1982).

[10] I. Kushiro, J. Geophys. Res. **81**, 6347 (1976).

[11] I. Kushiro, Earth Planet. Sci. Lett. **41**, 87 (1978).

[12] C. M. Scarfe, B. O. Mysen, and D. Virgo, in *Magmatic Processes: Physicochemical Principles*, B. O. Mysen, ed. (The Geochemical Society, University Park, PA 1987), p. 59.

[13] H. R. Shaw, J. Geophys. Res. **68**, 6337 (1963).

[14] I. Kushiro, H. S. Yoder, Jr., and B. O. Mysen, J. Geophys. Res. **81**, 6351 (1976).

[15] C. M. Scarfe, Carnegie Inst. Wash. Yearb. **80**, 336 (1981).

[16] T. Fujii and I. Kushiro, Carnegie Inst. Wash. Yearb. **76**, 419 (1977).

[17] T. Fujii and I. Kushiro, Carnegie Inst. Wash. Yearb. **76**, 461 (1977).

[18] M. Brearley, J. E. Dickinson, Jr., and C. M. Scarfe, Geochim. Cosmochim. Acta **50**, 2563 (1986).

[19] C. E. Lesher, R. L. Hervig, and D. Tinker, Geochim. Cosmochim. Acta **60**, 405 (1996).

[20] H. R. Shaw, Am. J. Sci. **272**, 870 (1972).

[21] B. T. Poe, P. F. McMillan, D. C. Rubie, S. Chakraborty, J. Yarger, and J. Diefenbacher, Science **276**, 1245 (1997).

[22] R. C. Liebermann and Y. Wang, in *High Pressure Research: Application to Earth & Planetary Sciences*, Y. Syono and M. H. Manghnani, eds. (American Geophysical Union, Washington, D.C., 1992), p. 19.

[23] M. Kanzaki, K. Kurita, T. Fujii, T. Kato, O. Shimomura, and S. Akimoto, in *High-Pressure Research in Mineral Physics: The Akimoto Volume*, M. H. Manghnani and Y. Syono, eds. (Terra Scientific, Tokyo, and American Geophysical Union, Washington, D.C., 1987), p. 195.

[24] G. E. LeBlanc and R. A. Secco, Geophys. Res. Lett. **23**, 213 (1996).

[25] D. C. Rubie, Phase Transitions **68**, 431–451 (1999).

[26] G. Slodzian, *Advances in Electronics and Electron Physics*, Suppl. 138 (Academic, New York, 1980), p. 1.

[27] J. Crank, *The Mathematics of Diffusion* (Clarendon, London, 1975).

[28] J. Diefenbacher, P. F. McMillan, and G. H. Wolf, J. Phys. Chem. B. **102**, 3002 (1998).

[29] D. Cranmer and D. R. Uhlmann, J. Geophys. Res. **86**, 7951 (1981).

[30] S. Mori, E. Ohtani, and A. Suzuki, Earth and Planet. Sci. Lett. **175**, 87 (2000).

[31] H. S. Waff, Geophys. Res. Lett. **2**, 193 (1975).

[32] J. F. Stebbins and P. McMillan, Am. Mineral. **74**, 965 (1989).

[33] J. F. Stebbins, Nature (London) **351**, 638 (1991).

[34] J. L. Yarger, K. H. Smith, R. A. Nieman, J. Diefenbacher, G. H. Wolf, B. T. Poe, and P. F. McMillan, Science **270**, 1964 (1995).

Chapter 6.3

Structural Characterization of Oxide Melts with Advanced X-Ray-Diffraction Methods

Yoshio Waseda and Kazumasa Sugiyama

X-ray-diffraction technique is one of the powerful tools for determining the atomic scale structures of oxide melts at high temperature. This chapter describes the conventional angle dispersive x-ray-diffraction (ADXD) results with some selected examples of pure oxide, binary silicate and borate melts. The usefulness of the energy dispersive x-ray-diffraction (EDXD) method coupled with a solid-state detecting system has been demonstrated by obtaining the results for the complex oxide melts of $NaAlSi_3O_8$ and $LiNbO_3$. The relatively new devised technique of anomalous x-ray scattering (AXS), which enables us to provide the environmental structural image around a specific element, is also discussed by using the local structure around Ge in germanate glasses.

6.3.1 Introduction

There is an increasing demand for understanding various properties of oxide melts, not only from metallurgical and petrological points of view but also from a new perspective of the crystal-growth technique through molten states. This has encouraged number of measurements of density, viscosity, surface tension, and electrical conductivity of oxide melts. To clarify their characteristic properties, a knowledge of the atomic-scale structures of oxide melts is essential, and particularly x-ray diffraction has been one of the most popular methods for the in situ structural analysis of materials since the first liquid result made by Debye in 1915 [1]. Recently, several modern x-ray-diffraction techniques have been developed for determining the fine local structure of oxides in both the molten and the glassy states. This chapter provides an extended discussion

H. Aoki et al. (eds), *Physics Meets Mineralogy* © 2000 Cambridge University Press.

of the structure of oxide melts at high temperatures obtained by several advanced diffraction techniques.

6.3.2   **Ordinary Angular-Dispersive X-Ray Diffraction**

The atomic-scale structure of oxide melts can be obtained by means of a high-temperature x-ray-diffraction technique [2, 3]. The description of the atomic-scale structure for disordered (noncrystalline) systems uses the radial distribution function (RDF). The RDF gives the probability of finding atoms from an origin as a function of the radial distance obtained by spherical and time averaging. The RDF result is one dimensional but it gives almost unique quantitative information that describes the atomic-scale structure in a disordered system.

The RDF analysis of melts and glasses by monochromatic x-rays [angular-dispersive x-ray diffraction (ADXD)] is now well developed and has been described together with the experimental details of x-ray intensity measurements at high temperatures [2–5]. Figure 6.3.1 shows a typical experimental setup of this conventional ADXD technique with a high-temperature chamber for a structural study of high-temperature melts (see, for example, Ref. [6]). This system consists of a θ–θ type goniometer in the vertical setting, in which an x-ray tube and a scintillation counter can rotate in opposite directions. The pyrolytic graphite monochromator in the diffracted beam path coupled with a pulse-height analyzer is frequently used for discriminating between Mo $K\alpha$ radiation and others such as Mo $K\beta$ and fluorescence from samples. This configuration makes it possible to measure the scattering intensities from the free surface of a melt sample. The high-temperature chamber mounted at the center of a goniometer axis

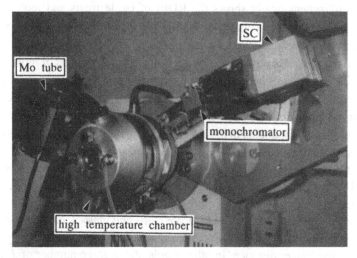

Figure 6.3.1 Overall view of an ADXD spectrometer with a high-temperature chamber for the structural study of a melt. SC, scintillation counter.

has a slot with a Be window through the water-cooled enclosure wall. This allows the passage of x rays and controls the atmosphere around melt samples. An ADXD profile is measured as a function of scattering angle 2θ by monochromatic x rays, and the ordinary measurement up to $2\theta = 120°$ with Mo $K\alpha$ radiation can cover a wave-vector region $Q = 4\pi \sin\theta/\lambda$ of ~150 nm^{-1}, where λ is the wavelength of the x rays. The observed scattering intensity $I_{obs}(Q)$ may be expressed by

$$I_{obs}(Q) = PAC[I_{coh}(Q) + I_{inc}(Q)], \tag{6.3.1}$$

where P is the polarization factor, A is the absorption factor, C is the geometric constant factor, including the so-called normalization constant, and $I_{coh}(Q)$ and $I_{inc}(Q)$ are the coherent and incoherent scattering intensities in electron units per atom, respectively. The factors P and A are dependent on scattering angle and the energy of the x rays as well as experimental conditions, and these are readily calculated by the analytical equations and the compiled absorption coefficient data [7]. After correcting for polarization and absorption, we can obtain $I_{coh}(Q)$ by the generalized Krogh–Moe–Norman method [2–5], including the consideration for Compton scattering [8–10]. Then the interference function $Qi(Q)$ can be obtained from the coherent scattering intensities in the following form:

$$Qi(Q) = Q[I_{coh}(Q) - \langle f^2 \rangle]/\langle f \rangle^2, \tag{6.3.2}$$

where $\langle f \rangle$ is the average atomic scattering factor and $\langle f^2 \rangle$ is average square atomic scattering factor. By the conventional Fourier transform of the function $Qi(Q)$, the ordinary RDF is readily calculated with the number density value of the sample ρ_0:

$$4\pi r^2 \rho(r) = 4\pi r^2 \rho_0 + \frac{2r}{\pi} \int_0^\infty Qi(Q) \sin Qr \, dQ. \tag{6.3.3}$$

Figure 6.3.2 schematically shows the RDFs of oxide melts and molten metals. Although the general features of these two RDFs are similar, the first peak for the oxide melt is almost completely resolved. Such a difference in the first peaks implies that the fundamental feature of the structure of oxide melts differs from that of molten metals. That is, oxide melts and glasses display a characteristically distinct local ordering within a narrow region accompanied by a complete loss of positional correlation at a few nearest-neighbor distance regions away from any origin. For this reason, one of the most important components in the structural study of oxide melts and glasses is to determine such distinct local ordering unit structures and their distribution.

Although no unique method for estimating the distance and the coordination number from the RDF results appears to exist at the present time, the interference function refining method is considered to be quite useful [3, 11]. This technique is based on the characteristic structural features of silicate melts and glasses, namely, the contrast between the narrow distribution of local ordering and a complete loss of positional correlation at the longer distance. In other words, the average number of j elements around i elements, N_{ij}, is separated by an average distance r_{ij}. The atomic distribution for $i-j$ pairs can be approximated by a discrete Gaussian-like distribution with a mean-square variation $2\sigma_{ij}$. The distribution for higher-neighbor correlations is likely to be

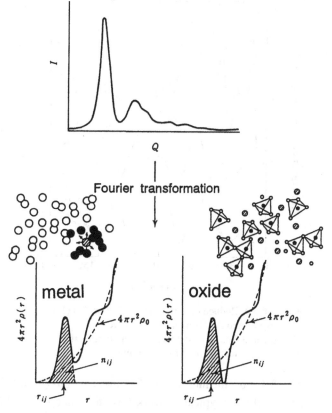

Figure 6.3.2 Typical RDFs and corresponding schematic structural features for molten metals and oxides.

given by a continuous distribution with the average number density of a system. This can be expressed by the following equation with respect to the reduced interference function $i(Q)$:

$$i(Q) = \sum_i \sum_j N_{ij} \, \exp(-\sigma_{ij} Q^2) f_i f_j \frac{\sin(Q r_{ij})}{(Q r_{ij})}$$

$$+ \sum_\alpha \sum_\beta \{\exp(-\sigma'_{\alpha\beta} Q^2) f_\alpha f_\beta 4\pi \rho_0 [Q r'_{\alpha\beta} \cos(Q r'_{\alpha\beta}) - \sin(Q r'_{\alpha\beta})]\} / Q^3.$$

$$(6.3.4)$$

The parameters $r'_{\alpha\beta}$ and $\sigma'_{\alpha\beta}$ in the boundary region are known not to be sharp [11]. The structural parameters for near-neighbor correlations are determined by a least-squares analysis so as to fit the experimental interference function $Qi(Q)$ by iteration. It should be stressed here that this method is not a unique mathematical procedure, but a useful semiempirical one for the resolution of the peaks in the RDF of disordered systems, accomplished by estimation of the fundamental local ordering units and the oxygen coordination number around a cation in oxide melts with variation of ±0.001 nm for r_{ij} and ±0.2 for N_{ij}.

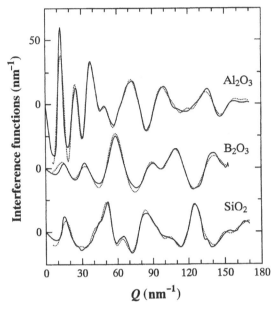

Figure 6.3.3 Reduced interference functions $Qi(Q)$ of SiO$_2$, B$_2$O$_3$, and Al$_2$O$_3$ melts. Solid curves, experimental data; dotted curves, calculation by the interference function refining technique.

Figures 6.3.3 and 6.3.4 show the interference function data $Qi(Q)$ and the resultant electron RDFs, respectively, for typical network-forming oxide melts of SiO$_2$ [12], B$_2$O$_3$ [13], and Al$_2$O$_3$ [14]. The solid curves in Fig. 6.3.3 are the experimental data, and the dotted curves correspond to the converged results of the least-squares refining technique starting from the respective crystalline structures of α-SiO$_2$, B$_2$O$_3$, and α-Al$_2$O$_3$. The solid curves of electron RDFs [3, 14] in Fig. 6.3.4 are the experimental data, and the dotted curves correspond to the theoretical predictions made with the pair function method proposed by Mozzi and Warren [15]. The structural parameters for the near-neighbor correlations in these simple oxide melts are summarized in Table 6.3.1. The agreement is satisfactory with respect to the structural parameters determined by the pair function method and the interference function refining technique. Thus it is suggested that the structural parameters determined by either method can be quite realistic. The structural parameters for the Si–O pair in SiO$_2$ melts indicate that each silicon is surrounded by four oxygens at a distance of 0.162 nm. These results imply that the SiO$_4$ tetrahedron is a realistic fundamental local ordering unit. Thus SiO$_2$ melts can be viewed as mixtures of small ordering unit structures of such SiO$_4$ tetrahedra, although their correlations decay rapidly at larger distances. Similarly the structural parameters for the first neighboring B–O pair in B$_2$O$_3$ melts indicate that each boron is surrounded by three oxygens at a distance of 0.138 nm so as to form fundamental structural units of BO$_3$ triangles. In the case of the Al$_2$O$_3$ melt, the atomic arrangement of AlO$_6$ octahedra with the averaged Al–O distance of 0.202 nm is easily confirmed as an unique fundamental structural unit.

Table 6.3.1. *Structural parameters of local ordering in* SiO_2, B_2O_3, *and* Al_2O_3 *melts*

Pair	Pair Function Method		Refining Technique	
	r_{ij} (nm)	N_{ij} (atom)	r_{ij} (nm)	N_{ij} (atom)
SiO_2 melt				
Si—O	0.162	3.8	0.162	3.4
O—O	0.265	5.6	0.263	5.5
Si—Si	0.312	3.9	0.312	3.7
B_2O_3 melt				
B—O	0.138	2.9	0.138	3.0
O—O	0.238	4.0	0.238	4.1
B—B	0.242	3.0	0.242	3.0
Al_2O_3 melt				
Al—O	0.202	5.4	0.202	5.6
O—O	0.278	5.0	0.282	6.2
Al—Al	0.286	2.0	0.287	2.3

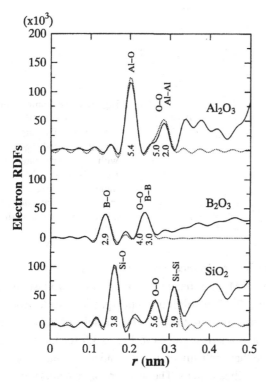

Figure 6.3.4 Comparison of the electron RDF estimated by the pair function method together with the experimental data for SiO_2, B_2O_3, and Al_2O_3 melts. Solid curves, experimental data; dotted curves, calculation by the pair function method.

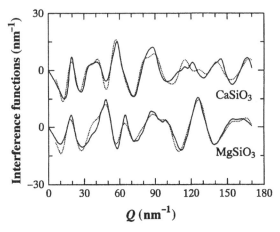

Figure 6.3.5 Interference functions $Qi(Q)$ of MgSiO$_3$
and CaSiO$_3$ melts. Solid curves, experimental data;
dotted curves, calculation by the interference function
refining technique.

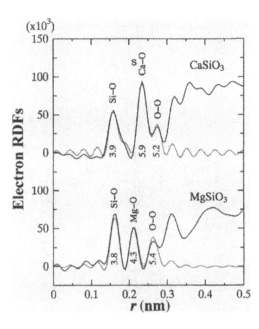

Figure 6.3.6 Comparison
of the electron RDF
estimated by the pair
function method together
with the experimental data
for MgSiO$_3$ and CaSiO$_3$
melts. Solid curves,
experimental data; dotted
curves, calculation by the
pair function method.

Figures 6.3.5 and 6.3.6 show the interference function data $Qi(Q)$ and the resultant
electron RDFs for metasilicates of MgSiO$_3$ (enstatite) and CaSiO$_3$ (wollastonite) (for
example see Refs. [12], [16], and [17]). The solid curves in Fig. 6.3.5 are the experimental data, and the dotted curves correspond to the converged results of the interference
function refining technique. The numerical values in Fig. 6.3.6 for the Si—O coordination numbers again indicate that each silicon is surrounded by four oxygens at a
distance of ~0.16 nm in these two silicate melts. The structural parameters with respect
to the near-neighbor correlations in these silicate melts are summarized in Table 6.3.2

Table 6.3.2. *Variations of the coordination number CaO—SiO₂ and MgO—SiO₂ determined by x-ray diffraction*

CaO (mol.%)	Temperature (°C)	Si—O	Ca—O	O—O	Si—Si
0	1750	3.8	0.0	5.6	3.9
34	1700	4.0	5.2	5.6	3.6
41	1600	3.8	5.4	5.4	3.4
45	1600	3.9	5.8	5.3	3.3
50	1600	3.9	5.9	5.2	3.1
57	1750	3.7	6.2	5.1	3.2

MgO (mol.%)	Temperature (°C)	Si—O	Mg—O	O—O	Si—Si
0	1750	3.8	0.0	5.6	3.9
44	1700	4.1	4.1	5.7	3.5
51	1700	3.9	4.3	5.4	3.3
56	1790	3.8	4.4	5.3	2.9

together with the results of some systematic structural studies of binary silicate melts [16, 17]. The essential points of these results are summarized below. The fundamental local ordering unit of SiO_4 tetrahedra has been confirmed from the systematic results of the $MgO-SiO_2$ and $CaO-SiO_2$ binary silicate melts over a wide composition range. An almost constant value of the oxygen coordination number of four around a silicon clearly indicates the formation of SiO_4 tetrahedra in binary silicate melts containing CaO and MgO over a wide concentration range, suggesting that the change in the fundamental local ordering units is insensitive to the change in melt composition. However, beyond the equimolar composition, the coordination number of Si–Si pairs, which corresponds to the linkage of SiO_4 tetrahedra, decreases from four to three in both binary silicate melts, as clearly seen in Table 6.3.2. This change of Si–Si pairs implies that the addition of CaO or MgO causes the breakdown of the network structure of pure SiO_2, resulting in loosely packed SiO_4 tetrahedral units. The slight decrease of O–O pairs could also be attributed to such disconnection of the network structure. On the other hand, the essential structural features of silicate melts are rather insensitive to temperature, at least within the temperature range investigated, although a peak broadening in the RDF occurs generally with an increase of temperature. The oxygen coordination number around alkaline-earth metals of Ca or Mg is found to depend essentially on the concentration, in which the coordination numbers for Ca–O and Mg–O are six in calcium silicate and four in magnesium silicate at the equimolar composition, respectively. It may be added that the oxygen coordination number is four for Li, six for Na, and eight for K in binary silicate melts [18]. These values are consistent with the ionic radii of these metallic ions [19, 20].

For the structure of the binary borate, Sugiyama et al. [14] have reported the structure of $xLi_2O + (1 - x)B_2O_3$ melts determined by x-ray diffraction at temperatures of ~100 K above the melting points. The results are summarized in Table 6.3.3. The number of oxygens around a boron in lithium borate melts slightly increases with the addition of Li_2O. This observation is not consistent with the local ordering found in crystal structures of this lithium borate system, in which the average nearest-neighbor

Table 6.3.3. *Structural parameters of liquid $xLi_2O + (1.0 - x)B_2O_3$ ($x = 0.0$, 0.2, 0.25, 0.33, 0.4, and 0.5)*

x	Temperature (K)	Density (mg/m^3)	B–O (nm) (atom)	Li–O (nm) (atom)	O–O (nm) (atom)
0.0	973	1.58	0.138(3.0)	—(—)	0.238(4.1)
0.2	1223	1.88	0.141(3.1)	0.203(3.6)*	0.242(4.3)
0.25	1263	1.92	0.140(3.1)	0.203(3.6)*	0.239(4.2)
0.33	1288	1.92	0.140(3.2)	0.203(3.6)*	0.239(4.2)
0.4	1223	1.94	0.140(3.2)	0.203(3.6)*	0.241(4.3)
0.5	1223	1.89	0.141(3.2)	0.203(3.6)	0.241(4.1)

*Fixed to the values of $0.5Li_2O + 0.5B_2O_3$.

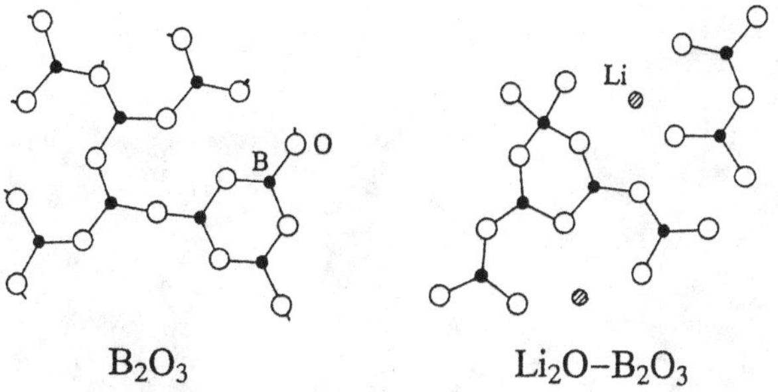

Figure 6.3.7 Schematic diagram for the possible change in coordination around boron in the molten $Li_2O-B_2O_3$.

correlations of $Li_2O_33B_2O_3$, $Li_2O_2B_2O_3$ and $Li_2OB_2O_3$ are 0.141 nm (3.3), 0.143 nm (3.5), and 0.137 nm (3.0), respectively. Nevertheless, a part of the coordination poly-hedra around boron appears to change into a tetrahedral coordination of oxygens at the expense of the trigonal coordination of pure B_2O_3 melt with the increment of the Li_2O component, as is schematically exemplified by the diagram of Fig. 6.3.7. Such a change in the local ordering structure in lithium borate melts is consistent with the conclusions obtained from nuclear magnetic resonance, Raman, and IR spectroscopy for lithium borate glasses [21–23]. It should be also noted that with respect to Li—O pairs, x-ray results are not definitive because of the poor resolution of the RDFs.

6.3.3 Energy-Dispersive X-Ray Diffraction

One of the most important requirements in the structural analysis of oxide melts is to determine accurately the local ordering structure and its distribution. For this purpose, energy-dispersive x-ray diffraction (EDXD), for which a solid-state detector (SSD) is used, may be one of the most powerful diffraction methods making it possible to obtain the high-resolution RDF from the structure factor with in a wide wave-vector region. As an example, the interference function over $Q = 200$ nm^{-1} can be measured when an x ray of $E = 40$ keV with a diffraction angle of 80° is used. This contrasts with the usual limit of 150 nm^{-1} observed by the conventional ADXD method with Mo $K\alpha$ radiation. The relative advantages and disadvantages of the ADXD and the EDXD methods have already been discussed in detail [3]. Nevertheless, some unfavorable problems, such as the quantitative determination of the primary beam spectrum and the correction for Compton scattering in the EDXD analysis, should be noted.

The EDXD method utilizes white x-ray radiation and a fixed diffraction angle. This is in contrast to the conventional ADXD method, which uses a monochromatic x ray and the angular scanning mode. The EDXD experiments were carried out in a labo-ratory setup consisting of a high-power constant-potential generator, a conventional x-ray tube with a W anode, and a $\theta-\theta$ vertical-type goniometer, as shown in Fig. 6.3.8 [24]. The x-ray source was usually operated at 60 kV and 30 mA. Under these operating conditions the continuous energy spectrum of x rays delivered by the source extends up

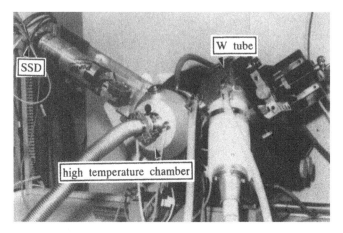

Figure 6.3.8 Overall view of an EDXD spectrometer with a
high-temperature chamber for the structural study of a melt. A SSD
of intrinsic Ge is used for obtaining the diffraction spectrum as a
function of x-ray energy.

to 60 keV. Because sharp L fluorescent lines of W appear at energies of approximately
10 keV and the intensity of x-ray photons of energy higher than 45 keV was very low,
only the energy region from $\sim$15 to 45 keV was considered in the usual analysis. The
scattered x-ray photons were detected by a portable Ge SSD attached to one of the
arms of the goniometer. The x-ray photons registered by the SSD were processed by a
multichannel pulse-height analyzer, and the full spectrum up to 60 keV was discriminated electrically with an interval of $\sim$50 eV/channel. A more detailed description of
the experimental EDXD setup can be found in Refs. [6] and [24]. All EDXD spectra
obtained are composed of coherently and incoherently scattered components from the
sample, and x-ray intensities follow the shape of the continuous energy spectrum of
the primary x-ray source. The relation between the registered fixed-diffraction-angle
EDXD intensities, $I_{obs}(E, \theta)$, and their individual constituents may be expressed as
follows [6, 24–28]:

$$I_{obs}(E, \theta) = C(E)[I_{coh}(E, \theta) + I_{inc}(E, E', \theta)], \tag{6.3.5}$$

$$I_{coh}(E, \theta) = A(E, E, \theta)P(E, E, \theta)I_p(E)I_{coh}(Q), \tag{6.3.6}$$

$$I_{inc}(E, E', \theta) = R(E, E')(dE'/dE)A(E, E', \theta)P(E, E', \theta)I_p(E')I_{inc}(Q'), \tag{6.3.7}$$

where C is a normalization constant and $I_{coh}(E, \theta)$ and $I_{inc}(E, E', \theta)$ are the coherent
and the incoherent components, respectively, of scattering intensities from a sample
as functions of x-ray energies and scattering angles. E' is the initial energy of the
incoming x-ray photon that is reduced to E after the Compton scattering [8–10].
$I_p(E)$ is the intensity profile of the primary x rays. $P(E, E', \theta)$ and $A(E, E', \theta)$ are
the polarization and the absorption factors, respectively. It should be noted that the
Breit–Dirac recoil factor $R(E, E')$ cancels out a correction factor of the incoherent
spectrum contraction (dE'/dE) [26]. The absorption factor $A(E, E', \theta)$ is readily

calculated by the analytical expressions coupled with the mass absorption coefficient as a function of x-ray energies [24, 29]. The polarization factor for white x rays can be written as follows:

$$P(E, E', \theta) = \frac{1}{2}\left(\frac{E}{E'} + \frac{E'}{E} - \sin^2 2\theta\right) + \frac{1}{2}\pi(E)\sin^2 2\theta, \qquad (6.3.8)$$

where $\pi(E)$ is a degree of polarization for the incident x rays defined by $[I_{pn}(E) - I_{pp}(E)]/[I_{pn}(E) + I_{pp}(E)]$ and $I_{pn}(E)$ and $I_{pp}(E)$ are the intensities of the normal and the parallel polarization components as functions of x-ray energies, respectively. Because the value of $\pi(E)$ is suggested to be less than 0.1 below the energy of $\sim 75\%$ of the incident electron energy, the polarization correction with respect to this factor has no significant effect on the EDXD analysis as far as the energy region below $\sim 75\%$ of the incident electron energy is applied [24, 26]. It is, however, worth mentioning that the value of $\pi(E)$ is equal to zero when the x-ray tube is tilted away at $45°$ from the diffraction plane, which contains incident and diffracted x-ray beams, as shown in Fig. 6.3.8.

$I_{coh}(Q)$ is the only structurally sensitive part of the total EDXD spectrum. To determine $I_{coh}(Q)$, the determination of the primary x-ray intensity $I_p(E)$ is the most essential process. This can be readily carried out by use of the characteristic features in x-ray scattering of disordered materials [24, 26, 27]. Because $I_{coh}(Q)$ will tend toward $\langle f^2 \rangle$ with the increment of the value for the wave vector Q, Eq. (6.3.5) can be approximated as the following expression in the large-Q region by use of $I_0(E)[= CI_p(E)]$:

$$\begin{aligned} I_{obs}(E, \theta) = I_0(E)\Bigg[&P(E, \theta)A(E, \theta)\langle f^2\rangle \\ &+ \frac{I_0(E')}{I_0(E)}P(E, E', \theta)A(E, E', \theta)I_{inc}(Q')\Bigg]. \end{aligned} \qquad (6.3.9)$$

Although the Compton shift becomes appreciable in the region with large values of E and 2θ, $I_0(E) = I_0(E')$ is assumed in the first iteration step for obtaining the first-order estimation of $I_{01}(E)$ from Eq. (6.3.9). In the second iteration, the approximation of $I_0(E')/I_0(E) = I_{01}(E')/I_{01}(E)$ allows us to obtain a more reliable primary x-ray spectrum, $I_{02}(E)$. Such calculating steps are iterated four or five times to obtain a reliable $I_0(E)$ profile, which enables us to explain the measured intensity from the sample.

With the estimated primary beam spectrum determined, it is now possible to calculate the pieces of a Faber–Ziman-type interference function $S(Q) = i(Q) - 1$ [30] from $I_{coh}(Q)$ at several fixed scattering angles, as follows:

$$S(Q) = I_{coh}(Q) - (\langle f^2\rangle - \langle f\rangle^2)/\langle f\rangle^2. \qquad (6.3.10)$$

By joining different segments of $S(Q)$, starting from the portion for the maximum 2θ run used to estimate the primary beam x-ray spectrum, we obtain the interference function in the wide-Q range. From the full $S(Q)$ profiles, the high-resolution RDF is readily calculated by the usual Fourier transform of Eq. (6.3.3).

Figure 6.3.9 shows the raw EDXD data for a $NaAlSi_3O_8$ (albite) melt at 1460 K [6]. Because the determination of the complete $S(Q)$ requires several segments at different

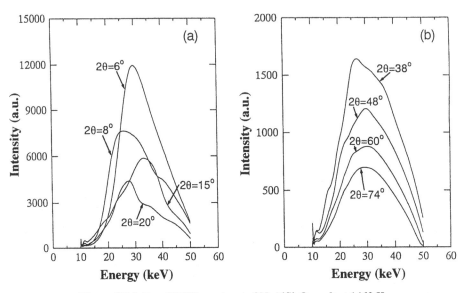

Figure 6.3.9 Raw EDXD spectrum of NaAlSi$_3$O$_8$ melt at 1460 K.

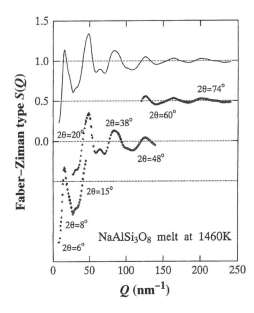

Figure 6.3.10 Segments of the Faber–Ziman-type interference function of NaAlSi$_3$O$_8$ melt at 1460 K.

2θ angles in the EDXD experiment, the angular settings of $2\theta = 6°, 8°, 10°, 15°, 20°,$ $38°, 48°, 60°,$ and $74°$ have to be chosen in such a way that any two consecutive 2θ runs overlap in the sufficient range in Q space. Through the data processing of the EDXD spectrum, the calculated pieces of $S(Q)$ for the molten NaAlSi$_3$O$_8$ were joined so as to obtain a full $S(Q)$ profile, as shown in Fig. 6.3.10. The resultant $Qi(Q)$ and RDF of NaAlSi$_3$O$_8$ melt together with those of the conventional ADXD results are given in Figs. 6.3.11 and 6.3.12, respectively. The interference function and RDF of

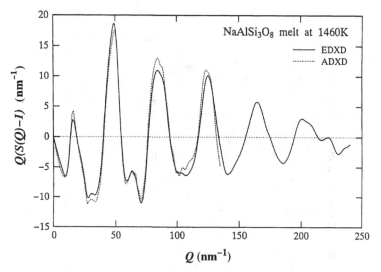

Figure 6.3.11 Reduced interference function $Qi(Q)$ of $NaAlSi_3O_8$ melt at 1460 K. The solid curve is the EDXD result, and the dotted curve is the ADXD result.

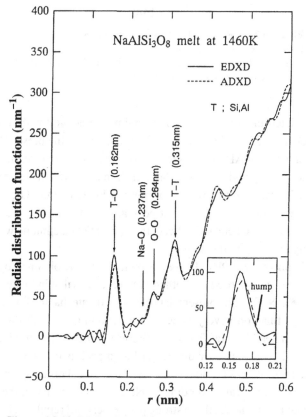

Figure 6.3.12 RDFs of $NaAlSi_3O_8$ melt at 1460 K. The solid curve is the EDXD result, and the dashed curve is the ADXD result. The arrows indicate the positions of some atomic pairs observed in crystalline $NaAlSi_3O_8$.

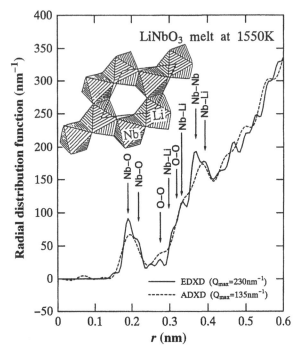

Figure 6.3.13 RDFs of $LiNbO_3$ melt at 1550 K. The arrows indicate the positions of some atomic pairs observed in crystalline $LiNbO_3$. The solid curve is the EDXD result, and the dashed curve is the ADXD result.

$NaAlSi_3O_8$ melt determined by the EDXD technique agree well with those obtained by the conventional ADXD technique. It is stressed here that the present RDF data of EDXD is superior to that of ADXD in its resolution because the interference function is determined in the large-Q region up to ~ 240 nm^{-1}. In particular, the first neighboring Na—O pair is well discriminated from a harmony of Si—O and O—O correlations. A slight hump at the larger r side of the first peak appears to indicate the contribution of Al—O pairs. These results clearly indicate that the EDXD technique works well for structural investigations of oxide melts.

Figure 6.3.13 shows the RDFs of $LiNbO_3$ melt at 1550 K obtained by the EDXD and the ADXD experiments [6, 31], together with the arrows that indicate some atomic pairs in crystalline $LiNbO_3$. The present data clearly indicate that the Fourier transform of the wide-$Qi(Q)$ function is very effective for obtaining the high-resolution RDF. Particularly, the first neighboring Nb—O correlation at ~ 0.20 nm is split into two peaks in the EDXD results, whereas only one rather broad peak is observed in the ADXD results. From the present EDXD and ADXD results, the local ordering unit of the melt is confirmed as the octahedrally coordinated NbO_6. The disparity in the first Nb—O atomic pairs apparently indicates that a strong [cation–cation] repulsion is still realized even in the melt structure at high temperatures. The disordered distribution of Nb and Li atoms could be inferred from the rather broad distribution of cation–cation atomic pairs from 0.3 to 0.5 nm. These features may imply the existence of relatively large clusters composed of NbO_6 octahedra, different from that of crystalline state.

The information about local ordering in the near-neighbor region obtained in these EDXD studies is found to be consistent with the previous conclusions based on results obtained by conventional ADXD methods. Nevertheless, these two examples of high-resolution RDFs obtained by the EDXD measurement are believed to be quantitatively accurate. Therefore the EDXD method is strongly favored as a way of obtaining the more fine local structural images of oxide melts.

6.3.4 Anomalous X-Ray-Scattering Method

Most oxide systems of petrological and metallurgical interest are usually not simple, and they contain more than two kinds of elements. Thus a single diffraction experiment with x rays, such as ADXD or EDXD, gives only a weighted sum of the atomic correlations for individual chemical constituents. For this reason, the near-neighbor atomic correlation or the local chemical environments around a specific element are essential for describing the quantitative structure in multicomponent oxide systems. When the so-called anomalous dispersion effect near the absorption edge of a constituent element is used [anomalous x-ray scattering (AXS)] [29, 32], we can obtain solutions to this problem by finding an accurate environmental structure around a specific element as a function of radial distance [29, 32]. The availability of synchrotron radiation has greatly improved both the acquisition and the quality of such data over those obtainable by conventional x-ray sources.

When the energy of incident x rays is close to an absorption edge of one of the constituent elements in a sample, a distinct energy dependence appears in the measured intensity because of the anomalous dispersion effect. This is interpreted by the resonance effect in which the oscillations of the corresponding K or L shell electrons are strongly disturbed [33]. In the vicinity of the absorption edge, where the anomalous dispersion phenomenon is significant, the atomic scattering factor $f(Q, E)$ should be described in the following form:

$$f(Q, E) = f^0(Q) + f'(E) + if''(E), \qquad (6.3.11)$$

where $f^0(Q)$ is the normal atomic scattering factor for x rays at energy far from any absorption edge and $f'(E)$ and $f''(E)$ are the real and the imaginary components of the anomalous dispersion. Figure 6.3.14 shows the energy dependence of anomalous dispersion terms with Fe metal as an example. The real part of f' indicates a sharp negative peak at the absorption edge, and its full width at half-maximum is typically 50 eV. It may also be noted that only the monotonic sharp energy dependence of f' is detected at the lower energy side of the absorption edge. These fundamental features of anomalous dispersion terms lead to the utilization of the x rays in the lower-energy side of the absorption edge for the general AXS method [29, 34]. Additionally, the experimental results of anomalous dispersion terms show reasonable agreement with

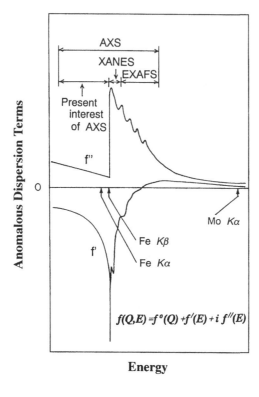

$$f(Q,E) = f^\circ(Q) + f'(E) + i\, f''(E)$$

Figure 6.3.14 Schematic diagram for anomalous dispersion terms of iron near the Fe K absorption edge.

the theoretical values determined by the Cromer–Liberman scheme at the lower-energy side of the absorption edge [35].

The variation in scattering intensity can be straightforwardly observed by use of incident x rays tuned near the vicinity of the absorption edge of a specific consitituent, for example, the element A. The detected energy dependence in intensity is attributable to only the change in the anomalous dispersion terms of element A. Because the variation of the imaginary term $f''(E)$ is known to be quite small for x rays on the lower-energy side of the absorption edge, the following useful relation can be readily obtained for multicomponent disordered systems [34, 36]:

$$\Delta I_A(Q) = \{I(Q, E_2) - \langle f^2(Q, E_2)\rangle\} - \{I(Q, E_1) - \langle f^2(Q, E_1)\rangle\}$$

$$= c_A[f'_A(E_2) - f'_A(E_1)] \int_0^\infty 4\pi r^2 \sum_{j=1}^{\text{Elements}} \mathrm{Re}[f_j(Q, E_1) + f_j(Q, E_2)]$$

$$\times [\rho_{Aj}(r) - \rho_{0j}]\frac{\sin(Qr)}{Qr}\, \mathrm{d}r, \qquad\qquad (6.3.12)$$

where $E_{\text{edge}} > E_1 > E_2$, c_j is the atomic fraction of the j element, $I(Q, E_i)$ is the coherent x-ray-scattering intensity in electron units per atom obtained in the usual way [2–5], Re indicates the real part of the scattering factors in the brackets, $\rho_{Aj}(r)$ corresponds to the radial density function of the j element around element A at a radial distance of r, and ρ_{0j} is the average number density of the j element in a system. Then the quantity of $\rho_A(r)$, which indicates the environmental structure around element A,

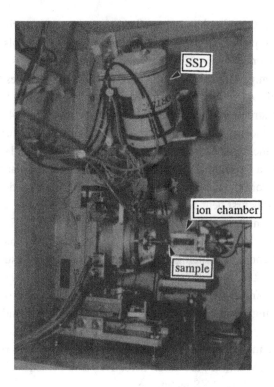

Figure 6.3.15 Overall view of an $\omega - 2\theta$ goniometer with a portable SSD at the Photon Factory, Institute of Materials Structure Science, Tsukuba, Japan, used for AXS measurements.

can be estimated by a Fourier transform of the quantity of $Q\Delta I_A(Q)$ [34, 36]:

$$4\pi r^2 \rho_A(r) = 4\pi r^2 \rho_0 + \frac{2r}{\pi} \int_0^\infty \frac{Q\Delta I_A(Q)\sin(Qr)}{c_A[f_A'(E_2) - f_A'(E_1)]W(Q)}\, dQ, \quad (6.3.13)$$

$$W(Q) = \sum_{j=1}^{\text{Elements}} c_j\, \text{Re}[f_j(Q, E_1) + f_j(Q, E_2)]. \quad (6.3.14)$$

Figure 6.3.15 shows an $\omega - 2\theta$ double-axis goniometer for AXS measurements at the Photon Factory of the Institute of Materials Structure Science (IMSS), Tsukuba, Japan. The white radiation from the synchrotron was monochromatized by double Si 111 crystals, and x rays in the vicinity of a relevant absorption edge are selected [37]. The incident x rays are monitored by an N_2 gas ion chamber, so as to maintain the constant incident intensity irradiating a sample. The fluorescent radiation from the sample arising mainly from the tail of the bandpass and higher harmonic diffraction of the monochromator is not negligible, and the separation of such fluorescent radiation from the scattering intensity is the crucial point in obtaining the sufficient reliability of the AXS measurements. Therefore the energy-sensitive SSD is recommended for use in the AXS measurements.

Structural analyses by the AXS method have been made on various ferrite and germanate glasses [38, 39] and aqueous solutions [40, 41]. Recently, a successful AXS measurement for determining partial structural functions of molten CuBr, which has a relatively small mass absorption coefficient, has been made in transmission geometry with a SiO_2 glass cell [42]. On the other hand, the reflection geometry from the free surface of the melt is strongly preferred in cases of molten oxides and metals with larger mass absorption coefficients. A new triple-axis goniometer apparatus, suitable for structural studies of high-temperature melts by the AXS method, has been designed as shown in Fig. 6.3.16, and the capability of this apparatus was demonstrated by obtaining the individual partial structural functions of molten $Bi_{30}Ga_{70}$ alloy [43]. To maintain the free surface of a molten sample, the diffractometer requires asymmetric reflection optics, in which the incident beam of the synchrotron radiation should irradiate the horizontal surface with a fixed, small glancing angle α. A monochromatic and horizontal x-ray beam tuned from the continuous synchrotron spectrum by a double-crystal Si 111 monochromator is first bent downward by an angle $2\phi(E)$ with a W/Si multilayer mirror. Then the bent beam is totally reflected by an angle $|2\phi(E) - \alpha|$ with a Pt-coated fused-quartz mirror, which is located at the center of the triple-axis goniometer, so as to keep the constant glancing angle against the sample free surface. It may be noted here that the left part of this facility with a high-temperature chamber is the $\omega - 2\theta$ double-axis goniometer shown in Fig. 6.3.15. This new facility promises to be powerful tool for obtaining the environmental RDFs of molten oxides, and such experimental challenges are now ongoing. Because no AXS study of high-temperature oxide melts, including silicate melts, is available at the present stage, the AXS results of several germanate glasses [39, 44, 45] are given here. This also provides a better understanding of the usefulness of the AXS technique.

Figure 6.3.17 shows the coherent intensity profiles of a GeO_2 glass measured with incident x rays of 10.805 and 11.080 keV that correspond respectively to 300 and 25 eV below the Ge K absorption edge (11.103 keV) together with that measured by Mo $K\alpha$ radiation [39, 44]. The bottom part of this figure indicates the differential intensity profile, corresponding to the environmental structure around Ge in the present GeO_2 glass. Figure 6.3.18 shows the interference function $Qi(Q)$ and the environmental interference function $Q\Delta i_{Ge}(Q)$. The environmental RDF curve obtained by the Fourier transform of $Q\Delta i_{Ge}(Q)$ in Fig. 6.3.18 is given in Fig. 6.3.19 together with the ordinary RDF curve that contains the average of the three Ge—O, O—O, and Ge—Ge partials. Comparing these two profiles, we find that the peak corresponding to the correlations of O—O pairs is absent in the environmental RDF for Ge. This is strong evidence that the present AXS measurement works well. The structural parameters of N_{ij}, r_{ij}, and σ_{ij} for the nearest-neighbor correlations were determined by the interference refining technique with Eq. (6.3.4) so as to reproduce the experimental interference functions $Qi(Q)$ and $Q\Delta i_{Ge}(Q)$ of Fig. 6.3.18. Because the two independent interference functions are used for the least-squares refinement for the structural parameters, the resolution of the structural parameters can be improved to a reasonable extent. The converged structural parameters are given in Table 6.3.4 together with those of the trigonal GeO_2, and both the interference functions and RDFs in Figs. 6.3.18 and 6.3.19, denoted by dots, are calculated from converged parameters. These data quantitatively indicate that each germanium is surrounded by four oxygens with a Ge—O distance of 0.174 nm and then the GeO_4 tetrahedron is a fundamental local

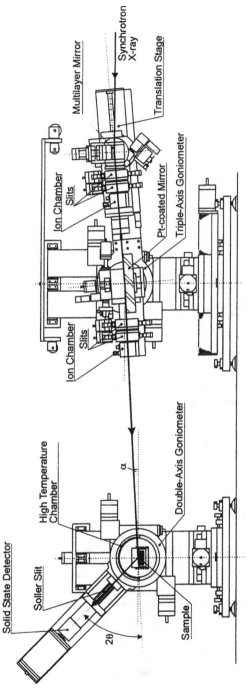

Figure 6.3.16 Experimental setting and its overview of AXS measurements for a melt in the asymmetrical reflection mode with synchrotron radiation. Left, the $\omega - 2\theta$ goniometer of Fig. 6.3.15 with a high-temperature chamber. Right, facility that bends the incidents x rays so as to keep the constant glancing angle α against the melt samples.

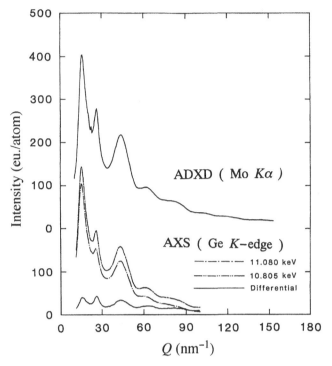

Figure 6.3.17 Bottom, intensity profiles of GeO_2 glass obtained by the AXS measurement at the Ge K edge, Top, intensity profile of GeO_2 glass obtained by the conventional ADXD measurement.

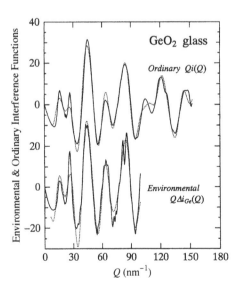

Figure 6.3.18 Bottom, environmental interference function $Q\Delta i_{Ge}(Q)$ for Ge of GeO_2 glass obtained by the AXS measurement at the Ge K edge, Top, ordinary interference function $Qi(Q)$ of the GeO_2 glass obtained by the conventional ADXD measurement.

unit structure in GeO_2 glass. As for the Ge—Ge pair, a germanium is surrounded by four germaniums at a distance of 0.237 nm. These features clearly indicate that GeO_4 tetrahedra link each other by sharing corners and form a three-dimensional network structure, similar to that of the SiO_2 glass.

Table 6.3.4. *Structural parameters of distance r (nm) and coordination number N (atom) in glassy and crystalline GeO₂*

	Glassy	GeO₂	Crystalline	GeO₂
Density(mg/m³)	3.64		4.21	
Ge—O	0.173	(4.1)	0.174	(4.0)
Ge—O	0.365	(4.1)	0.348	(6.0)
Ge—O	0.411	(9.8)	0.428	(14.0)
Ge—O	0.488	(3.4)	0.479	(4.0)
Ge—O	0.541	(6.0*)	0.540	(7.0)
Ge—Ge	0.317	(4.1)	0.315	(4.0)
Ge—Ge	0.453	(3.9)	0.447	(6.0)
Ge—Ge	0.497	(4.0)	0.499	(6.0)
Ge—Ge	0.535	(5.1*)	0.539	(6.0)
O—O	0.277	(6.4)	0.284	(6.0)
O—O	0.330	(5.1*)	0.337	(6.0)
O—O	0.434	(3.4*)	0.412	(4.0)
O—O	0.495	(13.6*)	0.507	(16.0)

*Fixed.

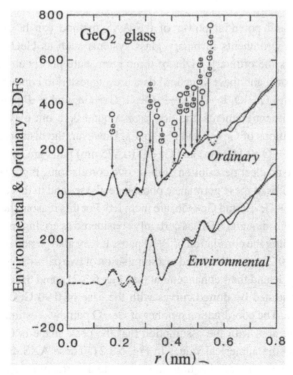

Figure 6.3.19 Environmental RDF $4\pi r^2 \rho_{Ge}(r)$ and ordinary RDF $4\pi r^2 \rho(r)$ calculated from the interference functions in Fig. 6.3.18.

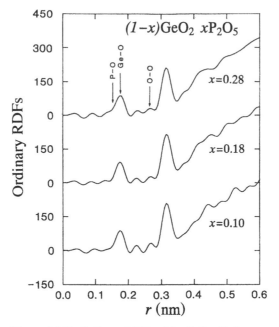

Figure 6.3.20 Ordinary RDFs of the GeO_2–P_2O_5 glasses obtained by the ADXD measurement by Mo $K\alpha$ radiation.

The usefulness and potential power of the AXS method can be demonstrated by obtaining the environments in binary glass systems such as GeO_2–P_2O_5 [45]. Figure 6.3.20 shows the ordinary RDFs of three germanate phosphate glasses with different compositions, and these structural data are suggested to contain the six partials corresponding to Ge–O, P–O, Ge–P, O–O, Ge–Ge, and P–P pairs. From the ionic radii of constituent elements, the first peak around 0.18 nm may be allocated to the mixed correlations of Ge–O and P–P pairs. However, the distances of tetrahedrally coordinated P–O (0.150 nm) and Ge–O (0.175 nm) pairs are so close that the ordinary RDFs give no clear description of these two correlations. Figure 6.3.21 gives the environmetal RDFs of these germanate phosphate glasses and in this case, only the three partials, Ge–O, Ge–P, and Ge–Ge, are included. For this reason, the first peak at ~0.18 nm in the environmental RDFs is certainly considered as arising from the Ge–O pair alone by excluding the correlation of P–O pairs. It may also be noted here that the peak appearing at ~0.24 nm is attributed to a summation of two pairs of correlation tails of Ge–O and Ge–Ge and their enhancement (see Fig. 6.3.21), and the corresponding contribution is illustrated by dotted curves with the case of 0.90 GeO_2–0.10 P_2O_5 glass as an example. The coordination number of Ge–O pairs was estimated from the corresponding peak area with the assumption that the peak shape is Gaussian. The results are given by the numerical values in Fig. 6.3.21. These AXS data imply that the increase in the oxygen coordination number around germanium induced by the addition of P_2O_5 component could be interpreted as the mixed state of the tetrahedral and the octahedral coordinations in germanate phosphate glasses. This is consistent with the results obtained by infrared absorption and Raman spectroscopy [46, 47].

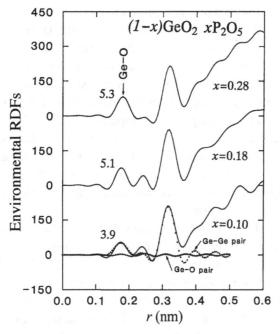

Figure 6.3.21 Environmental RDFs $4\pi r^2 \rho_{Ge}(r)$ of the GeO$_2$—P$_2$O$_5$ glasses. Dotted curves of 0.90 GeO$_2$—0.10 P$_2$O$_5$ glass are examples of calculations made with the contributions of Ge—O and Ge—Ge pairs.

More information on the AXS method, particularly its application to disordered systems, is found in Refs. [29] and [32]. Nevertheless, these examples of gernamate glasses are enough to shed light on the local chemical environmental structure around a specific element in a multicomponent system. For most of the elements, the change in the real component of the anomalous dispersion factor is typically 15%–25% of the normal atomic scattering factor at the K absorption edge, and it appears to be substantially larger (over 50%) at the L absorption edge. Thus the AXS method, coupled with an intense white x-ray source, is believed to be no longer simply a novel technique.

6.3.5 **Summary**

Structural characterization of high-temperature melts is essential for the understanding of many physical properties. For this purpose, angle-dispersive x-ray diffraction (ADXD), energy-dispersive x-ray diffraction (EDXD), and anomalous x-ray scattering (AXS) are undoubtedly useful and powerful experimental techniques for giving the unique structural information of radial distribution functions (RDFs).

In the case of the ADXD measurement, the resolution of the RDF appears to be limited to no better than 0.002 nm ($=\pi/Q_{max}$) by the truncation effect of the Fourier transform. However, this ADXD information is easy to obtain without any special equipment such as a SSD and/or synchrotron radiation and is enough for a discussion of the local ordering unit of a disordered structure. The EDXD method can determine the structure data in the wider wave-vector region over 200 nm^{-1}. This allows us to obtain a high-resolution RDF, and the resultant high-resolution RDF is quite useful for discussing the medium range structure in disordered materials as well as the local ordering structure in the near-neighbor region. This capability of the EDXD method has been well recognized in obtaining the anisotropic arrangement of the local structural unit found in the LiNbO$_3$ melt. However, some reservations should be stressed relative to the EDXD method, for example, the difficulty in the accurate determination of the energy distribution of the primary x-ray source. It is worth mentioning that this disadvantage is frequently solved by the fundamental structural information obtained by conventional ADXD measurements.

On the other hand, the AXS method can reduce difficulties for multicomponent disordered systems by obtaining the local chemical environment around a specific element. Of course, the resolution of the environmental RDF is again limited, and its careful interpretation is required when the measurable wave-vector region of the interference function is restricted mainly because of a relatively low-energy absorption edge of the desired element. Nevertheless, we believe that the AXS method, coupled with an intense white x-ray source such as synchrotron radiation, will become one of the most reliable and powerful methods for structural characterization of oxide melts in the near future.

Acknowledgments

The authors express their gratitude to Professor Yoshito Matsui for his valuable and critical suggestions regarding a systematic structural analysis of silicates melts. Shigeru Kimura of National Institute for Research in Inorganic Materials should be commended for the special promotion of EDXD analyses through Kimura Metamelt Project, ERATO, JRDC, in the period between 1992 and 1993. All of the AXS experiments described in this chapter were made as part of the AXS research projects supported by the staffs of the Photon Factory, IMSS, Tsukuba, Japan. The authors express their best thanks to Masahiro Nomura and Atsushi Koyama for their kind services and advice.

References

[1] P. Debye, Ann. Phys. **46**, 809 (1915).

[2] C. N. J. Wagner, H. Ocken, and M. L. Joshi, Z. Natureforsch. **20a**, 325 (1965).

[3] Y. Waseda, *The Structure of Non-Crystalline Materials, Liquid and Amorphous Solids* (McGraw-Hill, New York, 1980).

[4] B. E. Warren, *X-Ray Diffraction* (Addison-Wesley, New York, 1969).

[5] C. N. J. Wagner, in *Liquid Metals, Physics and Chemistry*, S. Z. Beer, ed.

(Marcel Dekker, New York, 1972).

[6] K. Sugiyama, T. Shingai, and Y. Waseda, Sci. Rep. Res. Inst. Tohoku Univ. A **42**, 231 (1995).

[7] *International Tables for Crystallography* (Kluwer Academic, Dordrecht, The Netherlands, 1992), Vol. C.

[8] D. T. Cromer and J. B. Mann, J. Chem. Phys. **47**, 1892 (1967).

[9] D. T. Cromer, J. Chem. Phys. **50**, 4857 (1969).

[10] V. H. Smith, Jr., A. J. Thakkar, and D. C. Chapman, Acta Crystallogr. A **31**, 391 (1975).

[11] A. H. Narten, J. Chem. Phys. **56**, 1905 (1972).

[12] K. Sugiyama, E. Matsubara, I. K. Suh, Y. Waseda, and J. M. Toguri, Sci. Rep. Res. Inst. Tohoku Univ. A **34**, 143 (1989).

[13] K. Sugiyama, K. Nomura, and S. Kimura, High Temp. Mater. Process. **14**, 131 (1995).

[14] Y. Waseda, K. Sugiyama, and J. M. Toguri, Z. Natureforsh. **50a**, 770 (1995).

[15] R. L. Mozzi and B. E. Warren, J. Appl. Crystallogr. **2**, 164 (1969).

[16] Y. Waseda and J. M. Toguri, in *Materials Science of the Earth's Interior*, F. Marumo, ed. (Terra Scientific, Tokyo, 1989), p. 37.

[17] Y. Waseda, Y. Shiraishi and J. M. Toguri, Mater. Trans. JIM **21**, 51 (1980).

[18] Y. Waseda and J. M. Toguri, *The Structure and Properties of Oxide Melts* (World Scientific, Singapore, 1998).

[19] L. Pauling, *The Nature of the Chemical Bond*, 3rd ed. (Cornell U. Press, Ithaca, NY, 1960).

[20] R. D. Shanon and C. T. Prewitt, Acta Crystallogr. B **25**, 925 (1969).

[21] Y. H. Yun and P. J. Bray, J. Non-Cryst. Solids **44**, 227 (1981).

[22] M. Tatsumisago, M. Takahashi, T. Minami, M. Tanaka, N. Umezaki, and N. Iwamoto, Yogyo-Kyokai-Shi **94**, 464 (1986).

[23] G. D. Chryssikos, E. I. Kamitsos, A. P. Patsis, M. S. Bitsis, and M. A. Karakassides, J. Non-Cryst. Solids **126**, 42 (1990).

[24] K. Sugiyama, V. Petkov, S. Takeda, and Y. Waseda, Sci. Rep. Res. Inst. Tohoku Univ. A **38**, 1 (1993).

[25] J. M. Prober and J. M. Schultz, J. Appl. Crystallogr. **8**, 405 (1975).

[26] T. Egami, J. Mater. Sci. **13**, 2587 (1978).

[27] C. N. J. Wagner, D. Lee, S. Tai, and L. Keller, Adv. X-Ray Anal. **24**, 245 (1981).

[28] S. Hosokawa, T. Matsuoka, and K. Tamura, J. Phys. Condens. Matter **3**, 4443 (1991).

[29] Y. Waseda, *Novel Application of Anomalous X-Ray Scattering for Structural Characterization of Disordered Materials* (Springer-Verlag, New York, 1984).

[30] T. E. Faber and J. M. Ziman, Philos. Mag. **11**, 153 (1965).

[31] K. Sugiyama, K. Nomura, Y. Waseda, P. Andonov, S. Kimura, and K. Shigematsu, Z. Naturforsch. **45a**, 1325 (1990).

[32] S. Ramaseshan and S. C. Abraham, eds., *Anomalous Scattering* (International Union of Crystallography, Munksgaard, Copenhagen, 1975).

[33] R. W. James, *The Optical Principles of the Diffraction of X-Rays* (Bell, London, 1954).

[34] Y. Waseda, IJIJ Int. **29**, 198 (1989).

[35] D. T. Cromer, and D. Liberman, J. Chem. Phys. **53**, 1891 (1970).

[36] Y. Waseda, E. Matsubara, and K. Sugiyama, Sci. Rep. Res. Inst. Tohoku Univ. A **34**, 1 (1988).

[37] M. Nomura and A. Koyama, User's Manual of BL6B and 7C at the Photon Factory, KEK Internal Rep. 93-1 (1993).

[38] E. Matsubara, K. Okuda, Y. Waseda, S. N. Okuno, and K. Inomata, Z. Naturforsch. **45a**, 1144 (1990).

[39] E. Matsubura, K. Harada, Y. Waseda, and M. Iwase, Z. Naturforsch. **43a**, 181 (1988).

[40] E. Matsubara and Y. Waseda, J. Phys. Cond. Matter **1**, 8575 (1989).

[41] E. Matsubara, K. Okuda, and Y.

Waseda, J. Phys. Cond. Matter **2**, 9133 (1990).

[42] M. Saito, C. Park, K. Omote, K. Sugiyama, and Y. Waseda, J. Phys. Soc. Jpn. **66**, 633 (1997).

[43] M. Saito, C. Park, K. Sugiyama, and Y. Waseda, J. Phys. Soc. Jpn. **66**, 3120 (1997).

[44] K. Sugiyama and E. Matsubara, High Temp. Mater. Process. **10**, 177 (1992).

[45] K. Sugiyama, Y. Waseda, and M. Ashizuka, Mater. Trans. JIM **32**, 1030 (1991).

[46] K. Takahashi, N. Mochida, H. Matsui, S. Takeuchi, and Y. Gohshi, Yogyo-Kyokai-Shi **81**, 482 (1976).

[47] N. Mochida, T. Sekiya, and A. Ohtsuka, Yogyo-Kyokai-Shi **96**, 271 (1988).

Chapter 6.4

Computer-Simulation Approach for the Prediction of Trace-Element Partitioning Between Crystal and Melt

Masami Kanzaki

A computer-simulation procedure for predicting the trace-element partition coefficient between crystal and melt is presented. It is a modified version of Nagasawa's model. The enthalpy change for trace-element partitioning between crystal and melt is approximated as the partial excess enthalpy of trace-element substitution in the crystal, which can be obtained by calculation of the enthalpy change of the crystal before and after substitution by means of molecular-dynamics simulation. As a test of the procedure, the partial excess enthalpies for trace cation substitution in KCl, diopside ($CaMgSi_2O_6$), and forsterite (Mg_2SiO_4) crystals were calculated. When the partial excess enthalpy is plotted against the size of the substituent cations, it has a minimum at the host cation position. With an increasing degree of misfit to the host ion site, the partial excess enthalpy increases. The calculation thus qualitatively reproduced the partitioning behavior observed in the Onuma (partition coefficient versus ionic radii) diagrams. The merit of this procedure is that it allows us to predict trace-element partitioning behavior even at extreme conditions solely from interatomic potentials.

6.4.1 Introduction

Partitioning of elements between minerals and melt (or fluid) is one of the most impor-tant factors in determining the elemental distribution in the Earth during its evolution. To understand geochemical processes by means of elemental distributions recorded in rocks, the partition coefficients of major and trace elements between minerals and

H. Aoki et al. (eds), *Physics Meets Mineralogy* © 2000 Cambridge University Press.

melt (or fluid) have been measured (e.g., Ref. [1]). However, much less effort has been applied to understanding the partitioning behavior from a microscopic perspective.

Matsui et al. [2] compiled the measured partition coefficients of major and trace elements between phenocrysts (crystal) and groundmasses (melt) in volcanic rocks. They noted that, when plotted against the ionic radius of elements (Onuma diagram), the partition coefficient for each isovalent ion series shows a parabolic shape in which the peak corresponds to the host ion position. From these observations, they suggested that the crystal structures of the phenocrysts are major factors in determining the partition coefficients.

Nagasawa [3] developed a microscopic model for the partition coefficient of trace elements between crystal and melt. In the model, the enthalpy change of partitioning is approximated as the partial excess enthalpy that is due to substitution of the host element by trace elements in the crystal. The latter is calculated from an elastic model as the elastic energy caused by putting a larger ion into a vacuum created by the removal of a host ion in an infinite elastic medium. The model explained the partitioning behavior observed in the Onuma diagrams described above quite well. Later, Blundy and Wood [4] extended Nagasawa's model by including the local elasticity around cation sites in the crystals. However, the applicability of these elastic models to the complex crystals at various pressure–temperature conditions is rather limited. For example, the effect of pressure on the partition is difficult to evaluate from these models. In addition, the effect of melt is ignored.

In this chapter, a modified version of the Nagasawa model is presented to overcome these limitations. Instead of calculations made with the elastic model, the enthalpies of host and trace-element-substituted crystals are directly calculated by means of a molecular-dynamics (MD) simulation technique. Application of the MD simulation method allows us to calculate the enthalpy of any crystal even at extreme pressure and temperature conditions. To test this method, partition coefficients (partial excess enthalpies) for trace monovalent cations in KCl and trace divalent cations in diopside ($CaMgSi_2O_6$) and forsterite (Mg_2SiO_4) have been calculated and are reported below.

6.4.2 Calculation Procedure

We follow the thermodynamic treatments of Nagasawa [3], except for the calculation of the partial excess enthalpy, as noted above. A major assumption in Nagasawa's model is the treatment of the melt as an ideal solution. Then the enthalpy change that is due to the exchange of trace elements between crystal and melt can be approximated as the partial excess enthalpy of the crystal [3]. As it is known that the excess enthalpies of melts are an order of magnitude smaller than those of solid solutions, this assumption is reasonable. In Nagasawa's original calculation, the partial excess enthalpy is obtained from the elastic deformation energy that is due to the substitution of a host ion by a trace element ion. In this study, the partial excess enthalpy of the crystal is calculated with a MD simulation.

For a MD simulation of crystals, the MXDORTO and MXDTRICL programs developed by Kawamura of Tokyo Institute of Technology are used [5]. The two-body potential used in this study is

$$V_{ij}(r_{ij}) = \frac{q_i q_j}{r_{ij}} + f_0(b_i + b_j)\exp\left[\frac{(a_i + a_j - r_{ij})}{(b_i + b_j)}\right] - \frac{c_i c_j}{r_{ij}^6}. \tag{6.4.1}$$

The potential V_{ij} between ions i and j separated by r_{ij} consists of three terms. The first term represents the Coulomb interaction between ions that have formal charges q_i and q_j. The second term models the repulsive interaction that is due to the overlap of electron clouds, where f_0 is a constant. The third term represents the van der Waals interaction where a_i, b_i, and c_i are potential parameters for ion i. Because the a_i parameter roughly corresponds to the ionic size, the effect of ionic radii on the partition can be studied by varying this parameter. To make a set of virtual elements with a wide range of effective ionic radii, a set of a_i and c_i parameters for each crystal is prepared; these are listed in Tables 6.4.1 and 6.4.2. The a_i and c_i parameters are determined by the interpolation and extrapolation of existing parameters for certain ions (e.g., Mg and Ca for diopside and forsterite). The mass and b_i parameter of the virtual elements are fixed as those of the host ion. Although these parameters are crude, our major goal here is to look for the general effect of ionic radii on the trace-element partitioning. In addition, it is apparent from the periodic table that there are not enough atoms that can be used to investigate the effect of ionic radii on partitioning. For example, there are few divalent cations that are smaller than Mg^{2+}. Therefore these virtual elements are used in this study. To calculate the partitioning behavior of certain elements relevant to the experimental data, more sophisticated empirical potentials or ones derived from first-principles calculation should be used.

Table 6.4.1. *Potential parameters of monovalent cations for KCl**

Ion	a_i	c_i
A1	1.260	10.00
A2	1.344	11.25
A3	1.428	12.50
A4	1.511	13.75
A5	1.595	15.00
A6	1.679	16.25
A7	1.763	17.50
A8	1.846	18.75
A9	1.930	20.00
Cl	1.950	30.00

*K corresponds to A5; b_i is 0.08 and 0.09 for the cations and Cl, respectively.

The calculation procedure of the enthalpy is similar to the study by Akamatsu et al. [6], and is shown below. The total number of atoms in a simulated cell was 512 for KCl, 640 for diopside, and 672 for forsterite. For the partial excess enthalpy calculation in KCl, 1 out of 256 K^+ ions in the cell is replaced with a trace-element ion X^+. Thus the composition of this substituted system is $K_{255/256}X_{1/256}Cl$. The enthalpies of the end members, KCl and XCl, are also calculated. The partial excess enthalpy at KCl composition can be approximated by

$$\frac{\partial H_{ex}}{\partial x} = H_{K_{255/256}X_{1/256}Cl} - \frac{255}{256}H_{KCl} - \frac{1}{256}H_{XCl}, \qquad (6.4.2)$$

where H_{KCl}, H_{XCl}, and $H_{K_{255/256}X_{1/256}Cl}$ are the enthalpies of KCl, XCl, and $H_{K_{255/256}X_{1/256}Cl}$, respectively, and are obtained by MD calculations. Similar equations are used for the M1 and M2 sites of diopside and forsterite.

For calculation of the enthalpies, the pure end members are calculated with 5000 steps for relaxing the initial structure at ambient pressure, followed by an additional 5000 steps for enthalpy sampling. The trace element is introduced into the pure system, and then the calculation is again performed with 5000 steps for relaxing the structure and a further 5000 steps for the sampling of the enthalpy.

Although the MD simulation should be performed at the temperature corresponding to that of crystal–melt equilibrium, a larger fluctuation of the partial excess enthalpy

Table 6.4.2. *Potential parameters of divalent cations for diopside and forsterite**

Ion	a_i	c_i
B1	0.957	0.0
B2	1.008	0.0
B3	1.059	0.0
B4	1.110	0.0
B5	1.161	2.0
B6	1.212	3.6
B7	1.263	5.2
B8	1.314	6.8
B9	1.365	8.4
B10	1.414	10.0
B11	1.465	11.6
B12	1.516	13.2
B13	1.567	14.8
B14	1.618	16.2
O	1.629	16.4
Si	1.012	0.0

*Mg and Ca ions correspond to B5 and B10, respectively; b_i is 0.080 and 0.085 for the cations and oxygen, respectively.

was noted at high temperature (as shown below). Therefore 300 K is selected for most of the enthalpy calculations, except for KCl. To evaluate the effect of pressure, the partial excess enthalpy of KCl is also calculated at 10 GPa.

6.4.3   Results

6.4.3.1   KCl

For KCl, K^+ ion is replaced with a monovalent trace ion in Table 6.4.1. The partial excess enthalpy of KCl as a function of the a_i parameter at 300 K and ambient pressure is shown in Fig. 6.4.1. A minimum in the partial excess enthalpy is located at the host K^+ ion position. The partial excess enthalpy shows a parabolic dependence on the a_i parameter. This is consistent with Nagasawa's elastic model for the partial excess enthalpy, in which the partial excess enthalpy is proportional to the square of the difference in ionic radius between the host and the substituted ions. Nagasawa [3] argued that his elastic energy formulation cannot be applied when the host ion is substituted by a smaller ion. However, the present study suggests that Nagasawa's formulation is valid for that case too. Figure 6.4.1 also shows the partial excess enthalpy at 300 K and 10 GPa. Application of pressure increased the partial excess enthalpy, as expected.

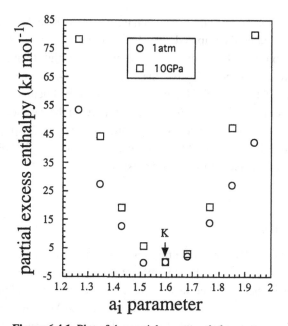

Figure 6.4.1 Plot of the partial excess enthalpy versus the a_i parameter calculated for KCl at 300 K. Open circles and squares show the data at ambient pressure and 10 GPa, respectively. The a_i parameter roughly corresponds to the ionic radii. The position for the host K^+ ion is shown by an arrow.

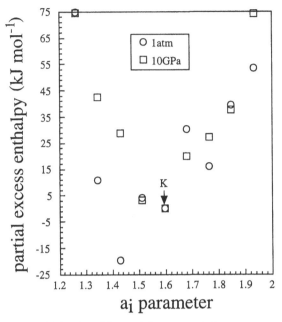

Figure 6.4.2 Plot of the partial excess enthalpy versus the a_i parameter calculated for KCl at 1000 K. The legend is the same as that of Fig. 6.4.1.

Figure 6.4.2 shows the partial excess enthalpy at 1000 K and ambient pressure. The data are more scattered compared with that at 300 K in Fig. 6.4.1. This is due to the larger fluctuation of enthalpy during the simulation at high temperature. Longer sampling times are necessary to improve the data. Figure 6.4.2 also shows the partial excess enthalpy at 1000 K and 10 GPa. It is apparent that the fluctuation is suppressed by the application of pressure. The effect of temperature on the partial excess enthalpy is not clear at ambient pressure because of the scatter of the data. The partial excess enthalpy becomes slightly smaller at 1000 K compared with 300 K at 10 GPa (Fig. 6.4.1). This trend opposes the effect of increasing pressure.

6.4.3.2 Forsterite

The partial excess enthalpies of substitution into the M1 and M2 sites of forsterite as functions of a_i parameter at 300 K and ambient pressure are shown in Fig. 6.4.3. In this case, Mg ions in these octahedral sites are replaced with the divalent trace ions in Table 6.4.2. Similar to the behavior of KCl, both the M1 and M2 sites show a parabolic dependence of the partial excess enthalpy on a_i. Because both sites are occupied by a Mg ion in forsterite, the behavior of the partial excess enthalpies for these sites is similar. However, the M2 site shows a slightly higher partial excess enthalpy than M1.

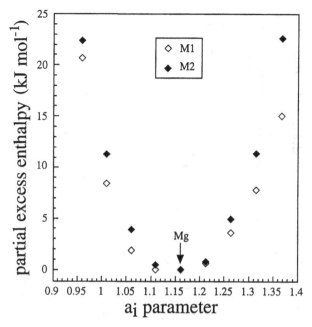

Figure 6.4.3 Plot of the partial excess enthalpy versus the
a_i parameter calculated for forsterite at 300 K. Open and filled
diamonds indicate the data for M1 and M2, respectively. The
position for the host Mg^{2+} ion for the M1 and M2 sites is shown.

During the MD simulation of forsterite with Mg in M1 or M2 sites completely
replaced with a smaller ion, sudden changes of volume and other properties, suggesting
a phase transition, were noted. This happens when Mg is replaced with B1 and B2 ions
in the M1 site, or B1, B2, and B3 ions in the M2 site. Inspection of the atomic positions
in these simulation cells reveals that they are displaced by those of forsterite. Further
characterization of the structure is underway. Figure 6.4.3 shows that this change
does not have a significant effect on the partial excess enthalpy, possibly because the
octahedral coordination in M1 and M2 sites is preserved during the transition.

6.4.3.3   Diopside

The partial excess enthalpies of M1 and M2 sites in diopside as functions of the a_i
parameter at 300 K and ambient pressure are shown in Fig. 6.4.4. In this crystal, Mg
and Ca ions, in M1 and M2 sites respectively, are replaced with divalent trace ions
(Table 6.4.2). Again, similar to the behavior observed in KCl and forsterite, both the
M1 and M2 sites show a parabolic dependence of the partial excess enthalpy on a_i.
Unlike forsterite, however, M1 and M2 sites are occupied by different cations with
different sizes, and the partial excess enthalpies for M1 and M2 sites show two separate
minima.

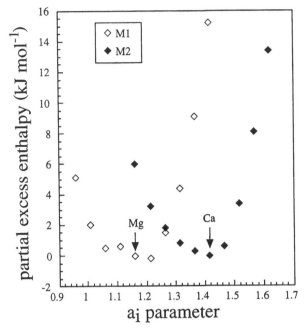

Figure 6.4.4 Plot of the partial excess enthalpy versus the a_i parameter calculated for diopside at 300 K. Open and filled diamonds indicate the data for M1 and M2, respectively. The positions for the host Mg^{2+} and Ca^{2+} ions for the M1 and M2 sites, respectively, are shown.

6.4.4   Discussion

The observed partition coefficients can be compared with the present partial excess enthalpies calculated above. The partition coefficient of trace element K is defined by [3]

$$\ln K = -\frac{\frac{\partial H_{ex}}{\partial x}}{RT}. \tag{6.4.3}$$

Therefore, when ln K is plotted against ionic radii as those in the Onuma (partition coefficient – ionic radii) diagrams in Ref. [2], they can be compared with upside-down pictures of the partial excess enthalpies shown in Figs. 6.4.1–6.4.4. The Onuma diagram for divalent ions in the olivine–groundmass system shows a single peak corresponding to the M1 and M2 sites in the olivine structure. This is consistent with the single minimum calculated for forsterite (Fig. 6.4.3). The observed K_{Ca}/K_{Mg} is ~0.05 for the olivine–groundmass partition [2]. Assuming an equilibrium temperature of olivine and magma of approximately 1400 K, the partial excess enthalpy can be estimated at ~35 kJ/mol with Eq. (6.4.3). The a_i of Ca corresponds to 1.414 in forsterite, and thus the partial excess enthalpy can be estimated from Fig. 6.4.3 to be ~30 kJ/mol. Considering the crude potentials used in this study, both partial excess enthalpies agree well.

The Onuma diagram of divalent cations for the augite–groundmass system shows two peaks corresponding to the M1 and M2 sites in this Ca-rich clinopyroxene structure. This agrees with the two minima calculated for diopside (Fig. 6.4.4). The partial excess enthalpy is rather asymmetric for diopside compared with that of KCl and forsterite. The origin of this asymmetry is not clear at the moment, but might be related to the local structure of octahedral sites in diopside.

In previous and present studies, the effect of melt is completely ignored in the calculation. Although the effect of melt is subtle, as is apparent from the Onuma diagram [2], it is expected to be significant under high pressure, because a melt is more compressible than the corresponding crystal. The effect of the melt is difficult to incorporate into Nagasawa's model, but can be readily evaluated in the present calculation scheme. The calculation will be essentially the same as that for the crystals discussed above. In this case, the enthalpy change of elemental substitution between crystal and melt can be obtained without calculating the partial excess enthalpy of the crystal. It is sometimes difficult to calculate the end member of a trace-element-substituted crystal because of structural changes, as is the case for forsterite. The inclusion of melt would circumvent such a problem. However, the local structure around the ions in the melt could be different from place to place and could also change with time. Therefore more sophisticated sampling of the enthalpy compared with that used for the crystals shown above is necessary to obtain reliable data.

Although the procedure discussed above is still primitive, it will potentially have many applications. The present approach is not limited to crystal–melt partitioning, but can also be used to calculate the partitioning between any combination of crystal, melt (fluid), and gas, as long as these phases are properly handled in the scheme of a MD simulation. The present simulation approach is also useful for predicting partitioning behavior under extreme conditions, such as those of the lower part of the lower mantle or the core, for which experimental measurements are difficult.

Acknowledgments

I have benefited greatly from Yoshito Matsui in understanding geochemistry and computer simulation. The idea for this study was developed in conversations with Yoshito. I am honored that the paper has been included in this volume.

References

[1] J. H. Jones, in *Rock Physics and Phase Relations, A Handbook of Physical Constants*, T. J. Ahrens, ed. (American Geophysical Union, Washington, D.C., 1995), pp. 73–104.

[2] Y. Matsui, N. Onuma, H. Nagasawa, H. Higuchi, and S. Banno, Bull. Soc. Fr. Mineral. Cristallogr. **100**, 315 (1977).

[3] H. Nagasawa, Science **152**, 767 (1966).

[4] J. Blundy and B. Wood, Nature (London) **372**, 452 (1994).

[5] K. Kawamura, MXDORTO, JCPE Newslett. **4**(3), 48 (1992).

[6] T. Akamatsu, M. Fukuhama, H. Nukui, and K. Kawamura, Mol. Sim. **12**, 431 (1994).

Subject Index

Materials Formula Index

Index of Contributors